Probiotics and Antimicrobial Effect

Probiotics and Antimicrobial Effect

Editor

Sabina Fijan

MDPI • Basel • Beijing • Wuhan • Barcelona • Belgrade • Manchester • Tokyo • Cluj • Tianjin

Editor
Sabina Fijan
Faculty of Health Sciences
University of Maribor
Maribor
Slovenia

Editorial Office
MDPI
St. Alban-Anlage 66
4052 Basel, Switzerland

This is a reprint of articles from the Special Issue published online in the open access journal *Microorganisms* (ISSN 2076-2607) (available at: www.mdpi.com/journal/microorganisms/special_issues/Probiotics_Antimicrobial).

For citation purposes, cite each article independently as indicated on the article page online and as indicated below:

LastName, A.A.; LastName, B.B.; LastName, C.C. Article Title. *Journal Name* **Year**, *Volume Number*, Page Range.

ISBN 978-3-0365-6933-8 (Hbk)
ISBN 978-3-0365-6932-1 (PDF)

Cover image courtesy of Sabina Fijan

Contents

About the Editor

Sabina Fijan

Sabina Fhas been employed at the University of Maribor since May 1999, first as a researcher at the Faculty of Mechanical Engineering on various EU projects in the 5th, 6th and 7th Framework programs of the European Commission. During this time, she has completed her Master thesis (2003) and Doctoral thesis (2005), followed by a postdoctoral project (2010) in the field of determining the influence of various antimicrobial procedures against pathogenic microorganisms.

Since 2011, she has been engaged as a university professor at the Faculty of Health Sciences, University of Maribor, Slovenia. She is the coordinator of the elective units: Beneficial foods and beneficial microbes (postgraduate students of nursing, course Preventative and Clinical Nutrition), Impact of beneficial microorganisms on health (postgraduate students of nursing, course Public health), Chosen chapters of Microbiology and Biochemistry (for postgraduate students of Bioinformatics) and Project acquisition, management and execution (undergraduate students of Nursing).

She has published research in several recognised international journals. Her research involves studying how pathogenic microorganisms are influenced by beneficial microorganisms, mainly probiotics.

Preface to "Probiotics and Antimicrobial Effect"

Dear Colleagues,

According to the definition accepted by the WHO and FAO in 2001 and the grammatical update conducted by the Panel of the International Scientific Association for Probiotics and Prebiotics in 2014, probiotics are defined as "live microorganisms that, when administered in adequate amounts, confer a health benefit on the host."The antimicrobial or antagonistic activity of probiotics is an important property that includes the production of antimicrobial compounds, competitive exclusion of pathogens, enhancement of the intestinal barrier function in resisting pathogens and others. There are many methods to ascertain probiotic properties, including various in vitro and in vivo methods. The in vitro methods include various modifications of the spot-on lawn assay, agar well diffusion assay (AWDA), co-culturing methods, usage of cell lines and others. The in vivo methods utilise animal models; however, their use is being restricted according to the European legislation OJ L136. The most important studies are double-blinded randomized placebo controlled clinical trials; however, these studies are difficult to perform as it is not easy to achieve uniform conditions. There is a clear need for more elaborate assays that would better represent the complex interactions between the probiotics and the final host. Our issue mainly focuses on the antimicrobial or antagonistic activity of probiotics and the methods to ascertain probiotic properties, including various in vitro and in vivo methods.

Sabina Fijan
Editor

microorganisms

MDPI

Editorial

Probiotics and Their Antimicrobial Effect

Sabina Fijan

Faculty of Health Sciences, University of Maribor, Žitna ulica 15, 2000 Maribor, Slovenia; sabina.fijan@um.si

This Special Issue of the journal *Microorganisms* highlights the importance of the antimicrobial effect of probiotics. According to the definition accepted by the World Health Organization and the Food and Agriculture Organization of the United Nations in 2001 and 2002 [1,2] as well as the grammatical update conducted by the Panel of the International Scientific Association for Probiotics and Prebiotics (ISAPP) in 2014 [3], probiotics are defined as "live microorganisms that, when administered in adequate amounts, confer a health benefit on the host." The antimicrobial or antagonistic activity of probiotics is an important property that includes the production of antimicrobial compounds, such as bacteriocins, competitive exclusion of pathogens, enhancement of the intestinal barrier function in resisting pathogens, as well as enhancing the immune system of the host in order to successfully combat pathogens [4,5]. There are many methods to ascertain antimicrobial probiotic properties, including various in vitro and in vivo methods. The in vitro methods include various modifications of the spot-on lawn assay, agar well diffusion assay (AWDA), co-culturing methods, usage of cell lines and others [6]. The in vivo methods utilize animal models; however, in favour of the protection of animals, alternative methods are being researched to replace all animal research according to the EU directive 2010/63/EU and its consolidated text EUR-Lex—02010L0063-20190626 from 2019 [7,8]. The most important studies on the efficacy of probiotic strains are robust and well-designed, double-blinded randomized placebo-controlled clinical trials that face their own challenges as it is not easy to achieve uniform conditions of participants to eliminate all other influences [6,9]. There is a clear need for more elaborate assays that would better represent the complex interactions between the probiotics and the final host.

The main common probiotics are members of the lactobacilli group, which has recently been divided into 25 genera [10] (including, but not limited to, certain strains of the following species: *Lactobacillus acidophilus, Lactobacillus delbrueckii* subsp. *bulgaricus, Lactobacillus gasseri, Lactobacillus crispatus, Lacticaseibacillus rhamnosus, Lacticaseibacillus casei, Lactiplantibacillus plantarum, Limosilactobacillus reuteri, Levilactobacillus brevis, Ligilactobacillus salivarius* and others), and *Bifidobacterium* genera (e.g., *Bifidobacterium animalis* subsp. *infantis, Bifidobacterium longum, Bifidobacterium bifidum,* and others). Furthermore, certain strains from other bacterial species (e.g., *Lactococcus lactis, Pediococcus mesenteroides, Enterococcus faecium, Streptococcus thermophilus, Bacillus subtilis, Bacillus coagulans, Clostridium butyricum, Escherichia coli*) and even certain strains from certain yeasts (e.g., *Saccharomyces cerevisiae* var. *boulardii*) qualify as probiotics [11]. Lactic acid bacteria constitute a diverse group of Gram-positive, non-spore-forming bacteria, involved in numerous fermentation processes that produce lactic acid from carbohydrates via the homofermentative or heterofermentative pathway [12,13]. The major representatives of this group are *Lactobacillus, Lactococcus, Streptococcus, Leuconostoc, Pediococcus, Enterococcus, Oenococcus* and *Weissella* genera [12]. The *Lactobacillus* genus as well as other lactic acid bacteria have many strains with well-known antimicrobial properties [14]. Cytokine production is also attributed to probiotic lactic acid bacteria, linked to their action in the gut-associated lymphoid tissue that influences host immunity by protecting the host from infections caused by pathogens as well as suppressing allergic symptoms and even cancer [15–17]. In the study by Yin and co-authors [15], it was found that the strain *Levilactobacillus brevis* JCM 1059 was most efficient in bacterial uptake by differentiated monocytic THP-1 cells, as well as in subsequent

Citation: Fijan, S. Probiotics and Their Antimicrobial Effect. *Microorganisms* **2023**, *11*, 528. https://doi.org/10.3390/microorganisms11020528

Received: 27 January 2023
Accepted: 15 February 2023
Published: 19 February 2023

interleukin-12 (IL-12) production. The review by Ahmed and co-authors [17] investigated the antimicrobial and anti-inflammatory effects of various *Weissella* species, and found that they are clinically treatable bacteria with emerging antimicrobial and probiotic benefits ranging from oral health, skin care, obesity, and inflammatory diseases to cancer.

Current research is focused on finding novel or next-generation probiotic strains with antimicrobial properties that can efficiently modulate the ecological taxa composition and functionality of the human microbiota in the gut and beyond. The most commonly used pathogens to assess the antimicrobial activity in the publication of this issue were from the following genera or species: *Pseudomonas* spp., *Klebsiella* spp., *Escherichia coli*, *Staphylococcus aureus*, *Streptococcus* spp., *Bacillus* spp. and *Salmonella* spp. [18–22]. In the study by Schifano and co-authors [22] several novel *Leuconostoc mesenteroides* strains (C1, C2, C3) and a *Weissella soli* strain (T4), isolated from carrots exhibited strong inhibition against common pathogens. Some strains of the *Bacillus* genus that fulfil the criteria of safety assessment and the status of qualified presumption of safety [23] have also shown an efficient phenotypic antimicrobial effect. Torres-Sánchez and co-authors [18] found that *Bacillus siamenensis*-like strains (rB1, rB3), isolated from the human gut microbiota, were most efficient in antimicrobial activity. Additionally, two potential probiotic strains: *Bacillus subtilis* CP9, isolated from a desert camel, in the study by Sudan and co-authors [20] and *Bacillus subtilis* Fa17.2, isolated from wild *Bromelia* sp. Flowers, in the study by Tenea and co-authors [21], exhibited antimicrobial activity. On the other hand, the strain *Bacillus coagulans* MTCC 5260 used in the study by Fijan and co-authors [19] exhibited only a slight antimicrobial effect against clinical wound pathogens, thus proving the importance of addressing strain specific properties [3,24].

Several multi-strain probiotics used in the study by Fijan and co-authors [19], such as OMNi-BiOTiC® dietary supplements, Bio-Kult® and NutriVital Ultra SB, exhibited more efficient antimicrobial action compared to single-strain probiotics, perhaps due to interactions in mixed microbial cultures are driven by metabolite exchanges and are dependent on symbiotic and sometimes competitive behaviours [20]. However, the same study [19] also found that various single strain lactobacilli with well-known antimicrobial properties, including *Lacticaseibacillus rhamnosus* LGG, *Lacticaseibacillus paracasei* Shirota, *Limosilactobacillus reuteri* DSM 17938 and *Lactiplantibacillus plantarum* subsp. *plantarum* DSM 2601 showed efficient antagonistic activity against clinical wound pathogens. The author concluded that perhaps an individualistic approach such as a 'probiogram' could be a possibility in the future as a method to find the most efficient targeted probiotic strains, cell-free supernatants, or neutralized cell-free supernatants that have the highest antagonistic effects against individual clinical wound pathogens.

The agar well diffusion assay using cell-free supernatants [18–22] was the most commonly used method to assess the antimicrobial efficiency of the selected probiotics and other beneficial microbes with probiotic properties. Cell-free supernatants are regarded as postbiotics if a beneficial effect to health is observed [25]; thus, efficient antimicrobial effects found in in vitro studies are the first step in establishing new postbiotics. Cell-free supernatants contain metabolites with antimicrobial properties such as bacteriocins, organic acids, including fatty acids, amphiphilic membrane active biosurfactants as well different metabolites with possible antimicrobial effects such as tryptophan-, polyamine-, glutathione- metabolites and others [19–21]. Organic acids may have potentiated the activity of other antimicrobial metabolites, which can trigger acidification and/or acid-mediated cell membrane variation to exert an apparent antagonistic effect [21]. Bacteriocins, such as nisins, lactacins, enterocins, colicins, etc., are ribosomal-synthesized peptides or proteins produced by bacterial strains with a strong ability to inhibit pathogenic bacteria and nanoencapsulation prevents proteolytic enzyme degradation and unwanted interactions with food components by enhancing food stability, as found by Shafique and co-authors [26]. Heat stability of antimicrobial substances is also an important trait when selecting bacteriocinogenic producer strains intended to be used as preservation agents in processed foods [21]. The agar spot and co-culturing assays were used in two publications of this issue [19,20]. Both methods investigate the antagonistic effect of viable probiotics against pathogens, where one measures the zone of inhibition of pathogen growth around the spotted probiotic and the other determines the cfu of the pathogen after incubation with the probiotic.

One route to treat or prevent infectious diseases caused by viruses is the use probiotics. Steyer and co-authors [27] conducted a literature review of randomised placebo controlled clinical studies on the antiviral properties against rotavirus gastroenteric infections in children and Hung and co-authors [28] conducted a literature review on the evidence of oral probiotics as a therapy for the gastrointestinal involvement in COVID-19 patients. Oral probiotics had been evidenced to improve gut health in achieving homeostasis by exhibiting their antiviral effects via the gut–lung axis [29] and patient with COVID-19 have significant changes in fecal microbiomes, characterized by the enrichment of opportunistic pathogens and the depletion of beneficial commensals, which is vastly associated with disease severity [28,30]. Registered clinical trials of probiotics in COVID-19, mainly with lactobacilli and mixtures of bifidobacteria and lactobacilli genera are ongoing and thus the preventive or therapeutic role of probiotics for such patients can be elucidated in the near future [28]. *Saccharomyces cerevisiae* var. *boulardii, Lacticaseibacillus rhamnosus* GG, and various multi-strain probiotics exhibited antiviral properties against rotavirus gastroenteric infections in children [27]. The underlying mechanism of the probiotics against rotavirus gastroenteric infections in children included immune enhancement and modulation of intestinal microbiota leading to the shortening of diarrhoea. Many factors influence the outcome of the clinical study, including: correct strain selection and dosage of probiotics, duration of treatment, quality of probiotics as well as the production process of probiotics [27]. More robust, well-designed clinical studies addressing all factors are warranted.

Overall, this Special Issue has brought together new studies on the antimicrobial effects of various novel probiotics from the *Weissella, Bacillus, Leuconostoc* and *Levilactobacillus* genera, as well as well-known probiotic food supplements. It also highlights successful applications of probiotics for different infectious diseases including rotaviral gastrointestinal infections, wound infections and even COVID-19.

Conflicts of Interest: The author declares no conflict of interest.

References

1. FAO/WHO. *Health and Nutritional Properties of Probiotics in Food including Powder Milk with Live Lactic Acid Bacteria*; Food and Agriculture Organization of the United Nations; World Health Organization: Cordoba, Argentina, 2001.
2. FAO/WHO. *Guidelines for the Evaluation of Probiotics in Food*; Report of a Joint FAO/WHO Working Group on Drafting Guidelines for the Evaluation of Probiotics in Food; Food and Agriculture Organization of the United Nations; World Health Organization: London, ON, Canada, 2002.
3. Hill, C.; Guarner, F.; Reid, G.; Gibson, G.R.; Merenstein, D.J.; Pot, B.; Morelli, L.; Canani, R.B.; Flint, H.J.; Salminen, S.; et al. Expert consensus document. The International Scientific Association for Probiotics and Prebiotics consensus statement on the scope and appropriate use of the term probiotic. *Nat. Rev. Gastroenterol. Hepatol.* **2014**, *11*, 506–514. [CrossRef] [PubMed]
4. Echegaray, N.; Yilmaz, B.; Sharma, H.; Kumar, M.; Pateiro, M.; Ozogul, F.; Lorenzo, J.M. A novel approach to Lactiplantibacillus plantarum: From probiotic properties to the omics insights. *Microbiol. Res.* **2023**, *268*, 127289. [CrossRef] [PubMed]
5. van Zyl, W.F.; Deane, S.M.; Dicks, L.M.T. Molecular insights into probiotic mechanisms of action employed against intestinal pathogenic bacteria. *Gut Microbes* **2020**, *12*, 1831339. [CrossRef] [PubMed]
6. Fijan, S. *Antimicrobial Effect of Probiotics against Common Pathogens*; InTech: London, UK, 2016. [CrossRef]
7. EUR-Lex. Consolidated text: Directive 2010/63/EU of the European Parliament and of the Council of 22 September 2010 on the protection of animals used for scientific purposes (Text with EEA relevance). *Off. J. Eur. Union* **2019**.
8. EUR-Lex. Directive 2010/63/EU of the European Parliament and of the Council of 22 September 2010 on the protection of animals used for scientific purposes Text with EEA relevance. *Off. J. Eur. Union* **2010**, L 276/33–L 276/79.
9. Binda, S.; Hill, C.; Johansen, E.; Obis, D.; Pot, B.; Sanders, M.E.; Tremblay, A.; Ouwehand, A.C. Criteria to Qualify Microorganisms as "Probiotic" in Foods and Dietary Supplements. *Front. Microbiol.* **2020**, *11*, 1662. [CrossRef] [PubMed]
10. Zheng, J.; Wittouck, S.; Salvetti, E.; Franz, C.; Harris, H.M.B.; Mattarelli, P.; O'Toole, P.W.; Pot, B.; Vandamme, P.; Walter, J.; et al. A taxonomic note on the genus Lactobacillus: Description of 23 novel genera, emended description of the genus Lactobacillus Beijerinck 1901, and union of Lactobacillaceae and Leuconostocaceae. *Int. J. Syst. Evol. Microbiol.* **2020**, *70*, 2782–2858. [CrossRef]

11. Fijan, S. Microorganisms with Claimed Probiotic Properties: An Overview of Recent Literature. *Int. J. Environ. Res. Public Health* **2014**, *11*, 4745–4767. [CrossRef]
12. Saha, U.B.; Saroj, S.D. Lactic acid bacteria: Prominent player in the fight against human pathogens. *Expert Rev. Anti-Infect. Ther.* **2022**, *20*, 1435–1453. [CrossRef]
13. Djadouni, F.; Kihal, M. Antimicrobial activity of lactic acid bacteria and the spectrum of their biopeptides against spoiling germs in foods. *Braz. Arch. Biol. Technol.* **2012**, *55*, 435–444. [CrossRef]
14. Arena, M.P.; Silvain, A.; Normanno, G.; Grieco, F.; Drider, D.; Spano, G.; Fiocco, D. Use of Lactobacillus plantarum Strains as a Bio-Control Strategy against Food-Borne Pathogenic Microorganisms. *Front. Microbiol.* **2016**, *7*, 464. [CrossRef]
15. Yin, T.; Zhang, X.; Iwatani, S.; Miyanaga, K.; Yamamoto, N. Uptake of Levilactobacillus brevis JCM 1059 by THP-1 Cells via Interaction between SlpB and CAP-1 Promotes Cytokine Production. *Microorganisms* **2023**, *11*, 247. [CrossRef]
16. Lebeer, S.; Vanderleyden, J.; De Keersmaecker, S.C. Host interactions of probiotic bacterial surface molecules: Comparison with commensals and pathogens. *Nat. Rev. Microbiol.* **2010**, *8*, 171–184. [CrossRef]
17. Ahmed, S.; Singh, S.; Singh, V.; Roberts, K.D.; Zaidi, A.; Rodriguez-Palacios, A. The Weissella Genus: Clinically Treatable Bacteria with Antimicrobial/Probiotic Effects on Inflammation and Cancer. *Microorganisms* **2022**, *10*, 2427. [CrossRef] [PubMed]
18. Torres-Sánchez, A.; Pardo-Cacho, J.; López-Moreno, A.; Ruiz-Moreno, Á.; Cerk, K.; Aguilera, M. Antimicrobial Effects of Potential Probiotics of Bacillus spp. Isolated from Human Microbiota: In Vitro and In Silico Methods. *Microorganisms* **2021**, *9*, 1615. [CrossRef]
19. Fijan, S.; Kocbek, P.; Steyer, A.; Vodičar, P.M.; Strauss, M. The Antimicrobial Effect of Various Single-Strain and Multi-Strain Probiotics, Dietary Supplements or Other Beneficial Microbes against Common Clinical Wound Pathogens. *Microorganisms* **2022**, *10*, 2518. [CrossRef] [PubMed]
20. Sudan, S.; Flick, R.; Nong, L.; Li, J. Potential Probiotic Bacillus subtilis Isolated from a Novel Niche Exhibits Broad Range Antibacterial Activity and Causes Virulence and Metabolic Dysregulation in Enterotoxic E. coli. *Microorganisms* **2021**, *9*, 1483. [CrossRef]
21. Tenea, G.N.; Gonzalez, G.L.; Moreno, J.L. Probiotic Characteristics and Antimicrobial Potential of a Native Bacillus subtilis Strain Fa17.2 Rescued from Wild Bromelia sp. Flowers. *Microorganisms* **2022**, *10*, 860. [CrossRef]
22. Schifano, E.; Tomassini, A.; Preziosi, A.; Montes, J.; Aureli, W.; Mancini, P.; Miccheli, A.; Uccelletti, D. Leuconostoc mesenteroides Strains Isolated from Carrots Show Probiotic Features. *Microorganisms* **2021**, *9*, 2290. [CrossRef] [PubMed]
23. Brodmann, T.; Endo, A.; Gueimonde, M.; Vinderola, G.; Kneifel, W.; de Vos, W.M.; Salminen, S.; Gómez-Gallego, C. Safety of Novel Microbes for Human Consumption: Practical Examples of Assessment in the European Union. *Front. Microbiol.* **2017**, *8*, 1725. [CrossRef]
24. Sanders, M.E.; Benson, A.; Lebeer, S.; Merenstein, D.J.; Klaenhammer, T.R. Shared mechanisms among probiotic taxa: Implications for general probiotic claims. *Curr. Opin. Biotechnol.* **2018**, *49*, 207–216. [CrossRef]
25. Salminen, S.; Collado, M.C.; Endo, A.; Hill, C.; Lebeer, S.; Quigley, E.M.M.; Sanders, M.E.; Shamir, R.; Swann, J.R.; Szajewska, H.; et al. The International Scientific Association of Probiotics and Prebiotics (ISAPP) consensus statement on the definition and scope of postbiotics. *Nat. Rev. Gastroenterol. Hepatol.* **2021**, *18*, 649–667. [CrossRef]
26. Shafique, B.; Ranjha, M.M.A.N.; Murtaza, M.A.; Walayat, N.; Nawaz, A.; Khalid, W.; Mahmood, S.; Nadeem, M.; Manzoor, M.F.; Ameer, K.; et al. Recent Trends and Applications of Nanoencapsulated Bacteriocins against Microbes in Food Quality and Safety. *Microorganisms* **2023**, *11*, 85. [CrossRef] [PubMed]
27. Steyer, A.; Mičetić-Turk, D.; Fijan, S. The Efficacy of Probiotics as Antiviral Agents for the Treatment of Rotavirus Gastrointestinal Infections in Children: An Updated Overview of Literature. *Microorganisms* **2022**, *10*, 2392. [CrossRef] [PubMed]
28. Hung, Y.-P.; Lee, C.-C.; Lee, J.-C.; Tsai, P.-J.; Ko, W.-C. Gut Dysbiosis during COVID-19 and Potential Effect of Probiotics. *Microorganisms* **2021**, *9*, 1605. [CrossRef]
29. Xia, Y.; Cao, J.; Wang, M.; Lu, M.; Chen, G.; Gao, F.; Liu, Z.; Zhang, D.; Ke, X.; Yi, M. Effects of Lactococcus lactis subsp. lactis JCM5805 on colonization dynamics of gut microbiota and regulation of immunity in early ontogenetic stages of tilapia. *Fish Shellfish. Immunol.* **2019**, *86*, 53–63. [CrossRef] [PubMed]
30. Zuo, T.; Zhang, F.; Lui, G.C.Y.; Yeoh, Y.K.; Li, A.Y.L.; Zhan, H.; Wan, Y.; Chung, A.C.K.; Cheung, C.P.; Chen, N.; et al. Alterations in Gut Microbiota of Patients With COVID-19 During Time of Hospitalization. *Gastroenterology* **2020**, *159*, 944–955. [CrossRef] [PubMed]

microorganisms

MDPI

Article

Uptake of *Levilactobacillus brevis* JCM 1059 by THP-1 Cells via Interaction between SlpB and CAP-1 Promotes Cytokine Production

Tingyu Yin [1,†], Xiaoxi Zhang [2,†], Shun Iwatani [1,3], Kazuhiko Miyanaga [4] and Naoyuki Yamamoto [1,*]

1 School of Life Science and Technology, Tokyo Institute of Technology, Yokohama 226-8501, Kanagawa, Japan
2 Department of Microbiology and Immunology, School of Medicine, Keio University, 35 Shinanomachi, Shinjuku, Tokyo 160-8582, Japan
3 Tsukuba Biotechnology Research Center, Astellas Pharma Inc., 5-2-3, Tokodai, Tsukuba-shi 300-2698, Ibaraki, Japan
4 Department of Infection and Immunity, School of Medicine, Jichi Medical University, 3311-1, Yakushiji, Shimotsuke-shi 329-0498, Tochigi, Japan
* Correspondence: n-yamamoto@bio.titech.ac.jp; Tel.: +81-45-924-5105
† These authors contributed equally to this work.

Abstract: Several probiotic lactic acid bacteria (LAB) exert immunomodulatory effects on the host. However, the reasons for the different effects of LAB have not been fully elucidated. To understand the different immunomodulatory effects of LAB, we evaluated the levels of critical molecules in differentiated monocytic THP-1 and dendritic cells (DCs) following the uptake of various LAB strains. *Lactobacillus helveticus* JCM 1120, *Lactobacillus acidophilus* JCM 1132, *Levilactobacillus brevis* JCM 1059, and *Lentilactobacillus kefiri* JCM 5818 showed significantly higher uptake among the 12 LAB species tested. The uptake of microbeads by THP-1 DC increased when coupled with the surface layer proteins (Slps) from the tested strains. SlpB was mainly observed in the *L. brevis* JCM 1059 Slps extract. The expected cell surface receptor for SlpB on THP-1 DC was purified using SlpB-coupled affinity resin and identified as adenylyl cyclase-associated protein 1 (CAP-1). SlpB binding to THP-1 DC decreased after the addition of anti-CAP-1 and anti-DC-SIGN antibodies but not after the addition of anti-macrophage-inducible C-type lectin (Mincle) antibody. These results suggest that SlpB on *L. brevis* JCM 1059 plays preferentially binds to CAP-1 on THP-1 DC and plays a crucial role in bacterial uptake by THP-1 cells as well as in subsequent interleukin-12 (IL-12) production.

Keywords: lactic acid bacteria; THP-1; dendritic cells; *Levilactobacillus brevis*; SlpB; CAP-1; IL-12

Citation: Yin, T.; Zhang, X.; Iwatani, S.; Miyanaga, K.; Yamamoto, N. Uptake of *Levilactobacillus brevis* JCM 1059 by THP-1 Cells via Interaction between SlpB and CAP-1 Promotes Cytokine Production. *Microorganisms* **2023**, *11*, 247. https://doi.org/10.3390/microorganisms11020247

Academic Editor: Sabina Fijan

Received: 19 December 2022
Revised: 9 January 2023
Accepted: 14 January 2023
Published: 18 January 2023

1. Introduction

Probiotics are bioactive microbes that have beneficial effects on the host and improve their intestinal microbial balance [1]. The human microbiota primarily includes bacteria that can profoundly influence health and disease. Several studies have shown that probiotics have various biological functions. They can improve intestinal morphology, maintain intestinal microbial balance, and improve host immunity [2,3]. Several probiotic lactic acid bacteria (LAB) exert immunomodulatory effects on the host, protecting from the infections caused by pathogenic bacteria and suppressing allergic symptoms [4]. *Lactobacillus* strains are known to be potent inducers of proinflammatory cytokines, such as interleukin-12 (IL-12) and tumour necrosis factor α (TNFα) in the gut [5,6]. The immunomodulatory effects of lactobacilli are closely linked to their uptake by the gut-associated lymphoid tissue (GALT) and ability to modulate mucosal immune responses. Sampling intestinal bacteria, such as probiotic LAB strains, by the mucosal epithelium is essential for initiating immune responses in the GALT. The uptake of LAB by the microfold cells (M cells) in the follicle-associated epithelium (FAE) is a crucial event for the activation of the immune cells, such as the antigen-presenting dendritic cells (DCs), within the lymphoid follicles of the GALT [7].

DCs play a crucial role in defence against many pathogens by inducing cellular immunity after pathogen recognition [8,9]. DCs are activated in response to intestinal microbes and mediate the differentiation of naïve T cells into T helper type 1 (Th1) and Th2 cells [10]. DCs recognise pathogens via cell surface receptors, such as toll-like receptors [11,12] and C-type lectins [13]. DC-specific ICAM-3-grabbing non-integrin (DC-SIGN) is a DC-specific type II transmembrane protein with a C-type lectin extracellular domain [14,15] that plays a crucial role in the first contact between DCs, and probiotic and pathogenic bacteria.

Microbial-associated molecular patterns (MAMPs) modulate multiple host immune responses. Lipopolysaccharide (LPS), lipoproteins (LP), peptidoglycan (PG), polysaccharide A (PSA), lipoteichoic acids (LTA), microbial RNA, and DNA have been reported as MAMPs components [16]. Recognition of pathogenic and commensal bacteria-derived MAMPs by host pattern recognition receptors (PRRs) leads to the induction of different host immune responses [17].

Enterobacteriaceae, such as *Escherichia coli* [18], *Shigella* spp., and *Salmonella* spp [19,20] have been reported to induce phagocytosis after binding to DC-SIGN on DC surfaces. Among LAB, the surface layer proteins (Slps) on *Lactobacillus acidophilus* [21,22], *Lentilactobacillus kefiri* [23], and *Lactobacillus helveticus* [22] have been shown to bind to the DC-SIGN receptor induced on DCs. Macrophage-inducible C-type lectin (Mincle), a C-type lectin on differentiated macrophages, was reported to be the receptor for the Slps from *Levilactobacillus brevis* [24]. However, the binding of Slps of *Lactobacillus* species with various sequences and isoelectric points [25] involved in these immunomodulatory properties remains unknown. Slps are readily released from *Lactobacillus* cells by treatment with chaotropic reagents, such as LiCl [26], due to their non-covalent ionic binding to the cell surface. Twelve kinds of strains commonly used as probiotic LAB species were randomly selected. The major Slps from various LAB strains were coupled with microbeads to determine their ability to support bacterial uptake. Additionally, the critical receptors on THP-1 DC, which are involved in LAB binding, were identified.

Here, we report a novel receptor protein called adenylyl cyclase-associated protein 1(CAP-1), which is expressed on THP-1 DC and has an affinity for *L. brevis* JCM 1059 SlpB. Finally, we outline the importance of SlpB-CAP-1 binding for cellular uptake and subsequent cytokine production.

2. Materials and Methods

2.1. Bacterial Strains and Fermentation

All LAB strains listed in the Table 1 were obtained from the Japan Collection of Microorganisms (JCM) and our culture collections. LAB strains were cultured in Man Rogosa Sharpe (MRS, Becton, Dickinson and Company, Sparks, MD, USA) medium at 30 °C or 37 °C for 20 h, as described in Table 1. Human originated THP-1 monocyte cells were obtained from RIKEN Cell Bank (JRCB0112).

2.2. Induction of DC-SIGN and CAP-1 Expression on Differentiated THP-1 Cells

To confirm the differentiation of THP-1 cells into DC (THP-1 DC), the cell surface expression of DC-SIGN was evaluated by flow cytometry (EC800, SONY) after adding anti-DC-SIGN (Novus Biologicals USA, Centennial, CO, USA) and anti-CAP-1 antibodies (Novus Biologicals USA, Centennial, CO, USA). THP-1 cells obtained from JCM were maintained in Roswell Park Memorial Institute 1640 (RPMI 1640) supplemented with 10% fetal bovine serum (FBS) at 37 °C in a 5% CO_2 humidified incubator. THP-1 cells were seeded in 24-well culture plates (3×10^5 cells/mL) and treated with 50 nM phorbol 2-myristate 13-acetate (PMA, Adipogen Life Science, Liestal, Switzerland) and 20 ng/mL IL-4 (Peprotech, Cranbury, NJ, USA) to promote the differentiation of THP-1 DCs. THP-1 cells were cultured with PMA for 1 d, PMA mixed with IL-4 for one day, PMA for two days, PMA mixed with IL-4 for two days, or PMA for two days, followed by IL-4 for two days at 37 °C.

Before harvesting the cells, they were incubated in 4% paraformaldehyde (PFA) for 10 min. The cells in each well were then detached by adding 1 mM ethylenediaminetetraacetic acid (EDTA)-PBS and transferred to a 1.5 mL plastic microtube. After washing the cells thrice with PBS, DC-SIGN and CAP-1 gene expression were monitored at 490/525 nm via flow cytometry. Anti-DC-SIGN or anti-CAP-1 antibody was added to the cells (5 ng/mL) before incubation at room temperature for 1 h. After washing the cells with PBS, they were mixed with anti-IgG-Alexa 488 (50 ng/mL, Thermo Fisher Scientific Company, Waltham, MA, USA) and incubated at 25 °C for 1 h.

Table 1. Lactic acid bacteria and culture incubation temperature.

Strain	Temperature (°C)	Origin
Lactobacillus helveticus JCM 1120	37	Cheese
Lactobacillus acidophilus JCM 1132	37	Human feces
Lactobacillus amylovorus JCM 1126	37	Cattle waste
Lacticaseibacillus rhamnosus JCM 1136	37	Unknown
Lactobacillus delbrueckii subsp. *bulgaricus* JCM 1002	37	Bulgarian yogurt
Levilactobacillus brevis JCM 1059	30	Human feces
Lactiplantibacillus plantarum JCM 1100	30	Unknown
Lentilactobacillus kefiri JCM 5818	30	Kefir grains
Lacticaseibacillus paracasei subsp. *paracasei* JCM 8130	30	Milk product
Lactiplantibacillus plantarum subsp. *plantarum* JCM 1149	30	Pickled cabbage
Lactococcus lactis subsp. *cremoris* NZ9000	30	Mutant of *L. lactis* MG1363
Lactococcus lactis subsp. *lactis* IL1403	30	Cheese

All LAB strains obtained from the Japan Collection of Microorganisms (JCM) and our culture collections were cultured in MRS medium at 30 °C or 37 °C for 20 h. Reported origin for each strain was showed.

2.3. Sulfo-Cyanine3 Labelling

Twelve types of LAB cells were labelled with sulfo-cyanine3 (Cy3) using a Cy3 Mono-reactive dye labelling kit (GE Healthcare Bio-Sciences KK, Tokyo, Japan), according to the manufacturer's instructions. Briefly, all LAB strains were cultured in 50 mL MRS medium at 30 °C or 37 °C for 20 h, harvested via centrifugation at 6000× *g* for 10 min, and then washed twice with 5 mL of 0.1 M $NaHCO_3$ (pH = 9.3). Cells were suspended in 500 µL of 1 M $NaHCO_3$ (pH 9.3) and mixed with the Cy3-labelling reagent at 25 °C for 1 h. The Cy3 labelled LAB were washed with PBS and used for the THP-1 DC uptake study.

2.4. Preparation of Slps Coupled Microbeads

Twelve LAB strains were washed twicw with PBS by centrifugation at 6000× *g* for 10 min to collect the pellet. Then the cells in the pellet were suspended in 1 M LiCl and washed once with 1 M LiCl. Then, the collected cells were suspended in 20 mL of 5M LiCl and the supernatant was collected after centrifugation and dialysed against 300 times the volume of 20 mM phosphate buffer (pH 6.8) and freeze-dried for storage at −30 °C until future use. To prepare the Slps and ovalbumin conjugated fluorescein isothiocyanate (^{FITC}OVA, Invitrogen) coupled microbeads, Slps (5 µg/mL) and ^{FITC}OVA (5 µg/mL) were incubated with the hydrophilic 1-ethyl-3-(3-dimethylaminopropyl) carbodiimide hydrochloride (GN) microbeads (Kamakura Techno-Science, Inc., Tokyo, Japan) pre-activated in 3.3% (*v*/*v*) glutaraldehyde at 30 °C for 1 h (Slps-microbeads). Subsequently, the Slps-microbeads were blocked by mixing with 40 nM glycine-PBS (phosphate buffered saline, pH 7.4, 0.8% NaCl) at 30 °C for 1 h.

2.5. Uptake of LAB and Slps-Microbeads by THP-1 DC

To prepare THP-1 DC, THP-1 cells were suspended in RPMI 1640 containing 10% FBS, 50 nM PMA, and 20 ng/mL IL-4 and seeded on a 24-well culture plate at a concentration of 1×10^6 cells/mL for two days at 37 °C. THP-1 DC were co-cultured with Cy3 labelled LAB cells (MOI = 10) for 2 h in 24-well culture plates with 500 µL RPMI 1640 containing 10% FBS. After cultivation, THP-1 DC were washed thrice with PBS and harvested by adding 10 mM EDTA containing PBS. Uptake of Cy3 labelled LAB and ^{FITC}OVA was measured using a SONY EC800 flow cytometer. Flow cytometric analyses were performed at 550/570 and 490/525 nm.

2.6. Cytokine Measurement

THP-1 DC were cultured in RPMI 1640 medium supplemented with 10% FBS, 100 U/mL penicillin, and 100 µg/mL streptomycin. The cells were then stimulated in 96-well culture plates with 50 nM PMA for 1 or 2 d, with or without IL-4 (final concentration of 20 ng/mL). Cells in the other two groups were incubated with PMA for 1 or 2 d and additional 1 or 2 d with IL-4, respectively. Heat-killed Lactobacillus cells were incubated with THP-1 DC (MOI = 10) at 37 °C for 24 h. After incubation, the culture supernatant was collected by centrifugation at 6000× *g* for 10 min. IL-10 and IL-12p40 (IL-12) in the supernatant were analyzed by using ELISA kits obtained from BioLegend Inc. (San Diego, CA, USA) and R&D systems (Minneapolis, MN, USA), respectively, with a plate reader (Thermo Fisher Scientific, Varioskan LUX SkanIt Software 4.0) in triplicates.

2.7. Purification of the SlpB Receptor on THP-1 Cells

To purify the SlpB receptor on THP-1 cells, purified SlpB from L. brevis JCM 1059 was covalently coupled with an affinity resin. The affinity resin was prepared by mixing 10 mg purified JCM 1059 SlpB with 1 mL of Profinity Epoxide (Bio-Rad Laboratories, Inc. Hercules, CA, USA). THP-1 DC (differentiated with 50 nM PMA and 20 ng/mL IL-4 for 1 d) were washed with PBS and harvested. Subsequently, their cell surface components were extracted with 0.1% Triton PBS (cell extract). The cell extract was centrifuged at 8000× *g* for 10 min to remove aggregates and cell debris. After centrifugation, the supernatant was mixed with the SlpB-resin and incubated at room temperature for 1 h. Next, the SlpB-resin was washed with 0.15 M NaCl PB containing 0.1% Triton-X 100. Proteins with an affinity for SlpB were eluted by washing with 0.1% triton containing 0.5 and 1 M NaCl PB.

2.8. Sodium Dodecyl Sulphate-Polyacrylamide Gel Electrophoresis and Western Blotting Analysis

Slps released from the 12 LAB with the help of 5 M LiCl were analysed using sodium dodecyl sulphate-10% polyacrylamide gel electrophoresis (SDS-10%PAGE). The released proteins were mixed with sample buffer (6 × 125 nM Tris-HCl, 4% SDS, 20% glycerol, 0.012% bromophenol blue, and 10% 2-mercaptoethanol) and heated for 5 min at 95 °C. Then, the Slps were analysed by SDS-10%PAGE according to the Laemmli method [27]. Protein bands in the gel were visualised by staining the gels with Coomassie Brilliant Blue (CBB). Protein Ladder One Plus (Nacalai Tesque, Kyoto, Japan) was used as the marker. Proteins in the gel were transferred to polyvinylidene difluoride (PVDF) membranes in Tris-Cl buffer, pH 8.3 (190 mM glycine, 5 mM Tris-Cl, 20% methanol), for 1 h, at 150 mA. Specific proteins were detected with antibodies against Mincle (NK MAX, Santa Ana, CA, USA), DC-SIGN (SAB, Greenbelt, MD, USA), and CAP-1 (Novus Biological. Centennial, CO, USA) after 1000-fold dilution and following bionylated anti-rabbit IgG after 1000-fold dilution and avidin-peroxidase reaction). Finally, the band was detected by adding of 4-Chloro-1-naphthol as a substrate (Wako, Japan).

2.9. Protein Identification

Proteome analysis was performed to identify the proteins. Protein bands were excised from SDS-10%PAGE gels after CBB staining. To remove the dye from the gel, a de-staining solution (30% acetonitrile, 50 mM NH_4HCO_3) was added and incubated for 30 min. Then, 60% acetonitrile and 20 mM NH_4HCO_3 were added to remove water from the gel. Next, 5% (*w*/*w*) trypsin (Promega Japan, Tokyo, Japan) was added to the dried gel and incubated at 37 °C for 12 h. Peptides released from the gel were analysed by mass spectrometry using an UltrafleXtreme TOF/TOF MS (Bruker Daltonics GmbH, Bremen, Germany) operating in positive reflection ion mode between m/z 0 and 5000 Da.

2.10. Binding of Glycan and SlpB to THP-1

THP-1 DCs were washed twice with cold PBS after differentiation and were fixed in 1% PFA for 10 min at 4 °C. Subsequently, ^{Cy3}SlpB was added (0.3 mg/mL) with or without galactose (Sigma-Aldrich, St. Louis, MO, USA) or mannose (Nacalai Tesque Inc., Kyoto, Japan) (each 0.5 mg/mL) and incubated for 1 h at 37 °C. THP-1 DCs were collected and subjected to flow cytometry to evaluate the binding of ^{Cy3}SlpB to THP-1 DC.

2.11. Binding of Deglycosylated SlpB to THP-1

A microwell plate was coated with anti-CAP-1 antibody (1 µg/mL) for 24 h at 4 °C, and THP-1 DC cell extract was incubated for 2 h at 25 °C. Then, ^{Cy3}SlpA or deglycosylated ^{Cy3}SlpB were added to each well, and the plates were incubated for 2 h at 25 °C. To remove the polysaccharides from SlpB, 4 µg of recombinant glycosidase (PNGase F PRIME^{TM}, N-Zyme Scientifics, Doylestown, PA, USA) was mixed with 20 µg of ^{Cy3}SlpB in PBS and incubated for 24 h at 37 °C. Cy3 fluorescence originating from ^{Cy3}SlpB was measured using a plate reader (Thermo Fisher Scientific, Varioskan LUX SkanIt Software 4.0).

2.12. Statistical Analysis

Statistical significance was analysed using GraphPad Prism software version 9.1. Statistically significant differences were was set at $p < 0.05$ by using one-way analysis of variance (ANOVA) followed by Duncan's test.

3. Results

3.1. Induction of DC-SIGN on THP-1 Cells

A previous study reported that SlpA binding was required for the basal level expression of DC-SIGN on THP-1 DC after PMA and IL-4 treatment [21,28,29]. Here, THP-1 cells were treated with a combination of PMA and IL-4 according to a previously report [29]. THP-1 cells showed a dendritic-like morphology (THP-1 DC) when PMA treatment (for 2 d) was followed by IL-4 treatment (for 2 d) (Figure 1A). The expression of DC-SIGN on THP-1 DC was then quantified as one of the differentiation markers by flow cytometry using anti-DC-SIGN antibody (Figure 1B). The results showed that administering PMA and PMA combined with IL-4 for 1 d was not effective in inducing DC-SIGN (induced less than 2%) expression (Figure 1C). Conversely, treatment with PMA for 2 d, PMA for 1 d and IL-4 for 1 d, or PMA combined with IL-4 for 2 d showed increased DC-SIGN expression on THP-1 cell surface (8.5, 9.2, and 10.3%, respectively) whereas the impact of additional IL-4 on DC-SIGN induction was not clear (Figure 1C). DC-SIGN levels on the cell surface were highest on THP-1 DC that were subjected to 2 days of PMA treatment, followed by IL-4 treatment for 2 days (23.3%) (Figure 1C).

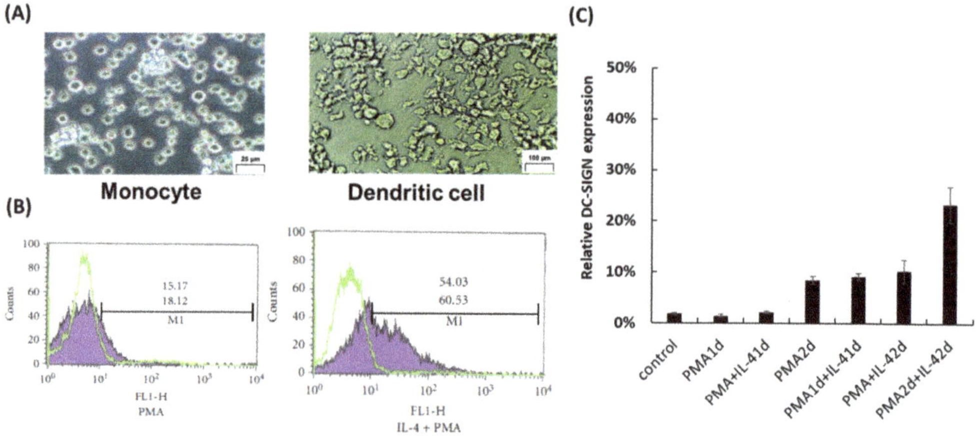

Figure 1. (**A**) Differentiation of monocytic THP-1 cells into dendritic-like THP-1 cells (THP-1 DC) by treatment with phorbol 2-myristate 13-acetate (PMA) and interleukin-4 (IL-4). (**B**) Expression of DC-specific ICAM-3-grabbing nonintegrin (DC-SIGN) on the surface of THP-1 DC was evaluated by flow cytometer using anti-DC-SIGN antibody. (**C**) Quantification of THP-1 cells expressing DC-SIGN on their cell surfaces after PMA and IL-4 treatment.

3.2. *Uptake of LAB by THP-1 DC and IL-12 Induction*

The SlpA on specific LAB strains is the key protein that binds to the DC-SIGN on THP-1 DC and plays a crucial role in the subsequent immune reaction. However, little is known about the other LAB cell surface proteins that may be capable of THP-1 DC binding. Twelve LAB strains were labelled with Cy3 and their uptake by THP-1 DC was analysed by flow cytometry. Significant differences were observed in the uptake ratios of the 12 LAB strains that were tested (Figure 2A). In particular, *L. helveticus* JCM 1120, *Lactobacillus delbrueckii* subsp. *bulgaricus* JCM 1002, *L. kefiri* JCM 5818, *L. acidophilus* JCM 1132, *Lactiplantibacillus plantarum* JCM 1100, and *L. brevis* JCM 1059 showed significantly higher uptake ratios than that of *Lacticaseibacillus paracasei* subsp. *paracasei* JCM 8130, which had the lowest uptake ratio (Figure 2A). Next, the Slps that were released from the LAB strains after treatment with the chaotropic reagent 5 M LiCl were coupled to ^{FITC}OVA conjugated microbeads and the importance of Slps in the uptake of the microbeads by THP-1 DC was evaluated. The microbeads coupled with Slps from various LAB strains showed different uptake ratios among the tested strains (Figure 2A). Microbeads-coupled with the Slps from *L. helveticus* JCM 1120, *L. acidophilus* JCM 1132, *L. brevis* JCM 1059, *L. plantarum* JCM 1100, and *L. kefiri* JCM 5818 showed potent uptake ratios, whereas decreased uptake ratios were observed for the microbeads that were coupled with the Slps from the LAB strains exhibiting reduced uptake (Figure 2A). This strongly suggests that the Slps released from *L. helveticus* JCM 1120, *L. acidophilus* JCM 1132, *L. brevis* JCM 1059, *L. plantarum* JCM 1100, and *L. kefiri* JCM 5818 contain the key components necessary for the binding of the bacteria to specific receptors on THP-1 DC. To understand the effect of LAB uptake on cytokine production, we monitored IL-12 production in THP-1 DC after treatment with 6 different LAB strains. Among the tested 6 strains, two types of *L. lactis* subspecies with lower uptake ratios showed lower IL-12 production, whereas *L. helveticus* JCM 1120, *L. acidophilus* JCM 1132, and *L. brevis* JCM 1059 with higher uptake ratios showed higher IL-12 production (Figure 2B). Therefore, the increased uptake of Slps by THP-1 DC may be important for the induction of IL-12 production in THP-1 DC. There were no significant differences in cytokine productions with bacterial cells collected at different growth times.

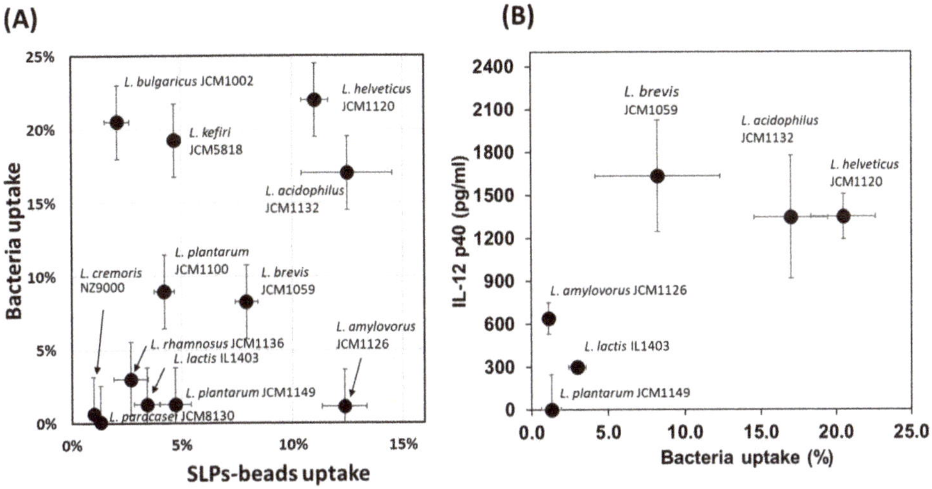

Figure 2. (**A**) Uptakes of Cy3 labelled surface layer protein A (^{Cy3}SlpA) lactic acid bacteria (LAB) and Slps-coupled microbeads by THP-1 DC. (**B**) Uptake of Cy3 labelled LAB (Bacterial uptake), and IL-12 production from THP-1 cells by treatments with various LAB strains.

3.3. Slps from Various LAB Strains and Binding to DC-SIGN

Previous studies have reported on the ability of the SlpA in the 5 M LiCl extract to bind the DC-SIGN receptor on THP-1 DC [21,30]. Thus far, no other Slps from the tested LAB strains have been described to have the ability to bind the DC-SIGN on THP-1 DC. Therefore, the Slps released in the 5 M LiCl extracts of various LAB strains were compared by SDS-10% PAGE analysis [27]. Expectedly, dense bands with molecular sizes of 47 and 45 kDa, corresponding to SlpA, were observed in *L. helveticus* JCM 1120 and *L. acidophilus* JCM 1132, respectively (Figure 3, lanes 8 and 9). In contrast, a major band with a molecular weight of 52 kDa was observed for *L. brevis* JCM 1059 (Figure 3, lane 6). The 52 kDa protein isolated from *L. brevis* JCM 1059 was identified as SlpB based on the proteome analysis after trypsin digestion of the excised gel by SDS-10%PAGE.

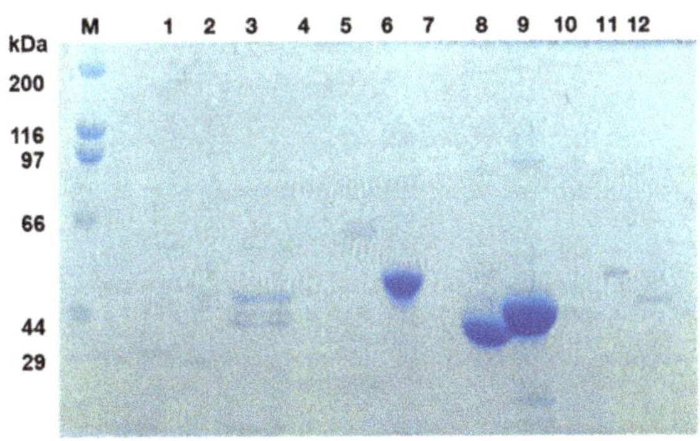

Figure 3. Sodium dodecyl sulphate-10% polyacrylamide gel electrophoresis (SDS-10%PAGE) analysis of 5 M LiCl released proteins. Lane 1: *L. plantarum* subsp. *plantarum* JCM 1149, lane 2: *L. lactic* subsp. *cremoris* NZ 9000, lane 3: *L. amylovorus* JCM 1126, lane 4: *L. paracasei* subsp. *paracasei* JCM 8130, lane 5: *L. kefiri* JCM 5818, lane 6: *L. brevis* JCM 1059, lane 7: *L. plantarum* JCM 1100, lane 8: *L. helveticus* JCM 1120, lane 9: *L. acidophilus* JCM 1132, lane 10: *L. plantarum* TIN- KL 001, lane 11: *L. lactis* subsp. *lactis* IL1403, lane 12: *L. rhamnosus* JCM 1136 and lane M: Marker proteins.

DC-SIGN on THP-1 DC is known to be the receptor for SlpAs and is crucial for the cellular uptake of *L. helveticus* and *L. acidophils* [21,22,29]. Mincle is also known as the receptor for the Slps on *L. brevis* [24]. However, the receptor for SlpB, which is induced on THP-1 DC and is necessary for the binding of *L. brevis* JCM 1059, has not been elucidated thus far. To confirm the ability of SlpA of *L. acidophilus* JCM 1132 and SlpB of *L. brevis* JCM 1059 to share the DC-SIGN receptor, ^{Cy3}SlpA and ^{Cy3}SlpB were prepared for competitive binding assays. As shown in Figure 4, the binding of ^{Cy3}SlpA SlpA to THP-1 DC was significantly inhibited by the addition of 10-fold of both, non-labelled SlpA and non-labelled SlpB. The binding of ^{Cy3}SlpA SlpB showed a reduced trend ($p = 0.069$) with the addition of 10-fold non-labelled SlpB, but not with non-labelled SlpA. These results suggest that SlpB on *L. brevis* JCM 1059 binds to both, DC-SIGN and other receptors on THP-1 cells.

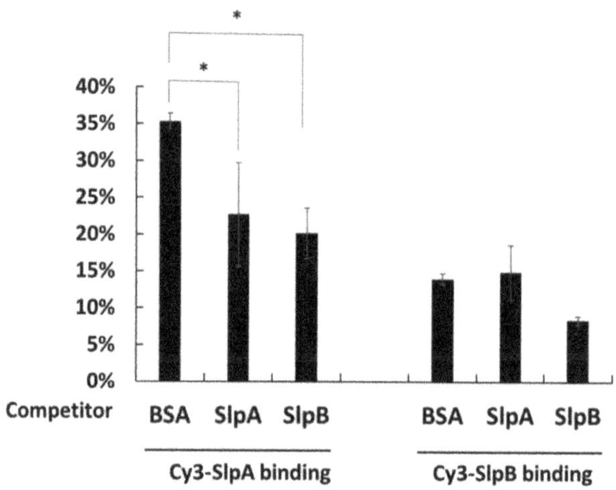

Figure 4. Bindings of ^{Cy3}SlpA prepared from *L. acidophilus* JCM 1132 and ^{Cy3}SlpB prepared from *L. brevis* JCM 1059 to THP-1 DC with or without 10 × non-labelled SlpA and SlpB. (Means ± SD. * $p < 0.05$).

3.4. Purification and Identification of the Receptor for SlpB from THP-1 Cells

Probable SlpB receptors in the THP-1 DC cell surface extract (THP-1 DC extract) were isolated with 0.1% Triton-PBS and applied to SlpB coupled Profinity Epoxide resins (SlpB-resin). The THP-1 DC extract was mixed with the SlpB-resin and the bound components were eluted with 0.5 and 1.0 M NaCl-PB after washing with 50 mM NaCl-PB. SDS-10%PAGE analysis showed the release of various sizes of components in 0.5 M elution, but a main protein with 57 kDa in 1 M NaCl-PB (Figure 5A). Proteome analysis indicated that the 57 kDa protein was CAP-1 [31,32]. To confirm the presence of CAP-1, and reported C-type lectin receptors, DC-SIGN, and Mincle in the THP-1 DC extract, western blotting was performed. A single band corresponding to 57 kDa was observed with anti-CAP-1 antibody for the affinity purified CAP-1 protein and for THP-1 cell extract (Figure 5A). The antibody against DC-SIGN showed a weak reaction towards some proteins including DC-SIGN corresponding reaction (arrow) as observed in previous study (29). However, no signal for Mincle was observed in THP-1 DC (Figure 5B). CAP-1 was observed on the whole cell surface of THP-1 DC when FITC-labelled anti-CAP-1 DC antibody was used (Figure 6A), as illustrated in Figure 6B.

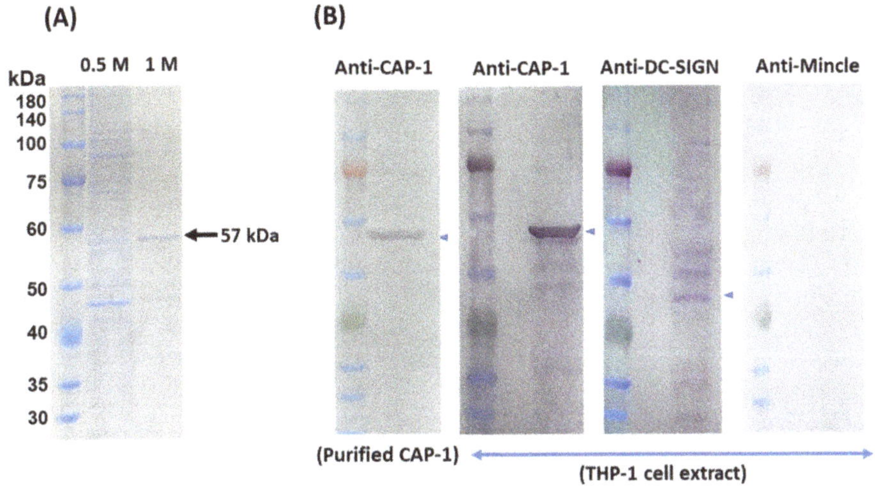

Figure 5. (**A**) SDS-10%PAGE of the affinity-purified 57 kDa receptor. (**B**) Western blotting analysis of affinity purified protein with anti-adenylyl cyclase-associated protein 1 (CAP-1) and crude extract of THP-1 cells with anti-CAP-1, anti-DC-SIGN, and anti-macrophage-inducible C-type lectin (Mincle) antibodies.

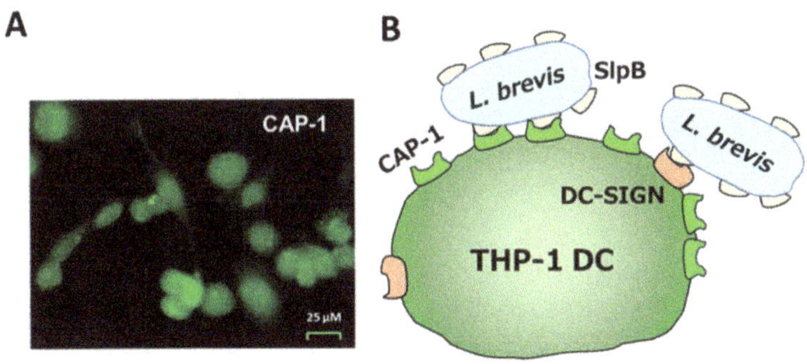

Figure 6. (**A**) Immunostaining of THP-1 DC with anti-CAP-1 antibody. CAP-1 on THP-1 DC was detected by incubation with FITC-labelled anti-CAP-1 antibody. (**B**) Illustration for *L. brevis* JCM 1059 binding of SlpB with CAP-1 and DC-SIGN.

3.5. Inhibition of SlpB Binding to THP-1 DC by Anti-CAP-1, Anti-DC-SIGN, and Anti-Mincle Antibodies

To confirm the preferential binding of SlpB to the receptors on THP-1 DC, we monitored the ability of ^{Cy3}SlpB to bind THP-1 DC in the presence of different antibodies. Addition of anti-CAP-1 antibody prevented the binding between ^{Cy3}SlpB and THP-1 DC (Figure 7A). The anti-DC-SIGN antibody also caused significant inhibition of the interaction between ^{Cy3}SlpB and THP-1 DC, whereas no significant inhibition was observed with the anti-Mincle antibody. The signal for DC-SIGN toward THP-1 DC extract was not strong in Western blotting (Figure 5); however, the specific binding of JCM 1059 with DC-SIGN was confirmed by the addition of anti-DC-SIGN antibody (Figure 7A). CAP-1 expression was induced in THP-1 DC after PMA and PMA/IL-4 treatment (Figure 7B).

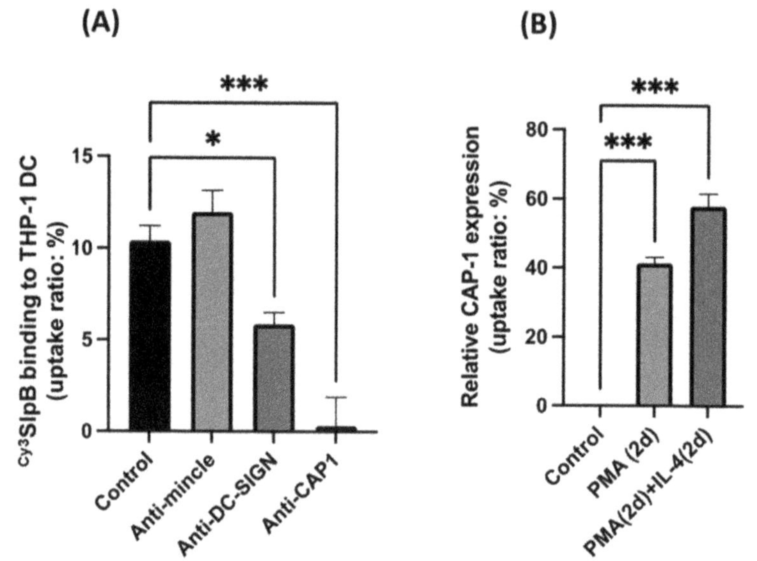

Figure 7. (**A**) ^{Cy3}SlpB was incubated with THP-1 DC with normal mouse serum (Control), anti-Mincle antibody, anti-DC-SIGN antibody, and anti-CAP-1 antibody for 1 h. (**B**) CAP-1 expression in THP-1 DC differentiated with PMA, and PMA and IL-4 (**B**). (Means ± SD. * $p < 0.05$; *** $p < 0.001$).

DC-SIGN is known for its ability to bind the carbohydrates in SlpA, and mannose prevents SlpA-DC-SIGN binding [22]. Therefore, we evaluated the effect of mannose and galactose on SlpB-THP-1 DC binding. As shown in Figure 8A, mannose and galactose strongly inhibited SlpB binding. Moreover, the interaction between ^{Cy3}SlpB and CAP-1 was significantly decreased after glycosidase treatment of ^{Cy3}SlpB (Figure 8B). Miner bands likely released from SlpB were observed after glycosidase treatment (Figure 8C). These results suggested that CAP-1 can recognise the carbohydrates on SlpB and function as a lectin-like protein. To confirm the existence of carbohydrates on SlpB, CAP-1 was captured on a microplate with anti-CAP-1 antibody and thereafter, ^{Cy3}SlpB was detected on the CAP-1 coated microplate. However, ^{Cy3}SlpB was not captured when ^{Cy3}SlpB was pre-treated with glycosidase (Figure 8B). These results indicate the presence of lectin-like activity of CAP-1 and are demonstrative of its interaction with the carbohydrates on SlpB.

3.6. Cytokine Production in THP-1 DC

To evaluate the impact of the SlpB from *L. brevis* JCM 1059 on bacterial uptake by THP-1 DC, we used Slp (mainly SlpB) removed *L. brevis* JCM 1059 and Slp associated JCM 1059 to monitor bacterial uptake and cytokine production. The uptake of *L. brevis* JCM 1059 was significantly reduced when Slp was removed from the cell surface (Figure 9). Furthermore, Slp associated *L. brevis* JCM 1059 induced robust IL-12 production, whereas Slp removed *L. brevis* JCM 1059 showed significantly reduced IL-12 production. The levels of IL-10 and IL-6 were relatively low. However, Slp associated *L. brevis* JCM 1059 induced IL-10 and IL-6 production. These results suggest that binding of SlpB in Slp fraction to the receptors on THP-1 DC may trigger *L. brevis* JCM 1059 uptake and cytokine (especially IL-12) production (Figure 9) during 24 h incubation.

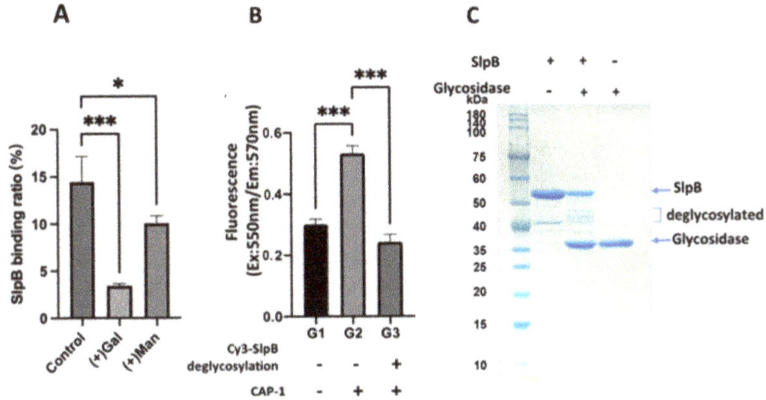

Figure 8. (**A**) ^{Cy3}SlpB and THP-1 DC binding with or without galactose (Gal) and mannose (Man). (**B**) Binding between CAP-1 and ^{Cy3}SlpB or glycosidase treated ^{Cy3}SlpB. (**C**) SDS-10%PAGE analysis of SlpB and deglycosylated SlpB. (Means ± SD. * $p < 0.05$; *** $p < 0.001$).

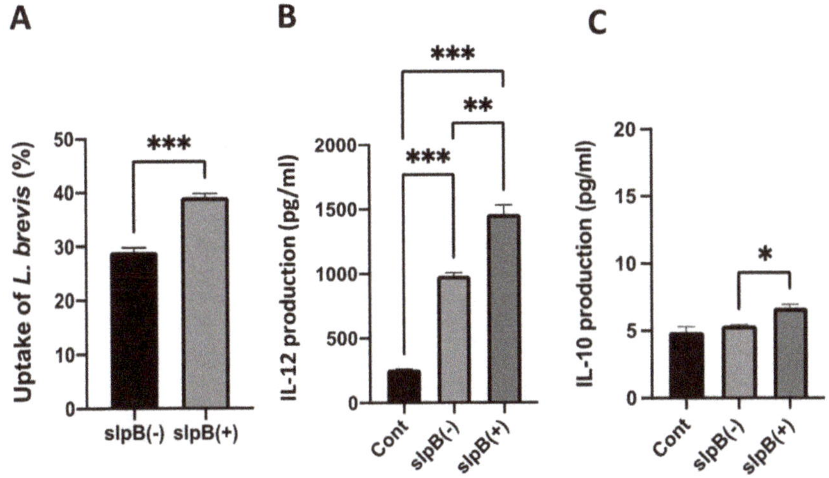

Figure 9. (**A**) Uptake of *L. brevis* JCM 1059 (slpB +) and LiCl treated JCM 1059 (slpB-) coating into THP-1 DC. (**B**) IL-12, and (**C**) IL-10 production by THP-1 DC after treatment with *L. brevis* with (+) or without (-) SlpB. (Means ± SD. * $p < 0.05$; ** $p < 0.01$, *** $p < 0.001$).

4. Discussion

Previous studies have highlighted the importance of the microbiota in the host immune system and other gut functions [33,34]. In the host gut immune system, little is known about the sampling of intestinal bacteria. Moreover, the Slps-receptor interactions between some *Lactobacillus* species, and DCs or macrophages facilitate bacterial uptake and induce the production of cytokines, such as IL-6, IL-10, IL-12, IL-17, and IL-23 [17,35,36]. Knocking out SlpA or the addition of a SlpA-specific antibody was shown to reduce the binding between *L. acidophilus* NCFM and THP-1 DC [21]. Furthermore, the interaction between the SlpA from *L. acidophilus* NCFM and DC-SIGN from THP-1 DC induced IL-10 and IL-12p70 production [21]. These results suggest that the frequent uptake of LAB by DC for 24 h may be essential for cytokine production. Most of previous studies mainly focused on C-type lectin receptor to identify receptors for SlpA [21] and SlpB [24] since C-type lectin localized on cell membrane of macrophage conserved carbohydrate recognition domains need for bacterial interactions [23]. In the present study, SlpB affinity purification was performed to screen SlpB receptors including non-C-type receptor and understand the role in the immunomodulatory responses and CAP-1 was identified as a novel SlpB receptor.

In the present study, we observed that among the tested LAB strains, *L. helveticus* JCM 1120, *L. acidophilus* JCM 1132, and *L. brevis* JCM 1059 had the highest uptake ratios, and could promote the productions of the proinflammatory (IL-12) and anti-inflammatory (IL-10) cytokines (Figure 2). Consistent with previous studies on the other *L. helveticus* and *L. acidophilus* strains, we identified SlpA to be the major Slps in the 5 M LiCl extracts from *L. helveticus* JCM 1120 and *L. acidophilus* JCM 1132 [21,29]. Furthermore, the present comparative study with various LAB strains is the first to demonstrate the importance of SlpB-CAP-1 binding for *L. brevis* JCM 1059 uptake and subsequent cytokine production by THP-1 DC. Previous studies have reported on the SlpB-dependent uptake of *L. kefiri* via its 3interaction with the Mincle receptor on macrophages [36]. However, Mincle was not expressed on the PMA/IL-4 differentiated THP-1 DC in the present study (Figure 5). SlpB binding to THP-1 DC was significantly reduced after the addition of anti-DC-SIGN antibodies. However, the decrease in SlpB binding was higher with the anti-CAP-1 antibody than that with the anti-DC-SIGN antibody. The higher production of CAP-1 than that of DC-SIGN (Figure 5) could be the main reason for the frequent access to THP-1 DC. Although CAP-1 was involved in the SlpB-dependent uptake of *L. brevis* JCM 1059 by THP-1 DC, the mechanism of binding and specificity remain unclear. Previously, CAP-1 was reported as a functional receptor for resistin expressed on monocyte and activated intracellular signalling pathway to modulate NF-κB-related inflammatory cytokines [31]. DC-SIGN is a C-type lectin receptor that contains the carbohydrate-recognition domain called Glu-Pro-Asn (EPN) [37] and has the potential to bind glucose-, mannose-, and N-acetylglucosamine-containing oligosaccharides [38]. There was no clear EPN sequence in CAP-1 sequence, but was low homology with DC-SIGN and Mincle (EPN-like sequence in Figure 10). Both galactose and mannose reduced SlpB binding to THP-1 DC (Figure 8A), indicative of the involvement of the galactose and mannose from SlpB in its binding with CAP-1. However, the ability of DC-SIGN and CAP-1 to recognise the different carbohydrates on SlpB remains unknown.

DC-SIGN 343 -WNRGEPNNVGEE- 354
MINCLE 165 -WDVGEPNNIATL- 176
CAP-1 329 -WRVENQENVSNL- 340

Figure 10. EPN sequence among DC-SIGN and Mincle, and similar sequence in CAP-1 Identical amino acids among 3 sequences were shown in red and between CAP-1 and Mincle of DC-SIGN were in blue.

DC-SIGN on THP-1 DC is crucial for its interaction with the SlpA on *L. helveticus* and *L. acidophilus* [21,30]. In the present study, we generated ^{Cy3}SlpA and ^{Cy3}SlpB to compare the ability of SlpB from *L. brevis* JCM 1059 and SlpA from *L. helveticus* JCM 1120 and *L. acidophilus* JCM1132 to interact with THP-1 DC. The binding between ^{Cy3}SlpA and the receptors on THP-1 DC was significantly inhibited by the addition of excess amounts of non-labelled SlpA or SlpB (Figures 4 and 7). In contrast, the interaction between ^{Cy3}SlpB and THP-1 DC was replaced by non-labelled SlpB but not by SlpA (Figure 7). SlpB has low homology with SlpA but shares the receptor DC-SIGN on THP-1 DC with the protein. Previous studies have predicted the carbohydrate-binding module sequences in the SlpB from *L. kefiri* that are involved in recognising DC-SIGN [23]. Carbohydrate-binding module-like sequences were also observed in the SlpB from *L. brevis* JCM 1059 (Figure 10).

The present study is the first to show that the ability of CAP-1 to bind the SlpB from *L. brevis* 1059 was greater than that of DC-SIGN. *L. brevis* can activate THP-1 cells after SlpB-dependent uptake by CAP-1. This contributes to our present knowledge of the immunomodulatory effect of SlpB-positive LAB on the gut immune system especially on gut-associated lymphoid tissues (GALT) and the host response.

5. Conclusions

Comparative studies with different 12 LAB species for the cytokine productions and the cell surface proteins revealed that SlpA on *L. helveticus* JCM 1120 and *L. acidophilus* JCM 1132 plays a crucial role for bacterial uptake by THP-1 DC. In contrast, SlpB on *L. brevis* JCM 1059 was a crucial to bind to THP-1 DC and following proinflammatory cytokine IL-12 production. SlpB receptor on THP-1 DC was purified by SlpB coated affinity resin and identified as CAP-1. CAP-1 expression on THP-1 DC was higher than that of DC-SIGN reported as the receptor for SlpB in WB analysis. Moreover, SlpB binding to THP-1 DC was completely inhibited by adding of anti-CAP-1 antibody and deglycosylation of SlpB, suggesting CAP-1 with interaction with carbohydrates on SlpB might be a major receptor for SlpB on THP-1 DC. Here, we identify a novel SlpB receptor CAP-1 on THP-1 DC which plays a crucial role in immunomodulatory effect of *L. brevis* in THP-1 cells.

Author Contributions: Conceptualization, S.I., K.M. and N.Y.; methodology, T.Y. and X.Z.; formal analysis, T.Y., X.Z. and S.I., validation, T.Y. and X.Z.; investigation, T.Y. and X.Z.; data curation, T.Y. and X.Z.; writing—original draft preparation, T.Y., S.I., K.M. and N.Y.; visualization, T.Y.; supervision, K.M. and N.Y.; writing-review & editing, N.Y. All authors have read and agreed to the published version of the manuscript.

Funding: This research received no external funding.

Institutional Review Board Statement: Not applicable.

Informed Consent Statement: Not applicable.

Data Availability Statement: Not applicable.

Conflicts of Interest: The authors declare no conflict of interest.

References

1. Gomaa, E.Z. Human gut microbiota/microbiome in health and diseases: A review. *Antonie Van Leeuwenhoek* **2020**, *113*, 2019–2040. [CrossRef] [PubMed]
2. Cortes-Perez, N.G.; de Moreno de LeBlanc, A.; Gomez-Gutierrez, J.; LeBlanc, J.G.; Bermúdez-Humarán, L.G. Probiotics and trained immunity. *Biomolecules* **2021**, *24*, 1402. [CrossRef] [PubMed]
3. Stavropoulou, E.; Bezirtzoglou, E. Probiotics in Medicine: A long debate. *Front. Immunol.* **2020**, *11*, 2192. [CrossRef]
4. Lebeer, S.; Vanderleyden, J.; De Keersmaecker, S.C. Host interactions of probiotic bacterial surface molecules: Comparison with commensals and pathogens. *Nat. Rev. Microbiol.* **2010**, *8*, 171–184. [CrossRef]
5. Shida, K.; Kiyoshima-Shibata, J.; Nagaoka, M.; Watanabe, K.; Nanno, M. Induction of interleukin-12 by *Lactobacillus* strains having a rigid cell wall resistant to intracellular digestion. *J. Dairy Sci.* **2006**, *89*, 3306–3317. [CrossRef]
6. Taverniti, V.; Guglielmetti, S. The immunomodulatory properties of probiotic microorganisms beyond their viability (ghost probiotics: Proposal of paraprobiotic concept). *Genes Nutr.* **2011**, *6*, 261–274. [CrossRef] [PubMed]
7. Oh, S.H.; Kim, S.H.; Jeon, J.H.; Kim, E.B.; Lee, N.K.; Beck, S.; Choi, Y.J.; Kang, S.K. Cytoplasmic expression of a model antigen with M Cell-Targeting moiety in lactic acid bacteria and implication of the mechanism as a mucosal vaccine via oral route. *Vaccine* **2021**, *39*, 4072–4081. [CrossRef]
8. Rahimi, N. C-type lectin CD209L/L-SIGN and CD209/DC-SIGN: Cell adhesion molecules turned to pathogen recognition receptors. *Biology* **2020**, *22*, 1. [CrossRef]
9. Feinberg, H.; Jégouzo, S.A.F.; Rex, M.J.; Drickamer, K.; Weis, W.I.; Taylor, M.E. Mechanism of pathogen recognition by human dectin-2. *J. Biol. Chem.* **2017**, *292*, 13402–13414. [CrossRef]
10. Butcher, M.J.; Zhu, J. Recent advances in understanding the Th1/Th2 effector choice. *Fac. Rev.* **2021**, *10*, 30. [CrossRef]
11. Underhill, D.M.; Ozinsky, A. Toll-like receptors: Key mediators of microbe detection. *Curr. Opin. Immunol.* **2002**, *14*, 103–110. [CrossRef] [PubMed]
12. Thoma-Uszynski, S.; Stenger, S.; Takeuchi, O.; Ochoa, M.T.; Engele, M.; Sieling, P.A.; Barnes, P.F.; Rollinghoff, M.; Bolcskei, P.L.; Wagner, M.; et al. Induction of direct antimicrobial activity through mammalian Toll-like receptors. *Science* **2001**, *291*, 1544–1547. [CrossRef] [PubMed]
13. Geijtenbeek, T.B.; Torensma, R.; van Vliet, S.J.; van Duijnhoven, G.C.; Adema, G.J.; van Kooyk, Y.; Figdor, C.G. Identification of DC-SIGN, a novel dendritic cell-specific ICAM-3 receptor that supports primary immune responses. *Cell* **2000**, *100*, 575–585. [CrossRef] [PubMed]
14. Akdis, M.; Verhagen, J.; Taylor, A.; Karamloo, F.; Karagiannidis, C.; Crameri, R.; Thunberg, S.; Deniz, G.; Valenta, R.; Fiebig, H.; et al. Immune responses in healthy and allergic individuals are characterized by a fine balance between allergen-specific T regulatory 1 and T helper 2 cells. *J. Exp. Med.* **2004**, *199*, 1567–1575. [CrossRef] [PubMed]

15. Thaiss, C.A.; Levy, M.; Suez, J.; Elinav, E. The interplay between the innate immune system and the microbiota. *Curr. Opin. Immunol.* **2014**, *26*, 41–48. [CrossRef]
16. Atarashi, K.; Tanoue, T.; Shima, T.; Imaoka, A.; Kuwahara, T.; Momose, Y.; Cheng, G.Y.; Yamasaki, S.; Saito, T.; Ohba, Y.; et al. Induction of colonic regulatory T cells by indigenous Clostridium species. *Science* **2011**, *331*, 337–341. [CrossRef] [PubMed]
17. Zhang, P.; Snyder, S.; Feng, P.; Azadi, P.; Zhang, S.; Bulgheresi, S.; Sanderson, K.E.; He, J.; Klena, J.; Chen, T. Role of N-acetylglucosamine within core lipopolysaccharide of several species of gram-negative bacteria in targeting the DC-SIGN (CD209). *J. Immunol.* **2006**, *177*, 4002–4011. [CrossRef]
18. Iyori, M.; Ohtani, M.; Hasebe, A.; Totsuka, Y.; Shibata, K.-I. A role of the Ca^{2+} binding site of DC-SIGN in the phagocytosis of *E. coli*. *Biochem. Biophys. Res. Commun.* **2008**, *377*, 367–372. [CrossRef]
19. Klena, J.; Zhang, P.; Schwartz, O.; Hull, S.; Chen, T. The core lipopolysaccharide of *Escherichia coli* is a ligand for the dendritic-cell-specific intercellular adhesion molecule nonintegrin CD209 receptor. *J. Bacteriol.* **2005**, *187*, 1710–1715. [CrossRef]
20. Takahara, K.; Yashima, Y.; Omatsu, Y.; Yoshida, H.; Kimura, Y.; Kang, Y.-S.; Steinman, R.M.; Gyu, C.P.; Inaba, K. Functional comparison of the mouse DC-SIGN, SIGNR1, SIGNR3 and Langerin, C-type lectins. *Int. Immunol.* **2004**, *16*, 819–829. [CrossRef]
21. Konstantinov, S.R.; Smidt, H.; de Vos, H.M.; Bruijns, S.C.; Singh, S.K.; Valence, F.; Molle, D.; Lortal, S.; Altermann, E.; Klaenhammer, T.R.; et al. S layer protein A of *Lactobacillus acidophilus* NCFM regulates immature dendritic cell and T cell functions. *Proc. Natl. Acad. Sci. USA* **2008**, *105*, 19474–19479. [CrossRef] [PubMed]
22. Martínez, M.G.; Prado, A.M.; Candurra, N.A.; Ruzal, S.M. S-layer proteins of *Lactobacillus acidophilus* inhibits JUNV infection. *Biochem. Biophys. Res. Commun.* **2012**, *422*, 590–595. [CrossRef] [PubMed]
23. Malamud, M.; Cavallero, G.J.; Casabuono, A.C.; Lepenies, B.; Serradell, M.L.A.; Couto, A.S. Immunostimulation by *Lactobacillus kefiri* S-layer proteins with distinct glycosylation patterns requires different lectin partners. *J. Biol. Chem.* **2020**, *295*, 14430–14444. [CrossRef] [PubMed]
24. Prado Acosta, M.; Goyette-Desjardins, G.; Scheffel, J.; Dudeck, A.; Ruland, J.; Lepenies, B.A. S-Layer from *Lactobacillus brevis* modulates antigen-presenting cell functions via the Mincle-Syk-Card9 axis. *Front. Immunol.* **2021**, *12*, 602067. [CrossRef] [PubMed]
25. Hynönen, U.; Palva, A. *Lactobacillus* surface layer proteins: Structure, function and applications. *Appl. Microbiol. Biotechnol.* **2013**, *97*, 5225–5243. [CrossRef] [PubMed]
26. Scolari, G.; Vescovo, M.; Zacconi, C.; Vescovi, F. Extraction and partial characterization of proteolytic activities from the cell surface of *Lactobacillus helveticus* Zuc2. *J. Dairy Sci.* **2006**, *89*, 3800–3809. [CrossRef]
27. Lemmli, U.K. Cleavage of structural proteins during the assembly of the head of bacteriophage T4. *Nature* **1970**, *227*, 680–685. [CrossRef]
28. Jin, C.; Wu, L.; Li, J.; Fang, M.; Cheng, L.; Wu, N. Multiple signaling pathways are involved in the interleukine-4 regulated expression of DC-SIGN in THP-1 cell line. *J. Biomed. Biotechnol.* **2012**, *2012*, 357060. [CrossRef]
29. Puig-Kröger, A.; Serrano-Gómez, D.; Caparrós, E.; Domínguez-Soto, A.; Relloso, M.; Colmenares, M.; Martínez-Muñoz, L.; Longo, N.; Sánchez-Sánchez, N.; Rincon, M.; et al. Regulated expression of the pathogen receptor dendritic cell-specific intercellular adhesion molecule 3 (ICAM-3)-grabbing nonintegrin in THP-1 human leukemic cells, monocytes, and macrophages. *J. Biol. Chem.* **2004**, *279*, 25680–25688. [CrossRef]
30. Prado Acosta, M.; Sandra, M.R.; Sandra, M.C. S-layer proteins from *Lactobacillus* sp. inhibit bacterial infection by blockage of DC-SIGN cell receptor. *Int. J. Biol. Macromol.* **2016**, *92*, 998–1005. [CrossRef]
31. Lee, S.; Lee, H.C.; Kwon, Y.W.; Lee, S.E.; Cho, Y.; Kim, J.; Lee, S.; Kim, J.Y.; Lee, J.; Yang, H.M.; et al. Adenylyl cyclase-associated protein 1 is a receptor for human resistin and mediates inflammatory actions of human monocytes. *Cell Metab.* **2014**, *19*, 484–497. [CrossRef]
32. Saba, L.A.; McComb, M.E.; Potts, D.L.; Costello, C.E.; Amar, S. Proteomic mapping of stimulus-specific signaling pathways involved in THP-1 cells exposed to *Porphyromonas gingivalis* or its purified components. *J. Proteome Res.* **2007**, *6*, 2211. [CrossRef]
33. Peterson, C.T.; Sharma, V.; Elmén, L.; Peterson, S.N. Immune homeostasis, dysbiosis and therapeutic modulation of the gut microbiota. *Clin. Exp. Immunol.* **2015**, *179*, 363–377. [CrossRef] [PubMed]
34. Lin, L.; Zhang, J. Role of intestinal microbiota and metabolites on gut homeostasis and human diseases. *BMC Immunol.* **2017**, *18*, 2. [CrossRef] [PubMed]
35. Gad, M.; Ravn, P.; Søborg, D.A.; Lund-Jensen, K.; Ouwehand, A.C.; Jensen, S.S. Regulation of the IL-10/IL-12 axis in human dendritic cells with probiotic bacteria. *FEMS Immunol. Med. Microbiol.* **2011**, *3*, 93–197. [CrossRef] [PubMed]
36. Malamud, M.; Carasi, P.; Assandri, M.H.; Freire, T.; Lepenies, B.; Serradell, M.L.Á. S-layer glycoprotein from *Lactobacillus kefiri* exerts its immunostimulatory activity through glycan recognition by Mincle. *Front. Immunol.* **2019**, *26*, 1422. [CrossRef]
37. Hatakeyama, T.; Kamiya, T.; Kusunoki, M.; Nakamura-Tsuruta, S.; Hirabayashi, J.; Goda, S.; Unno, H. Galactose recognition by a tetrameric C-type lectin, CEL-IV, containing the EPN carbohydrate recognition motif. *J. Biol. Chem.* **2011**, *286*, 10305–10315. [CrossRef]
38. Lee, R.T.; Hsu, T.-L.; Ku, S.; Hsieh, S.-L.; Wang, C.-H.; Lee, Y.C. Survey of immune-related, mannose/fucose-binding C-type lectin receptors reveals widely divergent sugar-binding specificities. *Glycobiology* **2011**, *21*, 512–520. [CrossRef]

microorganisms

MDPI

Review

Recent Trends and Applications of Nanoencapsulated Bacteriocins against Microbes in Food Quality and Safety

Bakhtawar Shafique [1,†], Muhammad Modassar Ali Nawaz Ranjha [1,†], Mian Anjum Murtaza [1], Noman Walayat [2], Asad Nawaz [3,4], Waseem Khalid [5], Shahid Mahmood [1], Muhammad Nadeem [1], Muhammad Faisal Manzoor [6], Kashif Ameer [1,*], Rana Muhammad Aadil [7,*] and Salam A. Ibrahim [8,*]

1 Institute of Food Science and Nutrition, University of Sargodha, Sargodha 40100, Pakistan
2 College of Food Science and Technology, Zhejiang University of Technology, Hangzhou 310014, China
3 Shenzhen Key Laboratory of Marine Microbiome Engineering, Institute for Advanced Study, Shenzhen University, Shenzhen 518060, China
4 Institute for Innovative Development of Food Industry, Shenzhen University, Shenzhen 518060, China
5 Department of Food Science, Government College University Faisalabad, Faisalabad 38000, Pakistan
6 Guangdong Provincial Key Laboratory of Intelligent Food Manufacturing, Foshan University, Foshan 528011, China
7 National Institute of Food Science and Technology, University of Agriculture, Faisalabad 38000, Pakistan
8 Food Microbiology and Biotechnology Laboratory, North Carolina Agricultural and Technical State University, Greensboro, NC 27411, USA
* Correspondence: kashifameer89@gmail.com (K.A.); muhammad.aadil@uaf.edu.pk (R.M.A.); ibrah001@ncat.edu (S.A.I.)
† These authors contributed equally to this work.

Abstract: Bacteriocins are ribosomal-synthesized peptides or proteins produced by bacterial strains and can inhibit pathogenic bacteria. Numerous factors influence the potential activity of bacteriocins in food matrices. For example, food additives usage, chemical composition, physical conditions of food, and sensitivity of proteolytic enzymes can constrain the application of bacteriocins as beneficial food preservatives. However, novel bacteriocin nanoencapsulation has appeared as an encouraging solution. In this review, we highlight the bacteriocins produced by Gram-negative bacteria and Gram-positive bacteria including lactic acid bacteria that have shown positive results as potential food preservatives. In addition, this review encompasses the major focus on bacteriocins encapsulation with nanotechnology to enhance the antimicrobial action of bacteriocins. Several strategies can be employed to encapsulate bacteriocins; however, the nanotechnological approach is one of the most effective strategies for avoiding limitations. Nanoparticles such as liposomes, chitosan, protein, and polysaccharides have been discussed to show their importance in the nanoencapsulation method. The nanoparticles are combined with bacteriocins to develop the nano-encapsulated bacteriocins from Gram-negative and Gram-positive bacteria including LAB. In food systems, nanoencapsulation enhances the stability and antimicrobial functionality of active peptides. This nanotechnological application provides a formulation of a broad range of antimicrobial peptides at the industry-scale level. Nano-formulated bacteriocins have been discussed along with examples to show a broader antimicrobial spectrum, increase bacteriocins' applicability, extend antimicrobial spectrum and enhance stability.

Keywords: bacteriocins; lactic acid bacteria; Gram-negative bacteria; Gram-positive bacteria; nanoencapsulation; antimicrobial activity

Citation: Shafique, B.; Ranjha, M.M.A.N.; Murtaza, M.A.; Walayat, N.; Nawaz, A.; Khalid, W.; Mahmood, S.; Nadeem, M.; Manzoor, M.F.; Ameer, K.; et al. Recent Trends and Applications of Nanoencapsulated Bacteriocins against Microbes in Food Quality and Safety. *Microorganisms* **2023**, *11*, 85. https://doi.org/10.3390/microorganisms11010085

Academic Editor: Sabina Fijan

Received: 27 November 2022
Revised: 19 December 2022
Accepted: 20 December 2022
Published: 28 December 2022

1. Introduction

Synthetic products based on chemicals are typically used to avoid spoilage and enhance the shelf life of foods. However, these chemicals have numerous adverse impacts on the health of humans. Compared with synthetic chemicals, naturally derived compounds are preferable and are applied by biopreservation [1,2]. Bacteriocins are antimicrobial peptides, therefore ribosomes are used to synthesize them. The main criterion for the classification of bacteriocins is their origin. Depending on whether they are produced by Gram-positive or Gram-negative bacteria, classes and subclasses of bacteriocins can be distinguished as shown in Figure 1. Lactic acid bacteria are utilized to produce bacteriocins that have the potential to retard the growth of pathogenic microorganisms [3]. Bacteriocins produced by lactic acid bacteria have been recognized as efficient at helping to maintain food safety and are generally recognized as safe (GRAS) [4,5]. Bacteriocins of lactic acid bacteria and their phylogenetically associated strains have a much broader antimicrobial activity spectrum [6].

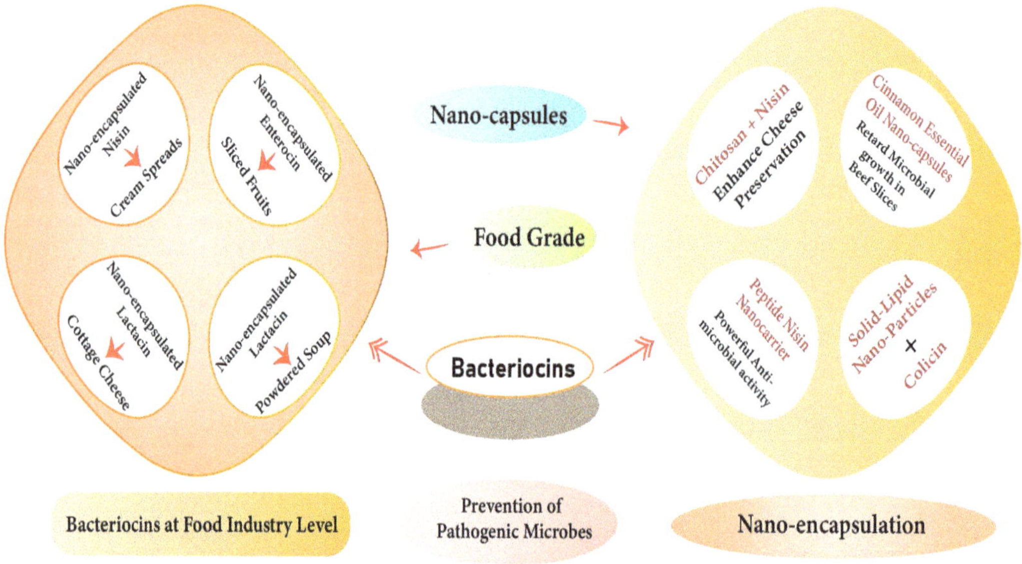

Figure 1. Graphical Abstract of Nano-encapsulated Bacteriocins in Food Safety.

Lactic acid bacteriocins are utilized in both the medical field and the food sector. Concerning food applications, they have been successfully applied as co-cultures or starters in experiments in the pilot study [7]. In the last decade, significant research has shown the effectiveness of bacteriocins in several branches of the food industry [8]. For example, one particular bacteriocin, nisin, is successfully used to improve the quality and storage period of milk and milk products [9,10]. Another study demonstrated that lactic acid bacteria and paracin with broad antimicrobial spectrum were successfully used to protect apple juice before bacterial contamination [11]. Enterocins are the broad-spectrum cyclic peptides against Gram-positive and Gram-negative bacteria, can be applied in milk products as well. Other products such as beef or fresh fish can also be protected with bacteriocins [12–14].

Metabolic products, bactericidal proteins, and antibiotic substances are produced by lactic acid bacteria. Lactic acid bacteria have the potential to inhibit numerous microorganisms in the food environment and exhibit vital antimicrobial characteristics related to food safety and preservation. In addition, strains of lactic acid bacteria provide health-enhancing capabilities regarding their potential in the medical sector. For instance, gastrointestinal pathogenic bacteria such as *Escherichia coli*, *Salmonella*, and *Helicobacter pylori* are mitigated by bacteriocins [15].

The functionality of bacteriocins produced by lactic acid bacteria is determined by many factors such as storage conditions, food matrix interaction, processing, pressure, temperature, and enzyme availability [16]. Effective nanoparticles are applied through nanoencapsulation to safeguard bacteriocins from degradation and deterioration. Nanoencapsulation is applied to bio-preservative bacteriocins to prevent enzyme degradation and to improve food product shelf life [17]. Nanoconjugates such as pediocin and nisin act as bacteriocins, and their applications provide unique packaging in food systems [18].

Gram-negative bacteriocins are used as proteins that efficiently play multi functions behavior to target pathogenic bacteria and bacterial species. Gram-positive bacteriocins are associated with advantageous properties such as the ability to deteriorate internal membranes and inhibit the growth of pathogenic microbes. Bacteriocins, either produced by Gram-negative or positive bacteria, always have the target of inhibiting the growth of other microorganisms in competition with the same environment. Hundreds of bacteriocins species have been discovered. The classification of bacteriocins is based on the sequence of amino acids, identified action mechanisms, and structure [19]. Starter cultures are used to produce bacteriocins, and exhibit potential applications in the dairy industry to protect fermented foods from the transmission of food pathogens [20]. Gram-positive bacteriocins are nano-encapsulated for application as bio-preservatives. Bacteriocin AS48 actively prevents pathogenic microbes such as *L. monocytogenes*, *S. aureus*, and *B. cereus* when used to prepare gelatin puddings, soy desserts, and baking cream. *L. monocytogenes* strains exist extensively in nature and are present in ready-to-eat meat. It has been recognized that Gram-positive bacteriocins inhibit or reduce the growth of *L. monocytogenes* during the processing of meat products [21].

Nanoencapsulation prevents proteolytic enzyme degradation and unwanted interactions with food components by enhancing food stability. Recently, several studies have revealed that bacteriocins encapsulation by nanoparticles improved the activity of peptides against multidrug-resistant bacteria and food spoilage microorganisms [22]. Anti-biofilms are manufactured by nanoencapsulation and have been regarded as a substantial mode for antimicrobial activity. Natural and synthetic nanoparticles are used in combination with bacteriocins to show improved efficacy in retarding the formation of biofilm and to assist in reducing antibiotic resistance [23]. Nanoparticles thus provide a wide range of characteristics to antimicrobial peptides such as bacteriocins with better functionality. These characteristics are physiological solution stability, a broad antibacterial spectrum, non-toxic nature, and ease of synthesis with less production cost and concentration [23].

Liposomes, silver nanoparticles, niosomes, nanovesicles, chitosan, solid lipid nanoparticles, phosphatidylcholine liposomes, and nanoliposomes are combined with Gram-negative and Gram-positive bacteriocins to develop nano-encapsulated bacteriocins for food systems. Nanoencapsulation promotes the antimicrobial action of bacteriocins to kill harmful bacteria, inhibits the interaction of pathogens directly with food substances, and ultimately extends the shelf life of food [24]. Nanoencapsulation is performed by several methods including nano-emulsification, nanoliposomes, electrospray, and formation of nanostructure with nanocrystals, nanostructure lipid carriers, and solid lipid nanoparticles [25].

This review addresses the importance of nanoencapsulation and highlights the bacteriocins of Gram-negative and Gram-positive bacteria and their potential application in various food industry systems. In addition, we review significant literature studies on the formulation of nano-encapsulated bacteriocins and the action mechanism of nano-encapsulated bacteriocins produced by lactic acid bacteria.

2. General Action Mechanism of Bacteriocins

Bacteriocins have been categorized into diverse classes, including Class I: lantibiotics that are stable in heat; Class II: non-lantibiotics (unmodified postranslationally); Subclass IIa: anti-Listeria-like and pediocin bacteriocins, Subclass IIb: two-peptide bacteriocins classified further into Subclass IIc: sec-dependent bacteriocins; Class III: non-lantibiotics that are labile to heat [26]. More bacteriocins have been separated and classified from lactic acid bacteria. Numerous bacteriocins have developed a position as effective antimicrobial agents due to their efficiency as food preservatives and provision of antagonistic impact to retard significant pathogens. The notable ones are nisin, pediocin, bulgarican, diplococcin, lactacins, acidophilin, plantaricins and helveticins [27].

The action mechanism of bacteriocins is focused on two diverse activity aspects, the physical interaction kinetics between susceptible cells and bacteriocin and certain biochemical lesions detection in target cells [28]. A wide variety of chemical structures permits bacteriocins to exhibit their impact on several important living cell functions (translation, transcription, biosynthesis of the cell wall, and replication, although most action is through the forming pores or membrane channels that disturb the potential energy [29]. It has been extensively postulated that the bacteriocin interacts with the target cell in two ways. The first way, which is perhaps reversible, resembles the bacteriocin's physical adsorption using receptors in the cell. At this stage, bacteriocins are emerged to remove the cell intact; meanwhile, no perpetual physiological damage could be done. The second way causes irreversible pathological alternations because of certain biochemical cell damage [30].

In nature, microorganisms contribute to various mechanisms for establishing protection and interaction. These mechanisms are linked with the bacteriocins peptides production with prolonged antimicrobial activity. Gram-negative bacteriocin protein establishes an effective approach to prevent multi-drug-resistant bacteria through the folding of proteins into bacteria and by the production of certain species of antibiotics [31]. Protein bacteriocins capture pathways of nutrient uptake to cause cell death and translocation of the cell envelope. Their importation is strengthened by the parasitizing of intermembrane complexes of protein joined to the motive proton force, which causes the delivery of a toxic domain within the cell. A plethora of biophysical, genetic, biochemical, and structural procedures have taken place to uncover the components of the cell envelope involved in the import of bacteriocin [32].

Colicins are antibacterial proteins that are produced by some intestinal strains of bacteria, primarily situated in chromosomally encoded plasmids. These huge proteins contain three domains such as the amino-terminal domain that facilitates the transport of the target cell's outer membrane, a receptor-binding domain that facilitates the transport within the periplasm, and a cytotoxic carboxy-terminal domain that shows the inhibitory impacts [33]. Three major action mechanisms for colicins have been described. These three mechanisms are the nuclease activity (e.g., target cell RNA/DNA), the formation of the pores that damage the integrity of the membrane, and murein synthesis inhibition [34]. *E. coli* is mainly responsible for the production of colicins which are bactericidal proteins that protect a colicinogenic plasmid. An immunity protein (imX or cxi), lysis proteins encoded gene and structural gene (cxa) are present in colicinogenic plasmids [32].

Colicin production regulation is facilitated by the response of SOS, which emerges to play a significant role in the host bacterial cell response to DNA damage. Three different action modes are provided by colicins: (1) the channels formation depends on voltage in the Gram-negative bacterial inner membrane, (2) the action of nuclease into the cytoplasm cells and (3) peptidoglycan degradation [32]. Genetic machinery available on chromosomes or plasmids is used to encode microcins. Microcins are categorized on the basis of three criteria: (1) the nature, localization, and presence of posttranslational modifications; (2) the organization of gene clusters; and (3) leader peptide sequences [35].

Bacteriocins display action by joining with the conforming receptor on the sensitive bacteria surface to retard bacterial growth. Bactericidal action mechanisms involve the peptidoglycanase type, RNase, and DNase function of the nuclease type and pore-developing type. Subtilisin A has an unmodified bacteriocin peptide structure and a low molecular weight. Colicin Ia and bacteriocin AS-48 have modified bacteriocin protein and peptide structures and low molecular weight as well. The strain is grown on the soft agar of sensitive bacteria LB to produce bacteriocin, and production zone occurs around the strains that produce bacteriocin. However, strains that do not produce bacteriocin have no inhibition zone [36]. The general bacteriocin mechanisms involve bacterial growth reduction comprising bacteriocin binding, bacteriocin translocation, enzyme activity modulation, and pore formation of the cytoplasmic membrane [37]. The general method of bacteriocins synthesis, anti-bacterial activity, and action mechanism of bacteriocins is shown in Figure 2.

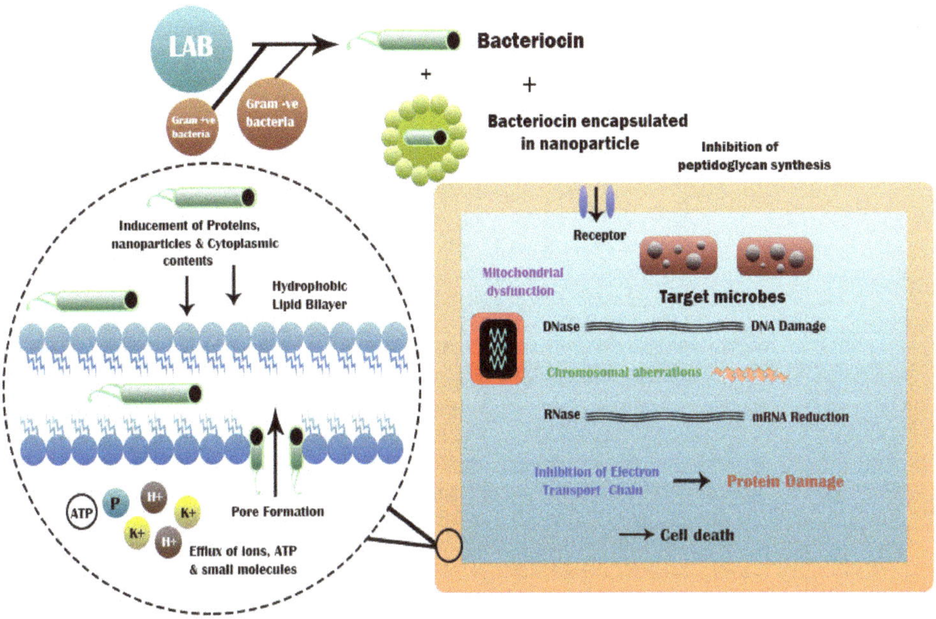

Figure 2. General method of bacteriocins synthesis, anti-bacterial activity, and action mechanism of bacteriocins [36,37].

3. Formulation of Nano-Encapsulated Bacteriocins

Nanoencapsulation is well defined as a novel technology to package constituents in minute assembly, with the usage of methods such as nano emulsification, nano structuration, and nanocomposite. It carries the final functionality of the product (containing limited core release) to maintain product quality expectations during storage [38]. Modification of nanomaterials can occur in the form of nanospheres, nanorods, nanoparticles, and nano frames and play their role in applications specifically with the means of biomedicine, electronic, solar energy conversion, environmental applications, water treatment, and some catalysis processes [39]. The principle of encapsulation includes core material or solid matrix usage to isolate bacterial cells, bioactive components, and other concerned agents from the environment. Nanoencapsulates are usually semi-permeable and spherical networks ranging in size from 10^{-9} m to 10^{-6} m [40,41].

3.1. Chitosan-Encapsulated Nisin

Nisin acts as a bacteriocin, also known as a food additive (E234) obtained from *Lactococcus lactis*, and exhibits antimicrobial properties and potential applications in food. Nisin imparts strong sporostatic and bactericidal actions to retard Gram-positive bacterial growth. Perhaps the bactericidal activity shown by nisin to inhibit Gram-negative bacterial growth is restricted due to nisin's inaccessibility to the plasma membrane [42]. A liposome is applied as a shell material by encapsulation but faces numerous disadvantages, comprising increased oxidative degradation susceptibility, increased phospholipids cost, and joined liposomes sedimentation during storage [43]. Nonionic surfactant niosomes or vesicles are better substances for shell material by encapsulation in the food area due to their low surfactant material cost compared to the more expensive liposome materials. Numerous nonionic surfactant classes such as glucosyl dialkyl ethers, polyglycerol alkyl ethers, polyoxyethylene alkyl esters, and ether and crown ether are added in niosomes preparation [44].

Nano-encapsulated nisin imparts significant actions to suppress the growth of *S. aureus* in raw milk for 24 h and pasteurized samples for 48 h as compared to free nisin, whose action cannot exceed the storage life of milk and remains 14 and 24 h, respectively [45]. Chitosan is a biodegradable, nontoxic copolymer that contains units of N-acetyle-D-glucosamine and D-glucosamine from the deacetylation of chitin in hot alkali availability. Chitosan is a nontoxic, biocompatible polymer that has the cohesive capability to act as an antimicrobial peptide and is developed as a nanoparticle-based vehicle. Chitosan also shows antimicrobial activity to retard numerous pathogenic and spoilage microorganisms from Gram-negative and Gram-positive bacteria, yeasts, and molds. The antimicrobial effect shown by chitosan is based on microorganism type, pH value, deacetylation degree, and molecular weight. Chitosan-encapsulated nisin is utilized to retard microbial growth and thus increase the shelf-life of food products [46–48].

3.2. *Liposome-Encapsulated Pediocin*

Nano-delivery systems containing carbohydrates, lipids, protein surfactants, and polymers have been modified to enhance and stabilize the biological activity of bacteriocins. For example, liposomes are comprised of spherical structures with phospholipids encompassing an aqueous medium through single or multiple bilayer membranes, and their size ranges from 10^{-9} to 10^{-6} nm. Liposomes are biodegradable and nontoxic agents appropriate for encapsulating both hydrophobic and hydrophilic substances [49]. The encapsulation of bacteriocins has been a significant development in utilizing natural antimicrobials in the area of food science. The activity of bacteriocins is influenced by various factors such as solubility changes, bacteriocins' charge, inactivation through proteases, and bacteriocins binding to food components. A liposome is potentially used to encapsulate pediocin to sustain its antimicrobial activity for a long duration. Silver nanoparticles have antimicrobial efficiency, which can be achieved and increased by encapsulation with antimicrobial agents, namely bacteriocins. Nanotechnology is a novel, fabricated, and a new platform to develop nano-structured substances possessing antimicrobial activities [50].

Pediococcus acidilactici ITV26 is potentially utilized to produce pediocin bacteriocin which is a bioconservative and exhibits antilisterial activity. Inactivation of pediocin may occur when available in foods in free form. The binding/interaction of proteolytic enzymes with a few food substances is an antimicrobial activity-affected factor. Therefore, encapsulation must be applied to protect these peptides and limit their release into liposomes. Researchers must express interest in Class II of bacteriocin, namely pediocin, because of its thermostability and strong action to prevent *Listeria monocytogenes* [51]. Liposomes contain more than one phospholipid bilayer and are known as spherical vesicles. Liposomes also possess active peptides and are recognized as bacteriocins by enclosing peptides between phospholipids bilayers and into the aqueous center due to their amphiphilic nature. Liposomal vesicles contain molecules of amphiphilic nature, and their structure can be described as molecules with a non-polar tail, a polar head, a concentric series, and two hydrophobic chains that are found on each molecule [52]. Class IIa bacteriocin contains pediocin in the gelatinous form to reduce *Listeria* infection from the consumption of hot dogs. Additionally, phosphatidylcholine nanovesicles are loaded with pediocin AcH possessing high antimicrobial activity, high efficiency for entrapment (80%), and high stability [53].

Nanoliposomes have diverse compositions and structures and are regarded as versatile tools for encapsulation technologies. Liposomes consist of spherical, closed structures and are made up of curved lipid bilayers. These lipid bilayers enclose the surrounding solvent in their interior part. Major liposomal substances are phospholipids containing amphiphilic molecules with a water-soluble section consisting of a hydrophilic head and a lipid-soluble section consisting of a hydrophobic tail which forms the composition of liposomes [54]. This phospholipid property provides distinctive benefits to liposomes, such as enclosing themselves in aqueous media and establishing a perfect carrier system. Liposomes provide applications in cosmetics, pharmaceuticals, and most importantly, food. Liposomes are extensively utilized as carrier systems based on lipids in the food area, predominantly antimicrobial formulations by liposomes. Phosphatidylcholine, lecithin, nanoliposomes; guar gum, and alginate capsules; plus guar gum and alginate hybrid capsules are combined with nanoliposomes to form nano liposomal encapsulated pediocin [55].

3.3. Solid Lipid Nano-Encapsulated Colicin

Colicins are sensitive to heat and protease enzymes with more molecular weight (30–80 kDa). They are known as bactericidal proteins produced by the strains of *E. coli* and have one colicinogenic plasmid. Colicins compounds are primarily used and studied as model systems to study the bacteriocin evolution functions and structures [56]. Synthesis of colicin is lethal for the production of cells because of lysis protein co-expression. Colicins are categorized into three major groups: degrading peptidoglycan, pore-forming, and nuclease based on the interaction mechanism with the target cell. The accomplishment of uptake of colicin due to the target cell is conducted by involved receptors in nutrients transport such as Fiu-bound iron, Tsx receptor nucleosides, cobalamin receptor BtuB vitamin B12, siderophore FepA-, FhuA-, and Cir- [57]. Colicins, lantibiocins, and enterocins are broad-spectrum bacteriocins that affect the larger bacterial genera group. Moreover, porin proteins are used by colicins to limit the passive diffusion of amino acids, phosphates, and sugars with the aid of the outer membrane [58].

Gram-negative bacteria have the potential to manufacture an extensive bacteriocins variety that is named specifically after the Klebsiella pneumoniae klebicins genus or species-producing bacteriocins such as *Serratia marcescens*-producing marcescins, *E. coli*-producing colicins, *Hafnia alvei*-producing alveicins, and *Enterobacter cloacae*-producing cloacins. Pseudomonads are generally associated with the production of pyocins [59]. Gram-negative bacteria mainly produce bacteriocins from Enterobacteriaceae. They are categorized into two major families such as colicins having more molecular mass (30–80 kDa) and microcins having less molecular mass (between 1 and 10 kDa) peptides [60]. SOS response regulon causes the mitigation in colicins production and emerges to play a key role in bacteria to DNA damage response. Microcins are hydrophobic peptides with increased stability and maximum stress production, mainly depletion of nutrients. Microcins and colicins are significantly available in *Escherichia coli*. However, many species of Gram-negative bacteria produce bacteriocin-like constituents [32]. Major bacteriocins of Gram-ve bacteria are colicins [33], pyocins [61] and microcins [62].

Colicin is a representative Gram-negative bacteriocin mainly produced by *Escherichia coli*. It is a protein with maximum molecular weight and it is used to reduce several Gram-negative bacteria. Mice-treated streptomycin is survived by *E. coli* rather than the non-colicin. The competitive benefit is gained, which permits the existence of the strain produced by bacteriocin. Multi-strain and even in multispecies, probiotics are superior and endorsed to produce bacteriocin [63,64]. Solid lipid nanoparticles (SLN) compose a suitable solid triglyceride core for the slow formulation of drug release. Colicins and nisin can be protected by SLN against degradation, increasing the anti-bacterial activity for the duration. Rather than free colicin and nisin, SLN enclosing colicin and nisin exhibited a great ability to retard *L. plantarum TISTR 850* for up to 15 days and *L. monocytogenes DMST 2871* for 20 days [65].

3.4. Nano-Encapsulated Microcins

The third type of bacteriocins produced by Gram-negative bacteria is microcins. Microcin is a bacteriocin that is synthesized by *E. coli* and has similarities with Gram-positive bacteriocins concerning thermal stability, protease resistance, and pH [66]. Microcins display dominant anti-bacterial activity and depend on subtle penetration mechanisms by the Gram-negative inner and outer bacterial membranes. Siderophore-microcins are involved in the binding of receptors to avoid the outer membrane in the transportation of iron. Cyclic microcin J25 is formed by the availability of an N-terminal macrolactam ring and utilizes the receptor of hydroxamate and intracellular protein SbmA membrane. Microcin C is synthesized as heptapeptide adenylate, requires external porins membrane and transporters of ABC membrane, and transforms into an adenylate that is a non-hydrolyzable aspartyl equivalent in the cytoplasm [67].

Microcin N (McnN) acts as antimicrobial peptides which have been examined for their capability to combat these foodstuff pathogens and selected for aquatic and human consumption. Bacteriocin McnN is synthesized by a non-pathogenic *E. coli* strain that shows activity to combat *Salmonella and E. coli* species [68]. Microcins from Class I contain Mirocin C7–C51, Microcin J25, and Microcin B. Microcin peptides from Class IIa are encoded by a plasmid including linkages of disulfide and require no modification of post-translation. Microcin peptides from Class IIb are chromosomally encoded peptides and undergo a general post-translation siderophore by C terminal modification [69]. Currently, antimicrobial peptides (AMPs) are known as host defense peptides (HDPs) and are interesting due to their potential for substitution in the novel strategies for resistance offered by bacteria to combat antibiotics in diseases and infections. However, several drawbacks are linked to antimicrobial peptides due to their lesser bioavailability, lesser solubility, and easy protease degradability that controls their antimicrobial usage. Vehicles used for AMPs delivery include polymers, micelles, nanoparticles, dendrimers, carbon nanotubes, and other system types which permit the AMPs to be used as a substitute for antibiotic treatment [70].

3.5. Nano-Encapsulated Lantibiotics

Gram-positive bacteria are categorized into four broad groups: lantibiotics, large proteins, non-modified small peptides, and cyclic peptides [71]. Gram-positive bacteriocins are classified into two major classes: lantibiotics are included in Class I while unmodified postranslationally small bacteriocins are included in Class II. Lantibiotics are known as peptides that include wide modifications of post-translational and possess methyllanthionine and/or lanthionine residues. In the previous literature, it has been elucidated that this ribosomally and postranslationally synthesized class of modified peptides (RiPPs) is considerably unusual [19].

Gram-positive bacteriocins establish an extensive anti-bacterial spectrum as compared to other bacteriocins. A peptidoglycan multilayered thick wall is generally attributed to providing a large spectrum rather than an outer membrane. Small peptide penetration is enabled by outer organization regardless of any binding of the receptor [72]. The third bacteriocins class formerly comprised bacteriolysins, currently recognized tailocins, and large anti-bacterial proteins of Gram-positive bacteria (up to 10 kDa). Lantibiotics include typical amino acids such as methylanthionine (MeLan), dehydrobutyrin (Dhb), D-alanine (D-Ala), lanthionine (Lan), and dehydroalanine (Dha) [73].

Lantibiotics are peptides that have a reduced molecular weight of approximately 5 kDa and include residues of methyllanthionine and/or lanthionine. Lantibiotics provide stability to the bacteriocin structure and resistance to protease action. Lantipeptides are classified into four classes depending on the biosynthesis specifics, in which two compounds possess anti-bacterial activity. Lantibiotics are categorized into three types depending on the structure features, such as AI, AII, and B equivalent linear bacteriocins, combined bacteriocins, and globular bacteriocins conformation [74].

AI-type lantibiotics contain microbisporicin, nisin, and epilancin 15×. Their anti-bacterial action depends on the reduction of cell wall production due to the N-terminal domain bacteriocin binding to lipid II, also known as peptidoglycan precursor. Moreover, the domain C-terminal plays a significant role in pores formation that causes potential membrane violation. Type-B lantibiotics include mersacidin, cinnamicin, and actagardin with a globular tertiary compact structure [75]. An isolated group contains two lantibiotic components that provide synergistic anti-bacterial action. The most-reported two components of lantibiotic are lactacin 3147 which includes type A1 β-peptide (LtnA2) and type B α-peptide (LtnA1). Lactacin 3147 shows anti-bacterial action and is examined in the targeted cell membrane by the formation of the pores. Lantibiotics exhibit strong activity and structural diversity in combatting Gram-positive pathogens. For instance, nisin has been utilized as an effective food preservative for the last 50 years. Numerous novel lantibiotics are presently undergoing clinical trials to examine their antimicrobial potential [76].

3.6. Nano-Encapsulated Peptides

The most interesting and stable form of bacteriocins are cyclic structure bacteriocins for practical applications. The compounds of this class are glycocins, lasso peptides, and peptide bonds with "head-to-tail" bacteriocins. Lasso peptides are named thus due to tertiary structure characteristics which are observed by the isopeptide bond formation between the N-terminal macrolactam amine ring and the aspartic acids or glutamic carboxylic acid residue at the 7, 8, or 9 positions of the C-terminal tail peptide sequence. Recently, the three lasso peptide bacteriocin structures have been synthesized by Gram-positive bacteria that have been categorized as follows: streptomonomycin from Streptomonospora alba, svicenin from Streptomyces sviceus, and lariatin A from Rhodococcus jostii [77].

The highly stable structures of lasso peptides are resistant to high temperatures and enzyme action. Properties of lasso peptides are lost during linearization; thus, it is not required to synthesize active bacteriocins chemically. At the receptors, every lasso peptide is target-specific according to the bacterial specie. As a result, streptomonomycin is a prevailing bacillus species inhibitor. It is supposed that the activity of the bacteria resulted in the interaction of the WalR protein for division and metabolism in the cell wall. Consequently, regardless of the other classes of bacteriocins that have comparable action mechanisms due to structural features, three-dimension structure of every peptide produces a definite signaling molecule [78]. Bioactive peptides by nanoencapsulation thus enhance bioavailability and defend stability during distribution, processing, and storage. The result is that consumers are presented with food with potential health benefits, and the stability of these peptides is also improved [79,80]. Studies on the formulation of nano-encapsulated bacteriocins through different nanoparticles are shown in Table 1.

Table 1. Reported Studies on the Formulation of Nano-encapsulated Bacteriocins.

LAB Bacteriocins	Purification of Bacteriocins	Nanoparticles	Applied Encapsulation Technique	Nanoencapsulation Conditions	Effective against Bacteria	Effect of Nanoformulation and Applied Technique	References
Nisin	Milk fat globule membrane (MFGM) phospholipids-based nanostructures	Rhamnolipids (RLs)	Ultrasonication-assisted self-assembly method	Sonicated for 30 min	Prevent *Escherichia coli* and *Listeria monocytogenes*	Enhanced cheese preservation to prevent the foodborne pathogens	[81]
Nisin	———	Cinnamon essential oil nanocapsules (CEO-NPs)	———	———	Retard the microbial growth and decreased lipid oxidation	Increased storage of beef slices for 15 days	[82]
Peptide nisin	Dissolution of 3 mg/mL nisin in acetic acid solution at pH 4.0 to obtain a stock solution of nisin	Nanocarrier based on polysaccharide with curcumin	Ultrafiltration tubes, Magnetic stirring	Stirring for 30 min at 25 °C with final 4.0 pH	14.00 mm on *B. subtilis* and 12.97 mm on *L. monocytogenes*	Nanocarriers were fabricated to provide multifunctional potential in the food and show powerful antimicrobial activity	[83]
Nisin	1 mg/mL nisin form Film-forming solutions (FFS) at 600 rpm by stirring at room temperature for 40 min	Nano-rhamnosomes	Field Emission Scanning Electron Microscope (FE-SEM)	10 kV accelerated voltage	Inhibit the *E. coli* and *L. monocytogenes* growth	Prolonged bioactive preservation of food by broad-spectrum antimicrobial activity to combat Gram-negative and Gram-positive foodborne pathogens	[84]
Nisin	Digestion of protein with trypsin 20 µg and incubation at 37 °C, dried for 16 h, and storage temp is −20 °C	Phosphatidylcholine liposomes	Sonication and Thin-film hydration method	Dried through thin film for 24 h in a desiccator and nisin solution in phosphate buffer (10 mm) addition at 100 µg/mL and sonicated for 3 min at 55 kHz	Liposome-encapsulated nisin decreases stresses and lowers the occurrence of *L. monocytogenes*	Liposome encapsulation might be an effective approach to prevent nisin resistance	[85]
Nisin	Nisin was dissolved in the water phase with soy oil (20 mL) and gelatin (1%)	Polyacrylate Sodium (PAAS) and polyvinyl alcohol (PVA)	Sonication, Response Surface Methodology, Electrospinning	Ultra-sonication time (15 min) and 15%, centrifugation (6000 rpm) at 4 °C for 5 min after freeze-drying for 48 h, at −50 °C	*Staphylococcus aureus* and *Escherichia coli* growth was prevented for 16 days	Nanofiber can potentially retard food microorganisms' activity in food and prolong the strawberry's shelf life	[86]
Plantaricin	Sodium sulfate method	Silver nanoparticles	———	———	Showed inhibitory activity towards *Listeria monocytogenes*	The stability period got increased from 5 days to 60 days	[87]
Bacteriocin	———	Au-zein-based nanomats	Electrospinning method	Samples were stored at (4 ± 1 °C)	~1 log CFU/g reduction of bacteria	Reduced the growth of mesophilic aerobic bacteria in skinless fish fillets	[88]
Nisin	———	Nanofibers (NP) with polyethylene (PE) packs	Electrospinning method	1.2 mL/h, 8 cm, and 20 kV	Total mesophilic bacteria from 5.03 to 3.52 log CFU/g, Lactic acid bacteria from 3.22 to 2.02 log CFU/g	Prevention of off-odor and reduction of microbial growth in rainbow trout fillets	[89]

4. Advantage of Antimicrobial Peptides at the Food Industry Scale

The potential application of nano-encapsulated antimicrobial peptides such as bacteriocins at the food industrial level needs to be encouraged. Several studies have demonstrated the possible use of lactic acid bacteria in the manufacturing of cheese. In this process, the active conversion of lactose to lactic acid produces bacteriocins to change the composition of complex cheese microflora and inhibition of pathogenic bacteria or adventitious spoilage. Particularly, the Joint Food and Agriculture Organization/World Health Organization Expert Committee on Food Additives in 1969 claimed that nisin is regarded as safe to be potentially used in food. In 1983, number E234 was allocated to this bacteriocin; it was added to the European food additive list and, finally, US Food and Drug Agency approved it to be used effectively in processed, pasteurized cheese spreads in 1988. Bacteriocins that are produced from lactic acid bacteria are generally utilized to preserve food and inhibit the spoilage and pathogenic microbes in food products and provide significant antimicrobial dimensions [90].

Enterocin AS-48 treatment deactivates *L. monocytogenes* cells inoculated on the slices and surface fruit and completely or partially disables *S. Aureus* in several sauces of fruit and vegetable [91]. A wild strain of *Streptococcus thermophilus ACA-DC 0001* was isolated from traditional products such as Greek yogurt used to produce Thermophilin ST-1 [92]. Thermophilin ST-1 has an inhibitory impact on numerous foodborne pathogens, food spoilage microorganisms, lactic acid bacteria, and on a few phytopathogens of Gram-negative bacteria including the following: *Enterococcus faecalis EF1, Xanthomonas campestris BPIC 1660, Listeria innocua BL 86/20, Erwinia rubrifasciens BPIC 1710, Pseudomonas syringae BPIC 1549* and *Staphylococcus aureus ATCC 29996*. Antimicrobial substances are sensitive to increased alkaline and acidic conditions, to proteolytic enzymes, specifically trypsin and pronase, heat-labile at 60 °C for 10 min, and exhibit a bactericidal action mode to prevent the *Lactococcus lactis ssp. cremoris CNRZ-117* indicator strain [92].

Lactacin has an active antimicrobial potential that can be utilized in food products such as powdered soup, baby milk, cottage cheese, and yogurt [57]. It has been reported that lactacin is non-immunogenic and nontoxic and mitigates the occurrence of infection. Lactacin is produced by lactic acid bacteria (*Lactococcus lactis 3147*), which is utilized significantly to prevent the development of several types of Gram-positive microorganisms such as *Pediococcus acidilactici, Clostridium botulinum, Listeria monocytogenes, Clostridium sporogenes, Staphylococcus aureus, Enterococcus faecalis,* and *Bacillus* spp. [93,94].

Nanoencapsulation of bacteriocins is the successful development through regulatory approval from initial biological experimentation and observation to commercial application. In recent years, it has stimulated a new approach in bacteriocins research for the industrial model and market innovations. It is a fact that without understanding the nature and mode of action, bacteriocins can be problematic in food. To reduce these disadvantages, nanoencapsulation can be an effective approach in various important commercial and imaginative applications. The core matrix and structurally enhanced nanocapsule probably have the potential in food applications, especially to encapsulate the antimicrobial peptides that are produced by food-grade LAB. Most importantly, it is more likely to meet regulatory approval specifically to their origin for their introduction into fermented foods without any purification or concentration [4].

5. Conclusions

This review highlighted the integral role, features, action mechanism and nanoencapsulation method of bacteriocins produced by lactic acid bacteria. Nowadays, food spoilage is a major concern in the food industry. The trend of applying natural and chemical-free preservatives has been increasing due to consumer concerns regarding the side effects of artificial preservatives as well as due to their chronic health effects. Bacteriocins are more efficient, anti-bacterial agents with fewer adverse effects. However, challenges and limitations are associated with bacteriocins usage as anti-bacterial agents or bio-preservatives in the food industry. Currently, available bacteriocin peptides have proven their efficacy, and various examples have been provided to ensure their commercial applications in the food sector. For example, nanoencapsulation acts as a suitable strategy to preserve biological activities and enhance the stability of nanoparticles when successfully applied to a food product. The efficient technique of nanoencapsulation uses nanomaterials that enhance the antimicrobial potential of bacteriocin. Moreover, these nanomaterials interact with bacteriocins to develop nano-formulated bacteriocins as well as establish the mechanism of action against targeted microorganisms which have been explained.

6. Future Recommendations

Although nano-encapsulated bacteriocins have been applied in the food industry efficiently, additional research effort is needed to encapsulate nutraceuticals and to explore them as nutraceutical carriers [49]. It is of great importance to conduct more studies on nano-encapsulated enterocin, lactacin, thermophilin, lacidin, sakacin, and bulgaricin [6]. Novel research approach is required to produce bacteriocins from genetically modified organisms and to develop suitable conditions for their application through nanoencapsulation methods [95]. Further studies are also needed to evaluate the in vivo efficiency and safety of peptides. Identifying the gene of new recombinant bacteriocins has become important by using polymerase chain reaction techniques to determine their applicability in food and medicine [36]. The bioavailability and stability of bacteriocins throughout the food chain during production, processing, distribution and storage is significant through nanoencapsulation method to modify the food with health benefits.

Funding: This research was funded, in part, by Grants (project Number NC.X337-5-21-170-1 and NC.X341-5-21-170-1) in addition to the 1890 capacity building project from the National Institute of Food and Agriculture (NIFA).

Acknowledgments: The authors would also like to acknowledge the support of the Agricultural Research Station at North Carolina Agricultural and Technical State University (Greensboro, NC 27411, USA). This research was funded, in part, by Grants (project Number NC.X337-5-21-170-1 and NC.X341-5-21-170-1) in addition to the 1890 capacity building project from the National Institute of Food and Agriculture (NIFA). Its contents are solely the responsibility of the authors and do not necessarily represent the official views of NIFA.

Conflicts of Interest: Authors declare no conflict of interest.

References

1. García, P.; Rodríguez, L.; Rodríguez, A.; Martínez, B. Food biopreservation: Promising strategies using bacteriocins, bacteriophages and endolysins. *Trends Food Sci. Technol.* **2010**, *21*, 373–382. [CrossRef]
2. Chowdhuri, A.; Iqbal, A.; Giasuddin, M.; Bhuiyan, A.A. Study on Isolation and Identification of *Salmonella* and *Escherichia coli* from Different Poultry Feeds of Savar Region of Dhaka, Bangladesh. *J. Sci. Res.* **2011**, *3*, 403–411. [CrossRef]
3. Yi, Y.; Li, P.; Zhao, F.; Zhang, T.; Shan, Y.; Wang, X.; Liu, B.; Chen, Y.; Zhao, X.; Lü, X. Current status and potentiality of class II bacteriocins from lactic acid bacteria: Structure, mode of action and applications in the food industry. *Trends Food Sci. Technol.* **2022**, *120*, 387–401. [CrossRef]
4. Cotter, P.D.; Hill, C.; Ross, R.P. Bacteriocins: Developing Innate Immunity For Food. *Nat. Rev. Microbiol.* **2005**, *3*, 777–788. [CrossRef]
5. Fidan, H.; Esatbeyoglu, T.; Simat, V.; Trif, M.; Tabanelli, G.; Kostka, T.; Montanari, C.; Ibrahim, S.A.; Özogul, F. Recent developments of lactic acid bacteria and their metabolites on foodborne pathogens and spoilage bacteria: Facts and gaps. *Food Biosci.* **2022**, *47*, 101741. [CrossRef]
6. Papagianni, M. Ribosomally synthesized peptides with antimicrobial properties: Biosynthesis, structure, function, and applications. *Biotechnol. Adv.* **2003**, *21*, 465–499. [CrossRef]
7. De Vuyst, L.; Leroy, F. Bacteriocins from Lactic Acid Bacteria: Production, Purification, and Food Applications Bacteriocins from Lactic Acid Bacteria: Production, Purification, and Food. *J. Mol. Microbiol. Biotechnol.* **2007**, *13*, 194–199. [CrossRef]
8. Cui, Y.; Luo, L.; Yi, X.; Lu, Y.; Yanglei, Y.; Liu, B.; Zhou, Y.; Lü, X. Mining, heterologous expression, purification, antibactericidal mechanism, and application of bacteriocins: A review. *Compr. Rev. Food Sci. Food Saf.* **2020**, *20*, 863–899. [CrossRef] [PubMed]
9. Al-holy, M.A.; Al-nabulsi, A.; Osaili, T.M.; Ayyash, M.M.; Shaker, R.R. Inactivation of *Listeria innocua* in brined white cheese by a combination of nisin and heat. *Food Control* **2012**, *23*, 48–53. [CrossRef]

10. Pimentel-filho, N.D.J.; Mantovani, C.; De Carvalho, F.; Dias, R.S.; Vanetti, M.C.D. Original article Efficacy of bovicin HC5 and nisin combination against *Listeria monocytogenes* and *Staphylococcus aureus* in fresh cheese. *Int. J. Food Sci. Technol.* **2014**, *49*, 416–422. [CrossRef]
11. Pei, J.; Yue, T.; Yuan, Y.; Dai, L. Activity of paracin C from lactic acid bacteria against *Alicyclobacillus* in apple juice: Application of a novelty bacteriocin. *J. Food Safety* **2017**, *37*, e12350. [CrossRef]
12. Ribeiro, S.C.; Ross, R.P.; Stanton, C. Characterization and Application of Antilisterial Enterocins on Model Fresh Cheese. *J. Food Prot.* **2017**, *80*, 1303–1316. [CrossRef] [PubMed]
13. Woraprayote, W.; Pumpuang, L.; Tosukhowong, A.; Roytrakul, S.; Perez, R.H.; Zendo, T.; Sonomoto, K.; Benjakul, S.; Visessanguan, W. Two putatively novel bacteriocins active against Gram-negative food borne pathogens produced by *Weissella hellenica* BCC 7293. *Food Control* **2015**, *55*, 176–184. [CrossRef]
14. Lu, Y.; Aizhan, R.; Yan, H.; Li, X.; Wang, X.; Yi, Y.; Shan, Y. Characterization, modes of action, and application of a novel broad-spectrum bacteriocin BM1300 produced by *Lactobacillus crustorum* MN047. *Braz. J. Food Microbiol.* **2020**, *51*, 2023–2048. [CrossRef] [PubMed]
15. Perez, R.H.; Zendo, T.; Sonomoto, K. Novel bacteriocins from lactic acid bacteria (LAB): Various structures and applications. *Microb. Cell Fact.* **2014**, *13*, S3. [CrossRef] [PubMed]
16. Lopes, N.A.; Brandelli, A. Nanostructures for delivery of natural antimicrobials in food. *Crit. Rev. Food Sci. Nutr.* **2018**, *58*, 2202–2212. [CrossRef]
17. Chandrakasan, G.; Rodríguez-Hernández, A.I.; del Rocío López-Cuellar, M.; Palma-Rodríguez, H.M.; Chavarría-Hernández, N. Bacteriocin encapsulation for food and pharmaceutical applications: Advances in the past 20 years. *Biotechnol. Lett.* **2019**, *41*, 453–469. [CrossRef]
18. Fahim, H.A.; Khairalla, A.S.; El-Gendy, A.O. Nanotechnology: A valuable strategy to improve bacteriocin formulations. *Front. Microbiol.* **2016**, *7*, 1385. [CrossRef]
19. Ge, J.; Kang, J.; Ping, W. Effect of Acetic Acid on Bacteriocin Production by Gram-Positive. *J. Microbiol. Biotechnol.* **2019**, *29*, 1341–1348. [CrossRef]
20. Chikindas, M.L.; Weeks, R.; Drider, D.; Chistyakov, V.A.; Dicks, L.M.T. Functions and emerging applications of bacteriocins. *Curr. Opin. Biotechnol.* **2018**, *49*, 23–28. [CrossRef]
21. Abdulkarim, I.H.; Mohammed, S.S.D.; Orukotan, A.A. Gene Identification for Bacteriocin Production by Lactic Acid Bacteria Isolated from Selected Fermented Foods. *Asian J. Biochem. Mol. Biol.* **2020**, *3*, 1–12. [CrossRef]
22. Brandelli, A. Nanostructures as Promising Tools for Delivery of Antimicrobial Peptides. *Mini-Rev. Med. Chem.* **2012**, *12*, 731–741. [CrossRef] [PubMed]
23. Naskar, A.; Kim, K. Potential Novel Food-Related and Biomedical Applications of Nanomaterials Combined with Bacteriocins. *Pharmaceutics* **2021**, *13*, 86. [CrossRef] [PubMed]
24. Lelis, C.A.; De Carvalho, A.P.A.; Conte Junior, C.A. Systematic Review on Nanoencapsulation Natural Antimicrobials in Foods: In Vitro versus In Situ Evaluation, Mechanisms of Action and Implications on Physical-Chemical Quality. *Int. J. Mol. Sci.* **2021**, *22*, 12055. [CrossRef] [PubMed]
25. Yousefi, M.; Narmani, A.; Jafari, S.M. Dendrimers as efficient nanocarriers for the protection and delivery of bioactive phytochemicals. *Adv. Colloid Interface Sci.* **2020**, *78*, 102125. [CrossRef]
26. Cintas, L.M.; Casaus, M.P.; Herranz, C.; Nes, I.F.; Hernández, P.E. Review: Bacteriocins of Lactic Acid Bacteria. *Food Sci. Technol. Int.* **2001**, *7*, 281–305. [CrossRef]
27. Savadogo, A.; Ouattara, A.C.; Bassole, H.I.; Traore, S.A. Bacteriocins and lactic acid bacteria—A minireview. *Afr. J. Biotechnol.* **2006**, *5*, 678–683.
28. Todorov, S.D.; Chikindas, M.L. Lactic Acid Bacteria Bacteriocins and their Impact on Human Health. In *Lactic Acid Bacteria;* CRC Press: Boca Raton, FL, USA, 2020; ISBN 9780429422591.
29. Es, B.; Bhat, S.G. Marine Bacteriocins: A Review Marine Bacteriocins: A Review. *J. Bacteriol. Mycol.* **2016**, *2*, 00040. [CrossRef]
30. Muñoz, M.; Jaramillo, D.; Melendez, P.; Alméciga-diaz, C.J. Native and Heterologous Production of Bacteriocins from Gram-Positive Microorganisms. *Recent Pat. Biotechnol.* **2011**, *5*, 199–211. [CrossRef]
31. Atanaskovic, I.; Kleanthous, C. Tools and Approaches for Dissecting Protein Bacteriocin Import in Gram-Negative Bacteria. *Front. Microbiol.* **2019**, *10*, 646. [CrossRef] [PubMed]
32. Rebuffat, S. Bacteriocins from Gram-Negative Bacteria: A Classification? In *Prokaryotic Antimicrobial Peptides;* Springer: New York, NY, USA, 2011; pp. 55–72. [CrossRef]
33. Cascales, E.; Buchanan, S.K.; Duche, D.; Kleanthous, C.; Lloube, R.; Postle, K.; Riley, M.; Slatin, S. Colicin Biology. *Microbiol. Mol. Biol. Rev.* **2007**, *71*, 158–229. [CrossRef]
34. Sylvie, D.; Beeckman, A.; Vanrompay, D.C.G. Bacterial Secretion Systems with an Emphasis on the Chlamydial Type III Secretion System. *Curr. Issues Mol. Biol.* **2005**, *12*, 17–42.
35. Gillor, O.; Kirkup, B.C.; Riley, M.A. Colicins and Microcins: The Next Generation Antimicrobials. *Adv. Appl. Microbiol.* **2004**, *54*, 129–146. [CrossRef]
36. Yang, S.; Lin, C.; Sung, C.T.; Fang, J. Antibacterial activities of bacteriocins: Application in foods and pharmaceuticals. *Front. Microbiol.* **2014**, *5*, 241. [CrossRef]
37. Bharti, V.; Singh, S.; Jain, N.; Ahirwal, L. Bacteriocin: A novel approach for preservation of food. *Int. J. Pharm. Pharm. Sci.* **2015**, *7*, 20–29.

38. Hildeliza, Q.B.; Chanona-pe, J.; Jose, L.S.M.; Gutie, G.F.; Jimene, A. Nanoencapsulation: A New Trend in Food Engineering Processing. *Food Eng. Rev.* **2010**, *2*, 39–50. [CrossRef]
39. Divya, K.; Jisha, M.S. Chitosan nanoparticles preparation and applications. *Environ. Chem. Lett.* **2018**, *16*, 101–112. [CrossRef]
40. Kailasapathy, K. Microencapsulation of Probiotic Bacteria: Technology and Potential Applications Further Reading. *Curr. Issues Intest. Microbiol.* **2002**, *3*, 39–48. [PubMed]
41. Namasivayam, S.K.R.; Samrat, K.; Debnath, S.; Jayaprakash, C. Biocompatible chitosan nanoparticles incorporated bacteriocin (CSNps-B) preparation for the controlled release and improved anti-bacterial activity against food borne pathogenic bacteria Listeria monocytogenes. *Res. J. Pharm. Biol. Chem. Sci.* **2015**, *6*, 625–631.
42. Juncioni, L.; Arauz, D.; Faustino, A.; Vessoni, T.C. Nisin biotechnological production and application: A review. *Trends Food Sci. Technol.* **2009**, *20*, 146–154. [CrossRef]
43. Hsu, S.D.; Breukink, E.; Tischenko, E.; Lutters, M.A.G.; De Kruijff, B.; Kaptein, R.; Bonvin, A.M.J.J.; Van Nuland, N.A.J. The nisin—Lipid II complex reveals a pyrophosphate cage that provides a blueprint for novel antibiotics. *Nat. Struct. Mol. Biol.* **2004**, *11*, 963–967. [CrossRef] [PubMed]
44. Sahin, N. Niosomes as nanocarrier systems. In *Nanomaterials and Nanosystems for Biomedical Applications*; Springer Science & Business Media: Berlin/Heidelberg, Germany, 2007; ISBN 9781402062896.
45. A Eldaly, E.; Mahmoud, F.A.; Abobakr, H.M. Preservative effect of chitosan coating on shelf life and sensory properties of chicken fillets during chilled storage. *J. Nutr. Food Secur.* **2018**, *3*, 139–148.
46. da Silva Malheiros, P.; Daroit, D.J.; Brandelli, A. Food applications of liposome-encapsulated antimicrobial peptides. *Trends Food Sci. Technol.* **2010**, *21*, 284–292. [CrossRef]
47. Bahrami, A.; Delshadi, R.; Jafari, S.M.; Williams, L. Nanoencapsulated nisin: An engineered natural antimicrobial system for the food industry. *Trends Food Sci. Technol.* **2019**, *94*, 20–31. [CrossRef]
48. Riesute, R.; Salomskiene, J.; Moreno, D.S.; Gustiene, S. Effect of yeasts on food quality and safety and possibilities of their inhibition. *Trends Food Sci. Technol.* **2021**, *108*, 1–10. [CrossRef]
49. Aditya, N.P.; Espinosa, Y.G.; Norton, I.T. Encapsulation systems for the delivery of hydrophilic nutraceuticals: Food application. *Biotechnol. Adv.* **2017**, *35*, 450–457. [CrossRef] [PubMed]
50. Campos, C.A.; Castro, M.P.; Rivas, F.P.; Schelegueda, L.I. Bacteriocins in food: Evaluation of the factors affecting their effectiveness. In *Microbial Pathogens and Strategies for Combating Them: Science, Technology and Education*; Formatex Research Center: Badajoz, Spain, 2013.
51. Kemperman, R.; Kuipers, A.; Karsens, H.; Nauta, A.; Kuipers, O.; Kok, J. Identification and characterization of two novel clostridial bacteriocins, circularin A and closticin 574. *Appl. Environ. Microbiol.* **2003**, *69*, 1589–1597. [CrossRef]
52. Laridi, R.; Kheadr, E.E.; Benech, R.O.; Vuillemard, J.C.; Lacroix, C.; Fliss, I. Liposome encapsulated nisin Z: Optimization, stability and release during milk fermentation. *Int. Dairy J.* **2003**, *13*, 325–336. [CrossRef]
53. de Mello, M.B.; da Silva Malheiros, P.; Brandelli, A.; Pesce da Silveira, N.; Jantzen, M.M.; de Souza da Motta, A. Characterization and antilisterial effect of phosphatidylcholine nanovesicles containing the antimicrobial peptide pediocin. *Probiotics Antimicrob. Proteins* **2013**, *5*, 43–50. [CrossRef] [PubMed]
54. Bouaraba, L.; Maherania, B.; Kheirolomoomb, A.; Hasana, M.; Aliakbarianc, B.; Lindera, M.; Arab-Tehrany, E. Influence of lecithin—Lipid composition on physico-chemical properties of nanoliposomes loaded with a hydrophobic molecule. *Colloids Surf. B Biointerfaces* **2014**, *115*, 197–204. [CrossRef] [PubMed]
55. Narsaiah, K.; Robin, S.N.J.; Whilson, A.; Mandge, H.; Manikantan, M.R.; Malik, R.K.; Vij, S. Pediocin-Loaded Nanoliposomes and Hybrid Alginate—Nanoliposome Delivery Systems for Slow Release of Pediocin. *J. Biosci.* **2012**, *3*, 37–42. [CrossRef]
56. Acedo, J.Z.; Chiorean, S.; Vederas, J.C.; van Belkum, M.J. The expanding structural variety among bacteriocins from Gram-positive bacteria Jeella. *FEMS Microbiol. Rev.* **2018**, *42*, 805–828. [CrossRef] [PubMed]
57. Alvarez-sieiro, P.; Montalbán-lópez, M.; Mu, D.; Kuipers, O.P. Bacteriocins of lactic acid bacteria: Extending the family. *Appl. Microbiol. Biotechnol.* **2016**, *100*, 2939–2951. [CrossRef] [PubMed]
58. Kalia, A.; Parshad, V.R. Novel trends to revolutionize preservation and packaging of fruits/fruit products: Microbiological and nanotechnological perspectives. *Crit. Rev. Food Sci. Nutr.* **2013**, *55*, 159–182, ISBN 9116124030. [CrossRef] [PubMed]
59. Brown, C.; Smith, K.; Brown, C.L.; Smith, K.; Mccaughey, L.; Walker, D. Colicin-like bacteriocins as novel therapeutic agents for the treatment of chronic biofilm-mediated infection. *Prog. Chall. Bacteriocins Res.* **2012**, *40*, 1549–1552. [CrossRef]
60. Chavan, M.A.; Riley, M.A. *Molecular Evolution of Bacteriocins in Gram-Negative Bacteria*; Springer: Berlin/Heidelberg, Germany, 2007; pp. 19–43.
61. Saha, S. A Tale of Two Tails: Characterization of R-Type and F-Type Pyocins of *Pseudomonas aeruginosa*. Ph.D. Thesis, University of Toronto, Toronto, ON, Canada, 2016.
62. Duquesne, S.; Destoumieux-Garzón, D.; Peduzzi, J.; Rebuffat, S. Microcins, gene-encoded antibacterial peptides from enterobacteria. *Nat. Prod. Rep.* **2007**, *24*, 708–734. [CrossRef] [PubMed]
63. Ponnachan, B.; Dash, N.; Jaybhay, P.; Tiwari, R. Reviewing isolation and production of probiotics. *Int. J. Med. Biomed. Stud.* **2021**, *5*. [CrossRef]
64. Mousavi Khaneghah, A.; Abhari, K.; Eş, I.; Soares, M.B.; Oliveira, R.B.A.; Hosseini, H.; Rezaei, M.; Balthazar, C.F.; Silva, R.; Cruz, A.G.; et al. Interactions between probiotics and pathogenic microorganisms in hosts and foods: A review. *Trends Food Sci. Technol.* **2020**, *95*, 205–218. [CrossRef]

65. Feng, L.; Mumper, R.J. A critical review of lipid-based nanoparticles for taxane delivery. *Cancer Lett.* **2012**, *334*, 157–175. [CrossRef]
66. Micenková, L.; Bosák, J.; Kucera, J.; Hrala, M.; Dolejšová, T. Colicin Z, a structurally and functionally novel colicin type that selectively kills enteroinvasive Escherichia coli and Shigella strains. *Nat. Res.* **2019**, *9*, 11127. [CrossRef]
67. Crone, W.J.K.; Vior, N.M.; Santos-aberturas, J.; Schmitz, L.G.; Leeper, F.J.; Truman, A.W. Zuschriften Dissecting Bottromycin Biosynthesis Using Comparative Untargeted Metabolomics Zuschriften. *Angew. Chem.* **2016**, *128*, 9791–9795. [CrossRef]
68. Gomaa, A.I.; Martinent, C.; Hammami, R.; Fliss, I. Dual Coating of Liposomes as Encapsulating Matrix of Antimicrobial Peptides: Development and Characterization. *Front. Chem.* **2017**, *5*, 103. [CrossRef]
69. Satchwell, K.L. Application of Microcin N and Tridecaptin A 1 to Control Bacterial Pathogens. Master's Thesis, University of Alberta, Edmonton, AB, USA, 2016.
70. Martin-serrano, Á.; Rafael, G.; Ortega, P.; De Mata, F.J. Nanosystems as Vehicles for the Delivery of Antimicrobial Peptides (AMPs). *Pharmaceutics* **2019**, *11*, 448. [CrossRef]
71. Heng, N.C.; Wescombe, P.A.; Burton, J.P.; Jack, R.W.; Tagg, J.R. The diversity of bacteriocins in Gram-positive bacteria. In *Bacteriocins*; Springer: Berlin/Heidelberg, Germany, 2007; ISBN 9783540366034.
72. Ansari, A. Bacteriocin from LAB for Medical and Health Applications. In *Beneficial Microorganisms in Medical and Health Applications*; Springer: Cham, Swizerland, 2015; pp. 199–221. [CrossRef]
73. Wencewicz, T.A.; Miller, M.J. Sideromycins as Pathogen-Targeted Antibiotics. In *Antibacterials*; Springer: Berlin/Heidelberg, Germany, 2017; pp. 151–183. [CrossRef]
74. Ekblad, B.; Kyriakou, P.K.; Oppega, C.; Nissen-meyer, J.; Kaznessis, Y.N. Structure—Function Analysis of the Two-Peptide Bacteriocin Plantaricin EF. *Biochemistry* **2016**, *55*, 5106–5116. [CrossRef] [PubMed]
75. Dias, N.; Augusta, I.M. A Review "Clean Labeling": Applications of Natural Ingredients in Bakery Products. *J. Food Nutr. Res.* **2018**, *6*, 285–294. [CrossRef]
76. Ovchinnikov, K.V.; Kristiansen, P.E.; Straume, D.; Jensen, M.S.; Aleksandrzak-piekarczyk, T.; Nes, I.F.; Diep, D.B.; Vederas, J.C. The Leaderless Bacteriocin Enterocin K1 Is Highly Potent against Enterococcus faecium: A Study on Structure, Target Spectrum and Receptor. *Front. Microbiol.* **2017**, *8*, 774. [CrossRef] [PubMed]
77. Yadav, S.C.; Kumari, A.; Yadav, R. Development of peptide and protein nanotherapeutics by nanoencapsulation and nanobioconjugation. *Peptides* **2011**, *32*, 173–187. [CrossRef] [PubMed]
78. Zimina, M.; Babich, O.; Prosekov, A.; Sukhikh, S.; Ivanova, S.; Shevchenko, M.; Noskova, S. Overview of Global Trends in Classification, Methods of Preparation and Application of Bacteriocins. *MDPI Antibiot.* **2020**, *9*, 553. [CrossRef] [PubMed]
79. Cabuk, B.; Okulklu, B.; Stanciuc, N.; Harsa, S.T. Nanoencapsulation of Biologically Active Peptides from Whey Proteins. *J. Nutr. Health Food Sci.* **2017**, *2*, 1–4.
80. Mosquera, M.; Giménez, B.; Mallmann, I.; Ferreira, J.; Montero, P.; Gómez-guillén, M.C.; Brandelli, A. Nanoencapsulation of an active peptidic fraction from sea bream scales collagen. *Food Chem.* **2014**, *156*, 144–150. [CrossRef]
81. Khalid, A.S.; Niaz, T.; Zarif, B.; Shabbir, S.; Noor, T.; Shahid, R.; Imran, M. Milk phospholipids-based nanostructures functionalized with rhamnolipids and bacteriocin: Intrinsic and synergistic antimicrobial activity for cheese preservation. *Food Biosci.* **2022**, *47*, 101442. [CrossRef]
82. Zhang, M.; Luo, W.; Yang, K.; Li, C. Effects of sodium alginate edible coating with cinnamon essential oil nanocapsules and Nisin on quality and shelf life of beef slices during refrigeration. *J. Food Prot.* **2022**, *85*, 896–905. [CrossRef] [PubMed]
83. Luo, L.; Wu, Y.; Liu, C.; Zou, Y.; Huang, L.; Liang, Y.; Ren, J.; Liu, Y.; Lin, Q. Elaboration and characterization of curcumin-loaded soy soluble polysaccharide (SSPS)-based nanocarriers mediated by antimicrobial peptide nisin. *Food Chem.* **2021**, *336*, 127669. [CrossRef] [PubMed]
84. Niaz, T.; Imran, M. Diffusion kinetics of nisin from composite coatings reinforced with nano-rhamnosomes. *J. Food Eng.* **2021**, *288*, 110143. [CrossRef]
85. Pinilla, C.M.B.; Stincone, P.; Brandelli, A. Proteomic analysis reveals differential responses of Listeria monocytogenes to free and nanoencapsulated nisin. *Int. J. Food Microbiol.* **2021**, *346*, 109170. [CrossRef] [PubMed]
86. Jiang, Y.; Ma, D.; Ji, T.; Sameen, D.E.; Ahmed, S.; Li, S. Long-Term Antibacterial E ff ect of Electrospun Polyvinyl Alcohol / Polyacrylate Sodium Nanofiber Containing Nisin-Loaded Nanoparticles. *Nanomaterials* **2020**, *10*, 1803. [CrossRef] [PubMed]
87. Amer, S.A.; Abushady, H.M.; Refay, R.M.; Mailam, M.A. Enhancement of the antibacterial potential of plantaricin by incorporation into silver nanoparticles. *J. Genet. Eng. Biotechnol.* **2021**, *19*, 13. [CrossRef]
88. Çetinkaya, T.; Ceylan, Z.; Meral, R.; Kılıçer, A.; Altay, F. A novel strategy for Au in food science: Nanoformulation in dielectric, sensory properties, and microbiological quality of fish meat. *Food Biosci.* **2021**, *41*, 101024. [CrossRef]
89. Oner, B.; Meral, R.; Ceylan, Z. Determination of some quality indices of rainbow trout fillets treated with nisin-loaded polyvinylalcohol-based nanofiber and packed with polyethylene package. *LWT* **2021**, *149*, 111854. [CrossRef]
90. Akbar, A.; Sadiq, M.B.; Ali, I.; Anwar, M.; Muhammad, J.; Shafee, M.; Ullah, S.; Qasim, S.; Ahmad, S.; Anal, A.K.; et al. Lactococcus lactis subsp. lactis isolated from fermented milk products and its antimicrobial potential. *CyTA-J. Food* **2019**, *17*, 214–220. [CrossRef]
91. José, M.; Burgos, G.; Pulido, R.P.; López, C. The Cyclic Antibacterial Peptide Enterocin AS-48: Isolation, Mode of Action, and Possible Food Applications. *Int. J. Mol. Sci.* **2014**, *15*, 22706–22727. [CrossRef]
92. Aktypis, A.; Kalantzopoulos, G. Purification and characterization of thermophilin ST-1, a novel bacteriocin produced by Streptococcus thermophilus ACA-DC 0001. *Le Lait* **2003**, *83*, 365–378. [CrossRef]

93. Kotenkova, E.; Ushakova, N.; Fedosov, E. The effect of lactating and some gastrointenstinal symbiotic bacteria on the growth of young rabbits in prepubertal period. *Muz. Olten. Craiova Oltenia. Stud. Şi Comunicări. Ştiinţele Nat.* **2010**, *26*, 1091–1096.
94. Andryukov, B.; Mikhailov, V.; Besednova, N. The Biotechnological Potential of Secondary Metabolites from Marine Bacteria. *J. Mar. Sci. Eng.* **2019**, *7*, 176. [CrossRef]
95. Balciunas, E.M.; Martinez, F.A.C.; Todorov, S.D.; de Melo Franco, B.D.G.; Converti, A.; de Souza Oliveira, R.P. Novel biotechnological applications of bacteriocins: A review. *Food Control* **2013**, *32*, 134–142. [CrossRef]

microorganisms

MDPI

Article

The Antimicrobial Effect of Various Single-Strain and Multi-Strain Probiotics, Dietary Supplements or Other Beneficial Microbes against Common Clinical Wound Pathogens

Sabina Fijan [1,*], Primož Kocbek [1], Andrej Steyer [2], Polona Maver Vodičar [3] and Maja Strauss [1]

[1] Faculty of Health Sciences, University of Maribor, Žitna ulica 15, 2000 Maribor, Slovenia
[2] Environment and Food, Division of Public Health Microbiology, National Laboratory of Health, Grablovičeva 44, 1000 Ljubljana, Slovenia
[3] Institute of Microbiology and Immunology, Faculty of Medicine, University of Ljubljana, Zaloška 4, 1000 Ljubljana, Slovenia
* Correspondence: sabina.fijan@um.si

Abstract: The skin is the largest organ in the human body and is colonized by a diverse microbiota that works in harmony to protect the skin. However, when skin damage occurs, the skin microbiota is also disrupted, and pathogens can invade the wound and cause infection. Probiotics or other beneficial microbes and their metabolites are one possible alternative treatment for combating skin pathogens via their antimicrobial effectiveness. The objective of our study was to evaluate the antimicrobial effect of seven multi-strain dietary supplements and eleven single-strain microbes that contain probiotics against 15 clinical wound pathogens using the agar spot assay, co-culturing assay, and agar well diffusion assay. We also conducted genera-specific and species-specific molecular methods to detect the DNA in the dietary supplements and single-strain beneficial microbes. We found that the multi-strain dietary supplements exhibited a statistically significant higher antagonistic effect against the challenge wound pathogens than the single-strain microbes and that lactobacilli-containing dietary supplements and single-strain microbes were significantly more efficient than the selected propionibacteria and bacilli. Differences in results between methods were also observed, possibly due to different mechanisms of action. Individual pathogens were susceptible to different dietary supplements or single-strain microbes. Perhaps an individual approach such as a 'probiogram' could be a possibility in the future as a method to find the most efficient targeted probiotic strains, cell-free supernatants, or neutralized cell-free supernatants that have the highest antagonistic effect against individual clinical wound pathogens.

Keywords: probiotics; beneficial microbes; wound pathogens; skin pathogens; agar spot; co-culturing; agar well diffusion; molecular methods; PCR

Citation: Fijan, S.; Kocbek, P.; Steyer, A.; Vodičar, P.M.; Strauss, M. The Antimicrobial Effect of Various Single-Strain and Multi-Strain Probiotics, Dietary Supplements or Other Beneficial Microbes against Common Clinical Wound Pathogens. *Microorganisms* **2022**, *10*, 2518. https://doi.org/10.3390/microorganisms10122518

Academic Editor: Alex Galanis

Received: 28 October 2022
Accepted: 13 December 2022
Published: 19 December 2022

1. Introduction

The skin is the largest organ in the human body and is colonized by diverse microbiota. Most of these microbes are harmless or even beneficial and serve as physical barriers, protecting our bodies from potential assaults by foreign organisms or toxic substances. The skin, therefore, prevents disruption of this balance caused by the invasion of pathogens due to skin damage because of illness, surgery, and burns [1,2]. Skin damage can be caused by a variety of different reasons such as trauma (including cuts, abrasions, chemical burns, fire burns, cold, heat, radiation, and surgery), or as a consequence of underlying illnesses such as diabetes [3]. The most common wound pathogens include biofilm-forming bacteria such as *Staphylococcus aureus*, *Pseudomonas aeruginosa*, *Enterococcus faecalis*, *Acinetobacter baumannii*, *Escherichia coli*, *Klebsiella pneumoniae*, *Enterobacter* spp., *Bacteroides* spp., *Peptostreptococcus* spp. [4–7]. Especially chronic wounds are a prominent health concern as they represent an important cause of morbidity and mortality and can significantly reduce the quality of life of patients due to delayed healing, inflammation process, and excessive scarring. They also result in enormous healthcare expenditures [6,8–10]. Wound debridement and the topical application of antibiotics or other antimicrobial substances are the conventional methods usually considered to eradicate wound infection. The main disadvantage of recurrent antibiotics used in the context of delayed wound healing and frequent hospitalizations is exacerbated by the rising risk of therapeutic resistance [3].

Probiotics that are by definition "live microorganisms that, when administered in adequate amounts, confer a health effect on the host" [11] aid in skin healing by stimulating the production of immune cells. They also exhibit antagonistic effects against pathogens via the competitive exclusion of pathogens [3,9,12]. Interestingly enough, the Organization for Economic Cooperation and Development (OECD) also states that probiotics are a promising alternative therapy to the topical use of antibiotics due to the increasing occurrence and transmission of antibiotic-resistant microorganisms [13]. A recent review found that exogenous and oral application of probiotics has shown a reduction in wound infections, especially when used as an adjuvant to antibiotic therapy, and therefore the potential use of probiotics in this field remains worthy of further studies, perhaps focused more on typical skin inhabitants as next-generation probiotics with high potential [9]. On the other hand, using postbiotics could be a safer adjuvant therapy for wound or skin infections as this would mean a safer version of applying metabolites of beneficial microbes without live cells as postbiotics are by definition a "preparation of inanimate microorganisms and/or their components that confers a health benefit on the host" [14].

Some probiotic strains or their cell-free supernatants, mainly from the lactobacilli group, which was recently divided into several genera [15], have shown strong antimicrobial potential against some common wound pathogens using in vitro studies [9]. The investigated probiotics include *Lactiplantibacillus plantarum* ATCC 10241 [16,17], *Limosilactobacillus fermentum* NCIMB 7230 [18], *Limosilactobacillus reuteri* SD2112 [19], *Lacticaseibacillus rhamnosus* GG [20], *Cutibacterium acnes* ATCC 6919 (previously known as *Propionibacterium acnes*) [21] as well as some multi-strain probiotics [22–26] and the investigated pathogens in these studies mainly include *Pseudomonas aeruginosa, Staphylococcus aureus,* and *Escherichia coli*. Animal studies have also shown that topical application of probiotics such as: *Lactiplantibacillus plantarum* ATCC 10241 [17,27–29], ATCC 8014 [30], USM8613 [31], *Limosilactobacillus fermentum* NCIMB 7230 [32], and *Cutibacterium acnes* ATCC 6919 [21] were efficient in reducing the pathogen load of skin wounds. The most important type of study to ascertain the efficacy of probiotics is clinical study. In fact, probiotics can only be named as such, if a beneficial effect is supported by at least one well-designed human clinical study [33]. Two human clinical studies have shown that topical application of probiotics reduced pathogen load [5,34] and some recent clinical studies have shown that probiotic consumption indirectly reduced pathogen load via improvement of immune function [9,35–39]. A recent study [40] also addressed the differences between the in vitro and in vivo effects of probiotics on the removal of pathogens using *Lactiplantibacillus plantarum* ATCC 8014 (PTCC 1058) in simulated wound fluid together with *Pseudomonas aeruginosa* and *Staphylococcus aureus* on and animal model. The authors found that the efficacy of probiotics in the presence of different wound pathogens was different and that further investigations are warranted.

To our knowledge, no study has investigated a wide range of single-strain and multiple-strain dietary supplements against a wide range of clinical wound pathogens. Therefore, the aim of our study was to evaluate the antimicrobial effect of eleven single-strain and seven multiple-strain probiotic dietary supplements or other beneficial microbes and their efficiency against fifteen clinical wound pathogens using three methods: agar-spot assay, co-culturing assay, and an agar well diffusion assay, and to statistically compare all results.

2. Materials and Methods

2.1. Microbial Strains and Clinical Isolates

Eleven single-strain and seven multi-strain dietary supplements or other beneficial microbes noted in Tables 1 and 2 were used.

Table 1. Multi-strain dietary supplements containing various probiotic strains.

Label	Supplement	Lactobacilli [1]	Bifidobacteria	Other Bacteria or Fungi
MS1	OMNi-BiOTiC® HETOX, Institut Allergosan, Austria	*Lacticaseibacillus casei* W56 *Lactobacillus acidophilus* W37 *Levilactobacillus brevis* W63 *Lactobacillus salivarius* W24	*Bifidobacterium lactis* W52 *Bifidobacterium bifidum* W23	*Lactococcus lactis* W58 *Lactococcus lactis* W19
MS 2	OMNi-BiOTiC® STRESS Repair, Institut Allergosan, Austria	*Lacticaseibacillus casei* W56 *Lactobacillus acidophilus* W22 *Lacticaseibacillus paracasei* W20 *Lactiplantibacillus plantarum* W62 *Ligilactobacillus salivarius* W24	*Bifidobacterium lactis* W52 *Bifidobacterium lactis* W51 *Bifidobacterium bifidum* W23	*Lactococcus lactis* W19
MS 3	OMNi-BiOTiC® 6, Institut Allergosan, Austria	*Lactobacillus acidophilus* W55 *Ligilactobacillus salivarius* W57 *Lacticaseibacillus casei* W56	*Bifidobacterium animalis* W53	*Enterococcus faecium* W54 *Lactococcus lactis* W58
MS 4	OMNi-BiOTiC® FLORA plus+, Institut Allergosan, Austria	*Lactobacillus crispatus* LBV88 *Lacticaseibacillus rhamnosus* LBV96 *Lactobacillus gasseri* LBV150N *Lactobacillus jensenii* LBV116	/	/
MS 5	OMNi-BiOTiC® Activ, Institut Allergosan, Austria	*Lacticaseibacillus casei* W56 *Lactobacillus acidophilus* W37, *Levilactobacillus brevis* W63, *Ligilactobacillus salivarius* W24	*Bifidobacterium lactis* W52, *Bifidobacterium longum* W108, *Bifidobacterium breve* W25, *Bifidobacterium lactis* W51, *Bifidobacterium bifidum* W23	*Lactococcus lactis* W58, *Lactococcus lactis* W19,
MS 6	NutriVital Ultra SB, NutriVital Ply Ltd., Australia	*Lactobacillus acidophilus* La14 *Lactiplantibacillus plantarum* Lp-115	*Bifidobacterium animalis* subsp. *lactis* BI-04	*Saccharomyces cerevisiae* var. *boulardii*
MS 7	(Bio-Kult®), Protexin Lopsen Head, UK,	*Lacticaseibacillus casei* PXN 37, *Lactiplantibacillus plantarum* PXN 47, *Lacticaseibacillus rhamnosus* PXN 54, *Lactobacillus acidophilus* PXN 35, *Lactobacillus delbrueckii* subsp. *bulgaricus* PXN 39, *Lactobacillus helveticus* PXN 45, *Ligilactobacillus salivarius* PXN 57	*Bifidobacterium bifidum* PXN 23, *Bifidobacterium breve* PXN 25, *Bifidobacterium longum* PXN 30, *Bifidobacterium infantis* PXN 27	*Bacillus subtilis* PXN 21, *Lactococcus lactis* subsp. *lactis* PXN 63, *Streptococcus thermophilus* PXN 66

[1] The *Lactobacillus* genus has been recently divided into novel genera [15], therefore the novel genera have been used.

Table 2. Single-strain dietary supplements and other beneficial microbes.

Label	Supplement	Strains
SS01	Waya® LGG® forte, Medis GmbH, Austria	*Lacticaseibacillus rhamnosus* LGG
SS02	Yakult®, Yakult Honsha Co, Ltd., Yakult Europe, Italy	*Lacticaseibacillus paracasei* Shirota
SS03	BioGaia®, TwoPac AB, Sweden	*Limosilactobacillus reuteri* DSM 17938
SS04	German Collection of Microorganisms and Cell Cultures GmbH	*Lactiplantibacillus plantarum* subsp. *plantarum* DSM 2601
SS05	Probactiol® senior, Metagenics Italia S.r.l., Italia	*Bifidobacterium animalis* subsp. *lactis* HN019
SS06	Baby Linbi®, Lek Pharmaceutical company d.d., Slovenia	*Bifidobacterium animalis* subsp. *lactis* BB-12
SS07	ProLife® sporogenes, Zeta Farmaceutici, S.p.a., Italy	*Bacillus coagulans* MTCC 5260
SS08	German Collection of Microorganisms and Cell Cultures GmbH	*Propionibacterium freudenreichii subsp. freudenreichii* DSM 20271
SS09	German Collection of Microorganisms and Cell Cultures GmbH	*Acidipropionibacterium acidipropionici* DSM 20272
SS10	Optim PropioniBacter, Laboratoire Optim, Bionoto sprl, Belgium	*Propionibacterium freudenreichii* subsp. *shermanii*
SS11	SB Probiotic, Blooms, Phytologic Pty Ltd., Australia	*Saccharomyces cerevisiae* var. *boulardii*

As noted in Tables 1 and 2, all multi-strain dietary supplements MS1 to MS7 and single-strain supplements SS1 to SS4 contain different strains of lactobacilli. Eight strains of the modified *Lactobacillus* genus (MS1 to MS7), seven strains of the *Lacticaseibacillus* genus (MS1, MS2, MS3, MS4, MS5, MS7, SS01, SS02), three strains of *Ligilactobacillus salivarius* (MS1, MS2, MS3, MS5, MS7), three strains of *Lactiplantibacillus plantarum* (MS2, MS6, MS7, SS04), one strain of *Levilactobacillus brevis* (MS1, MS5) and one strain of *Limosilactobacillus reuteri* in SS03. The bifidobacteria genus is included in eight samples: MS1, MS2, MS3, MS5, MS6, MS7, SS05, and SS06. All eight samples include strains of the species *Bifidobacterium animalis* that contains two subspecies: *B. animalis* subsp. *lactis* and *B. animalis* subsp. *animalis.* Two strains of *Bifidobacterium breve* W25, PXN 25 and two strains of *Bifidobacterium longum* W108, PXN 30 in MS5, and MS7, one strain of *Bifidobacterium bifidum* W23 in MS1, MS2, MS5, and MS7 and one strain of *Bifidobacterium infantis* in MS7. Three *lactococci* strains were included in MS1, MS2, MS3, MS5, and MS7. Three propionibacteria strains are included in SS08, SS09, and SS10, [41]. One strain of each of the following bacteria are also included: *Enterococcus faecium* in MS3, *Bacillus subtilis* PXN 21 in MS7, *Bacillus coagulans* MTCC 5260 in SS07, and *Streptococcus thermophilus* in PXN 66 MS7. *Saccharomyces cerevisiae* var. *boulardii* is included in MS6 and SS11. The clinical pathogens were selected from the bacterial strains isolated from the wound samples received at the Institute of Microbiology and immunology at the Faculty of Medicine, University of Ljubljana, Slovenia in 2021. The genera/or species and origin are noted in Table 3.

Table 3. Clinical pathogenic isolates and their origin.

Label	Pathogen	Origin
1 2	*Staphylococcus aureus*	Patient with leg ulcer infection Patient with diabetic ulcer infection
3 4	*Pseudomonas aeruginosa*	Patient with inguinal infection after cardio intervention Patient with gastrostomy site infection
5 6	*Enterococcus faecalis*	Patient with infection at central venous catheterization Patient with sternal wound infection
7 8	*Escherichia coli*	Patient with surgical wound infection and dehiscence Patient with surgical wound infection
9 10	*Klebsiella pneumoniae*	Patient with sternal wound infection Patient with surgical wound infection
11 12	*Enterobacter* spp.	Patient with leg ulcer infection Patient with inguinal infection after cardio intervention
13	*Acinetobacter* spp.	Patient with bedsore (pressure ulcer) infection
14 15	*Bacteroides* spp.	Patient with perianal infection Patient with bedsore (pressure ulcer) infection

Clinical strains *Staphylococcus aureus, Pseudomonas aeruginosa, Enterococcus faecalis, Escherichia coli, Klebsiella pneumoniae, Enterobacter* spp., *Acinetobacter* spp. and *Bacteroides* spp. noted in Table 3 were collected from various different skin or wound infections, including skin ulcers, diabetic ulcers, pressure ulcers, inguinal infections, infections at jejunostomy, infections at central venous catheterization, sternal infection, wound dehiscence, surgical wound infection, perianal infection. All pathogens were identified using conventional microbiological methods in the medical diagnostics laboratory Institute of Microbiology and Immunology at the Faculty of Medicine, Ljubljana, Slovenia. All pathogens and probiotics except for those containing lactobacilli were cultured in tryptic soy broth (Fluka, 51228) as overnight cultures, incubated at 35 °C. All lactobacilli containing probiotics or other beneficial microbes were cultured in De Man, Rogosa, and Sharpe broth (Merck, 1.10661) as overnight cultures, incubated at 35 °C.

2.2. Molecular Methods for the Detection of Probiotic Strains

For the detection of bacterial and fungal strains of probiotics and other beneficial microbes used in our assays, genus-specific and species-specific PCR primers were used as shown in Table 4.

Table 4. Primer pairs of selected microbial genera or species.

Microorganism		Primer Pairs (5′–3′)	Product Size	Reference
		Lactobacilli spp.		
Lactobacilli spp.	LbLMA1-rev R-16-1	CTC AAA ACT AAA CAA AGT TTC CTT GTA CAC ACC GCC CGT C	220 bp	Dubernet et al., 2002 [42]
Lacticaseibacillus rhamnosus	Rham 1 RhamR	GTC GAA CGA GTT CTG ATT ATT G GAA CCA TGC GGT TCT TGG AT	158 bp	Sul et al., 2007 [43]
Lactobacillus acidophilus	LacidoF LacidoR	CAC TTC GGT GAT GAC GTT GG CGA TGC AGT TCC TCG GTT AAG C	575 bp	
Lacticaseibacillus casei	PrI CasII	CAG ACT GAA AGT CTG ACG G GCG ATG CGA ATT TCT TTT TC	200 bp	Walter et al., 2000 [44]
Limosilactobacillus reuteri	Lfpr Reu	GCC GCC TAA GGT GGG ACA GAT AAC ACT CAA GGA TTG TCT GA	350 bp	
Lactobacillus gasseri	Lgas-3 Lgas-2	AGC GAC CGA GAG AGA GAG A TGC TAT CGC TTC AAG TGC TT	360 bp	Takahashi et al., 2006 [45]
Lactiplantibacillus plantarum	LplanF LplanR	CGA GAC AGC AAT TCC TGC ACT CG CCT CAG AAA CAG TCC GGT TGA C	176 bp	Gaspar et al., 2019 [46]
		Bifidobacteria spp.		
Bifidobacterium spp.	Bif164F Bif601R	GGG TGG TAA TGC CGG ATG TAA GCC ATG GAC TTT CAC ACC	453 bp	Bernhard et al., 2000 [47]
Bifidobacterium bifidum	BifF BifR	ATT TGA GCC ACT GTC TGG TG CAT CCG GGA ACG TCG GGA AA	431 bp	Sul et al., 2007 [43]
Bifidobacterium longum	BiflongF BiflongR	TTC CAG TTG ATC GCA TGG TC GGG AAG CCG TAT CTC TAC GA	831 bp	
Bifidobacterium animalis	Bani-tF Bani-tR	TCA CGA CAA GTG GGT TGC CA GTT GAT CGG CAG CTT GCC G	178 bp	Sheu et al., 2010 [48]
		Other bacteria and fungi		
Lactococcus spp.	L1 L2	AAC TCT GTT GTT AGA G ATC TCT AGG AAT AGC AC	570 bp	Deasy et al., 2000 [49]
Propionibacterium spp.	PB1 PB2	AGT GGC GAA GGC GGT TCT CTG GA TGG GGT CGA GTT GCA GAC CCC AAT	865 bp	Rossi et al., 1999 [50]
Bacillus coagulans	BC1-F BC1-R	ACA GGG CTT TCA GAT ACC CG CGG GGA TCC GTC CAT CAA AA	990 bp	Majeed et al., 2017 2017 [51]
Bacillus subtilis	Bsub5F Bsub5R	AAG TCG AGC GGA CAG ATG G CCA GTT TCC AAT GAC CCT CCC C	595 bp	Wattiau et al., 2001 [52]
Enterococcus faecium	EM1F EM1R	TTG AGG CAG ACC AGA TTG ACG TAT GAC AGC GAC TCC GAT TCC	658 bp	Cheng et al., 1997 [53]
Saccharomyces cerevisiae	SC1 SC2	AAC GGT GAG AGA TTT CTG TGC AGC TGG CAG TAT TCC CAC AG	1170 bp	Mitterdorfer et al., 2002 [54]

Bacterial and fungal genomic DNA was extracted from the suspension of microorganisms using PrepMan Ultra Sample Preparation Reagent (Applied Biosystems) in accordance with the manufacturer's instructions. Amplification was carried out in a thermal cycler (S Labcycler, Sensoquest, Germany), applying the cycling conditions as presented in Table 5. The reaction mixture (50 µL) contained 2.5 U of Taq DNA Polymerase (Qiagen, Germany), 0.5 µM of each primer, 0.2 mM of each deoxyribonucleotide triphosphate, 1.5 mM of

10× reaction buffer, and different concentrations of $MgCl_2$ 2.5 mM $MgCl_2$ for *Lacticaseibacillus casei, Limosilactobacillus reuteri, Lactobacillus gasseri, Lactiplantibacillus plantarum, Bacillus subtilis*, 2 mM $MgCl_2$ for *Lacticaseibacillus rhamnosus, Lactobacillus acidophilus, Bifidobacterium bifidum, Bifidobacterium longum*, 1.5 mM $MgCl_2$ for *Bifidobacterium animalis, Bacillus coagulans, Lactococcus* genus, *Saccharomyces cerevisiae*, 1 mM $MgCl_2$ for *Propionibacterium* genus, and *Enterococcus faecium* and approx. 10 to 100 ng of bacterial or fungal DNA. In the case of single strains, a lower concentration of template was used to avoid inhibition of the reaction.

Table 5. Cycling parameters for polymerase chain reaction programs of selected microbes.

PCR Program	Denaturation [1]	Annealing	Extension	No. of Cycles	Reference/Modified Program
Lactobacilli spp.	30 s at 94 °C	30 s at 55 °C	30 s at 72 °C	20	Dubernet, et al., 2002 [42]
Lacticaseibacillus casei, Limosilalactobacillus reuteri				30	Walter et al., 2000 [44]
Lacticaseibacillus rhamnosus, Lactobacillus acidophilus, Bifidobacterium longum	30 s at 94 °C	30 s at 63 °C	30 s at 72 °C	30	Sul, et al., 2007 [43]
Lactobacillus gasseri	30 s at 94 °C	120 s at 65 °C	120 s at 72 °C	35	Takahashi et al., 2006 [45]
Lactiplantibacillus plantarum	15 s at 94 °C	30 s at 60 °C	60 s at 72 °C	40	Gaspar et al., 2019 [46]
Bifidobacterium spp.	30 s at 94 °C	60 s at 53 °C	120 s at 72 °C	35	Bernhard et al., 2000 [47]
Bifidobacterium bifidum	30 s at 94 °C	45 s at 57 °C	30 s at 72 °C	35	Modified in this study
Bifidobacterium animalis	35 s at 94 °C	35 s at 60 °C	40 s at 72 °C	35	Sheu et al., 2010 [48]
Propionibacterium spp.	30 s at 94 °C	15 s at 70 °C	60 s at 72 °C	40	Rossi et al., 1999 [50]
Bacillus coagulans	30 s at 94 °C	30 s at 60 °C	60 s at 72 °C	30	Majeed et al., 2017 [51]
Bacillus subtilis	30 s at 94 °C	120 s at 65 °C	120 s at 72 °C	30	Wattiau et al., 2001 [52]
Enterococcus faecium	60 s at 94 °C	60 s at 54 °C	60 s at 72 °C	40	Fijan et al., 2018 [55]
Lactococcus spp.	60 s at 94 °C	60 s at 50 °C	60 s at 72 °C	30	Modified in this study
Saccharomyces cerevisiae	60 s at 94 °C	60 s at 50 °C	60 s at 72 °C	30	Mitterdorfer et al., 2000 [54]

[1] Initial denaturation and final extension are 15 min at 95 °C and 7 min at 72 °C respectively for all amplifications.

Aliquots of the amplified products were subjected to electrophoresis (100 V, 45 min) in 1.5% agarose gel in TBE buffer (89 mM Tris base, 89 mM boric acid, 2 mM EDTA, pH 8.0). Gels were stained with 8 µL of Syber Green I and visualized under UV light (312 nm).

2.3. Agar Spot Assay

The antimicrobial effect of the chosen single-strain and multi-strain dietary supplements, probiotics, or other beneficial microbes against common skin or wound pathogens was determined using the modified agar spot assay [55–57]. Briefly, each probiotic overnight culture with a final concentration (10^8 cfu/mL) was inoculated as spots onto the following media: De Man, Rogosa, and Sharpe agar (Millipore, 1.10660) for all multi-strain probiotics and SS01-SS04, TOS-propionate agar (Sigma-Aldrich, 43314) for SS05 and SS06, Mannitol Egg Yolk Polymyxin agar (Merck, 1.05267) for SS07, Clostridium perfrigens agar (Liofilchen, 610207) for SS08-SS10, and Sabouraud glucose agar (BioMerieux, AEB152202) for SS11.

The plates were dried for 30 min at room temperature. All De Man, Rogosa, and Sharpe agar plates were then incubated anaerobically at 35 °C for 24 h using anaerobic jars together with a Genbag anaerobic pouch. Other agar plates were incubated aerobically for 24 h. All plates were then overlaid with 7 mL of soft agar (15g tryptic soy bujon (Fluka, 51228)/500 mL and 2g agarose (Fluka, 51228)/500 mL) inoculated with overnight cultures of the pathogens (with final concentration 7 log cfu/mL) and incubated at 35 °C for 48 h. Figure 1a shows a scheme of the agar spot assay.

After 48 h of incubation, measurements of inhibition zones around the colonies were measured using a ruler. The diameter of the zone of inhibition measuring from both sides of the clear zone around the colony was measured. The result also included 6–7 mm of the colony. Zones of more than 20 mm, between 10 and 20 mm, and less than 10 mm were considered as strong (3+), intermediate (2+), and low inhibitions (+), respectively. This is a modified scale [57], similar to those proposed by Davis and Stout [58]. If no zone was detected the result was reported as less than 6 mm. This assay was performed in triplicate. The mean of the zones of inhibition as well as the standard deviation SD were calculated.

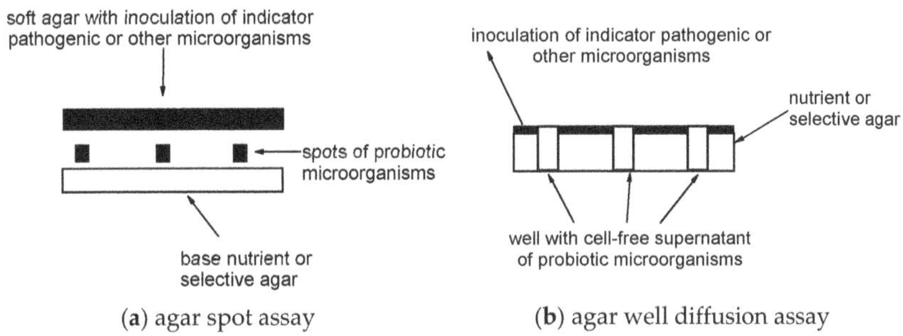

Figure 1. Scheme of the agar spot assay and the agar well diffusion assay. Adapted from Fijan, 2016 [59].

2.4. Co-Culturing for Microbial Competition Assay

The co-culturing for microbial competition assay of the pathogens and the chosen single-strain and multi-strain dietary supplements, probiotics, or other beneficial microbes was conducted similarly to Tranberg and co-authors [60] as follows: aliquots of 1 mL of an overnight culture of probiotics and 1 mL of the overnight culture of the clinical wound pathogens were inoculated into 500 mL sterile tubes with fresh broth containing 1 mL tryptic soy broth (Fluka, 51228) and 1 mL De Man, Rogosa and Sharpe broth (Merck, 1.10661). As controls, 1 mL overnight cultures of pathogens were grown in 1 mL tryptic soy broth and 1 mL De Man, Rogosa, and Sharpe broth. All samples were incubated overnight at 35 °C for 24 h.

After 24 h of incubation, colonies of surviving pathogens were counted using serial dilutions, ranging from 10^1 to 10^8. The following selective media were used for clinical isolates: Baird-Parker agar (Biolife, 4011162) for *Staphylococcus aureus* isolates, cetrimide agar (22470, Fluka) for *Pseudomonas aeruginosa* isolates, kanamycin esculin azide agar (Biolife, 4015522) for *Enterococcus faecium* isolates, violet red bile glucose agar (Fluka, 70189) for *Escherichia coli* and *Enterobacter* spp. isolates, HiCrome Klebsiella selective agar (Fluka, 90925) for *Klebsiella pneumoniae* isolates, MacConkey agar without salt (Sigma Aldrich, 51405) for *Acinetobacter* spp. isolates and bile esculin agar (Sigma Aldrich, 48300) for *Bacteroides* spp. isolates. All selective media were then incubated aerobically at 35 °C for 24 or 48 h according to the manufacturer's recommendation except for bile esculin agar for *Bacteroides* spp. isolates which were incubated anaerobically at 35 °C for 24 h using anaerobic jars together with a Genbag anaerobic pouch.

The reduction and log step reduction were then calculated as follows:

$$\%R = \frac{cfu_{pa} - cfu_{pa+pro}}{cfu_{pa}} \times 100$$

$$\log_{10} R = log\frac{cfu_{pa+pro}}{cfu_{pa}}$$

where: %R is the percent of reduction of the pathogen, $\log_{10}$R is the log step reduction, cfu_{pa} is the cfu of the pathogen after incubation and cfu_{pa+pro} is the cfu of the pathogen after incubation of the pathogen together with the probiotic. A log step reduction of more than six was considered strong inhibition as it corresponds to a 99.9999% reduction in the case of initial concentration of 10^6 cfu/mL. Between 3 and 6 was considered intermediate inhibition and less than 3 was considered low inhibition. Two separate experiments were conducted, and the average was calculated for each sample.

2.5. *Agar Well Diffusion Assay*

A slightly modified method of the agar well diffusion assay for the inhibition of pathogens by cell-free supernatants of chosen single-strain and multi-strain dietary supplements, probiotics, or other beneficial microbes by Holder and Boyce [61] was used. Briefly, overnight cultures of pathogens were confluently streaked onto Müller Hinton agar (BioLife, 4017402) plates with sterile cotton swabs and the plates were left to dry for 30 min. Wells (5 mm in diameter) were cut using 1000 µL sterile pipette tips. Cell-free supernatants of overnight cultures of chosen probiotics and other beneficial microbes were prepared by sedimentation of cells with centrifuging (4000× *g* for 10 min). The cell-free supernatant was filtered through a 0.22 µm pore size syringe filter. Half of each cell-free supernatant was used directly by inoculating 800 µL into the wells. The other half was neutralized using NaOH and adjusted to pH = 7 to achieve a neutralized cell-free supernatant that was inoculated into the remaining wells. Figure 1b shows a scheme of the agar well diffusion assay.

The antibacterial effect was determined by measuring the diameter of the zone of inhibition around the wells. Again, zones of more than 20 mm, between 11 and 20 mm, and less than 10 mm were considered strong (3+), intermediate (2+), and low inhibitions (+), respectively. The mean of the radii measuring from the edges of the colonies to the edges of the clear zones was calculated as well as the standard deviation SD. This assay was also performed in triplicate.

After 48 h of incubation, measurements of inhibition zones around the wells were measured using a ruler. The diameter of the zone of inhibition measuring from both sides of the clear zone around the well was measured. The result also included 6 mm diameters of the wells. Zones of more than 20 mm, between 10 and 20 mm, and less than 10 mm were considered as strong (3+), intermediate (2+), and low inhibitions (+), respectively according to the modified scale by Shokryazdan and co-authors [57]. If no zone was detected, the result was reported as less than 6 mm. This assay was performed in triplicate. The mean of the zones of inhibition as well as the standard deviation SD were calculated.

2.6. *Statistics*

The mean zones of inhibition were presented as 95% Confidence Intervals (CI) comparing, agar spot assay, co-culturing, and agar well diffusion assay respectively, explored and evaluated with appropriate statistical as needed for various probiotics groupings, such as single-/multi-strain, species. Student *t*-test was used to compare single-/multi-strains. One-way ANOVA test with post-hoc HSD comparing mean zone was used for multiple probiotic groups and two-factor ANOVA was used to compare agar well diffusion assay interaction with various groups. Assumptions of those tests were also checked. The statistical analysis was performed in the statistical program R (version 4.2.1).

3. Results

3.1. *Identification of Species and Genera of Microbial Strains Using Molecular Detection Methods*

The results of the polymerase chain reactions (PCR) using genera-specific and species-specific primer pairs for multi-strain probiotic dietary supplements and single-strain probiotics and other beneficial microbes are noted in Tables 6 and 7, respectively.

Table 6. Presence of conducted genera-specific and species-specific PCR products of multi-strain probiotics.

Sample	Confirmed Lactobacilli		Confirmed Bifidobacteria		Confirmed Other Bacteria or Fungi	
	Genus-Specific PCR	**Species-Specific PCR**	**Genus-Specific PCR**	**Species-Specific PCR**	**Genus-Specific PCR**	**Species-Specific PCR**
MS1	Lactobacilli [1] spp.	*L. acidophilus, L. casei*	*Bifidobacterium*	*B. animalis, B. bifidum*	*Lactococcus*	/
MS2	Lactobacilli spp.	*L. acidophilus, L. casei, L. plantarum*	*Bifidobacterium*	*B. animalis, B. bifidum*	*Lactococcus*	/
MS3	Lactobacilli spp.	*L. acidophilus, L. casei*	*Bifidobacterium*	*B. animalis*	*Lactococcus*	*E. faecium*
MS4	Lactobacilli spp.	*L. gasseri, L. rhamnosus*	/	/	/	/
MS5	Lactobacilli spp.	*L. acidophilus, L. casei*	*Bifidobacterium*	*B. animalis, B. bifidum, B. longum*	*Lactococcus*	/
MS6	Lactobacilli spp.	*L. acidophilus, L. plantarum*	*Bifidobacterium*	*B. animalis*	/	*Saccharomyces cerevisiae*
MS7	Lactobacilli spp.	*L. acidophilus, L. casei, L. plantarum, L. rhamnosus*	*Bifidobacterium*	*B. animalis, B. bifidum, B. longum*	*Lactococcus*	*B. subtilis*

[1] The *Lactobacillus* genus has been recently divided into novel genera [15], therefore the novel genera have been used.

Table 7. Presence of conducted genera-specific and species-specific PCR products of single strain microbes.

	Confirmed Genus-Specific PCR	Confirmed Species-Specific PCR
SS01	Lactobacilli spp.[1]	*Lacticaseibacillus rhamnosus*
SS02	Lactobacilli spp.	*Lacticaseibacillus casei*
SS03	Lactobacilli spp.	*Limosilactobacillus reuteri*
SS04	Lactobacilli spp.	*Lactiplantibacillus plantarum*
SS05	*Bifidobacterium* genus	*Bifidobacterium animalis*
SS06	*Bifidobacterium* spp.	*Bifidobacterium animalis*
SS07	(Not conducted)	*Bacillus coagulans*
SS08	*Propionibacterium* spp.	(Not conducted)
SS09	*Propionibacterium* spp.	(Not conducted)
SS10	*Propionibacterium* spp.	(Not conducted)
SS11	(Not conducted)	*Saccharomyces boulardii*

[1] The *Lactobacillus* genus has been recently divided into novel genera [15], therefore the novel genera have been used.

The PCR primer pairs LbLMA1-rev and R-16-1 that targets the nucleotide sequence of the spacer between the 16S and 23S rRNA genes in all lactobacilli genera confirmed by a positive band at 220 bp was found for all seven multi-strain probiotics and all single-strain samples that contained lactobacilli (SS1 to SS4). Species-specific PCR using primer pairs noted in Table 4 was run for the following lactobacilli species: *Lactobacillus acidophilus, Lactobacillus gasseri, Lacticaseibacillus casei, Lactocaseibacillus rhamnosus, Lactiplantibacillus plantarum* and *Limosilactobacillus reuteri*. Species-specific DNA fragments were found for *Lactobacillus acidophilus* in MS1, MS2, MS3, MS5, MS6, MS7, *Lactobacillus gasseri* in MS4, *Lacticaseibacillus casei* in MS1, MS2, MS3, MS5, MS7, and SS02, *Lactocaseibacillus rhamnosus* in MS4, MS7, and SS01, *Lactiplantibacillus plantarum* in MS2, MS6, MS7, and SS04 and *Limosilactobacillus reuteri* in SS03.

The genus *Bifidobacterium* using the primer pairs Bif164F and Bif601R for amplifying the 16S ribosomal rRNA fragments confirmed by a positive band at 453 bp was also confirmed for all bifidobacterial containing samples (MS1, MS2, MS3, MS5, MS6, MS7, SS05, SS06). Species-specific PCR using primer pairs noted in Table 4 was run for the following bifidobacterial species: *Bifidobacterium animalis*, *Bifidobacterium bifidum*, and *Bifidobacterium longum*. Species-specific DNA fragments were found for *Bifidobacterium animalis* in MS1, MS2, MS3, MS5, MS6, MS7, SS5, and SS6, *Bifidobacterium bifidum* in MS1, MS2, MS5, and MS7 and *Bifidobacterium longum* in MS5 and MS7. The genera *Lactococcus* and *Propionibacterium* were confirmed by primer pairs noted in Table 4 for MS1, MS2, MS3, MS5, and MS7 and SS08, as well as SS09 and SS10, respectively. Species-specific DNA fragments were also found for the bacteria *Enterococcus faecium* (MS3), *Bacillus subtilis* (MS7), and *Bacillus coagulans* (SS07). *Saccharomyces cerevisiae* species-specific DNA fragments were found (MS6, SS11) thus confirming the presence of *Saccharomyces cerevisiae* subsp. *boulardii*.

3.2. Agar Spot Assay

The evaluation of the mean zone of inhibition and standard deviation of the agar spot assay for all investigated probiotics and other beneficial microbes against the clinical pathogens, isolated from various skin wounds are noted in Table 8. All results of the zone of inhibition and standard deviation are noted in Supplementary Table S1.

Table 8. Evaluation of the antagonistic effect of various probiotics and other beneficial microbes against clinical skin pathogens using the agar spot assay.

	Evaluation of Zone of Inhibition Using the Agar Spot Assay *														
	Sa1	**Sa2**	**Pa3**	**Pa4**	**Ef5**	**Ef6**	**Ec7**	**Ec8**	**Kp9**	**Kp10**	**Ent11**	**Ent12**	**Ac13**	**Bac14**	**Bac15**
MS1	3+	3+	3+	3+	3+	2+	3+	2+	2+	2+	2+	2+	3+	3+	3+
MS2	3+	3+	3+	3+	3+	2+	3+	3+	3+	3+	3+	3+	3+	3+	3+
MS3	3+	3+	3+	3+	3+	3+	3+	3+	3+	3+	3+	2+	3+	3+	3+
MS4	3+	3+	3+	3+	3+	3+	3+	3+	3+	3+	2+	2+	2+	3+	3+
MS5	3+	2+	3+	3+	3+	3+	3+	3+	3+	3+	2+	3+	3+	3+	3+
MS6	3+	3+	3+	3+	3+	3+	3+	3+	3+	3+	2+	3+	3+	3+	3+
MS7	3+	3+	2+	3+	3+	3+	2+	3+	3+	3+	2+	3+	3+	3+	3+
SS01	3+	3+	3+	3+	3+	3+	3+	3+	3+	3+	3+	2+	3+	3+	3+
SS02	3+	3+	3+	3+	3+	3+	3+	3+	3+	2+	2+	2+	3+	3+	3+
SS03	2+	3+	3+	3+	3+	3+	3+	3+	3+	3+	2+	2+	3+	3+	3+
SS04	3+	3+	3+	3+	3+	3+	3+	3+	3+	3+	2+	3+	3+	3+	3+
SS05	2+	2+	2+	2+	2+	2+	2+	3+	3+	2+	3+	3+	2+	2+	3+
SS06	2+	2+	2+	2+	2+	2+	2+	3+	3+	3+	3+	3+	2+	3+	2+
SS07	+	2+	2+	2+	+	+	+	+	+	+	+	+	+	+	+
SS08	+	+	+	+	+	+	+	+	+	+	+	+	+	+	+
SS09	+	+	+	+	+	+	+	+	+	+	+	+	+	+	+
SS10	+	+	+	+	+	+	+	+	+	+	+	+	+	+	+
SS11	2+	2+	3+	3+	2+	3+	3+	2+	2+	2+	+	+	2+	2+	2+

* More than 20 mm was considered strong inhibition (3+), between 11 and 20 mm was considered intermediate inhibition (2+), and less than 10 mm was considered low inhibition (+). The diameter of the colony is included. If no zone of inhibition was detected the result was reported as <6 mm.

As obvious from Table 8, all multiple-strain probiotics and single-strain probiotics SS01to SS04 (including *Lacticaseibacillus paracasei* Shirota, *Limosilactobacillus reuteri* DSM 17938, *Lacticaseibacillus rhamnosus* GG, and *Lactiplantibacillus plantarum* DSM 2601) were successful against most clinical wound pathogens as strong inhibition (the zone of inhibition was more than 20 mm) was found in most of the assays. On the other hand, the single strain probiotics *Bacillus coagulans* MTCC 5260 (SS07), *Propionibacterium freudenreichii* DSM 20271 (SS08), *Propionibacterium propionici* DSM 20272 (SS09), and *Propionibacterium freudenreichii* susp. *shermanii* (SS10) exhibited only low inhibition (the zone of inhibition was less than 10 mm). Intermediate average inhibition (zone of inhibition was between 10 and 20 mm) was found for both single-strain bifidobacteria: *Bifidobacterium lactis* HN019 (SS05),

Bifidobacterium lactis BB12 (SS06), and the single-strain fungi *Saccharomyces boulardii* (SS11). The average zone of inhibition of all probiotic strains against individual clinical pathogens was intermediate for most strains and even above 20 mm for one strain of *Staphylococcus aureus*, *Pseudomonas aeruginosa*, *Escherichia coli*, and *Acinetobacter* and both clinical strains of the *Bacteroides* genus thus indicating that no specific pathogen stood out or was more resistant to the antimicrobial effect of the chosen probiotics.

The means of the inhibition zone of probiotics against wound pathogens with 95% CI are noted in Figure 2. The *Propionibacterium* strains and the *Bacillus* strain (SS07-SS10) seem to have smaller mean zones of inhibition and all multi-strain probiotics seem to have a larger zone of inhibition against all challenge wound pathogens. Checking the mean zone of inhibition against all wound pathogens for the various probiotics we observed statistical differences ($F(17.252) = 40.5$, $p < 0.001$).

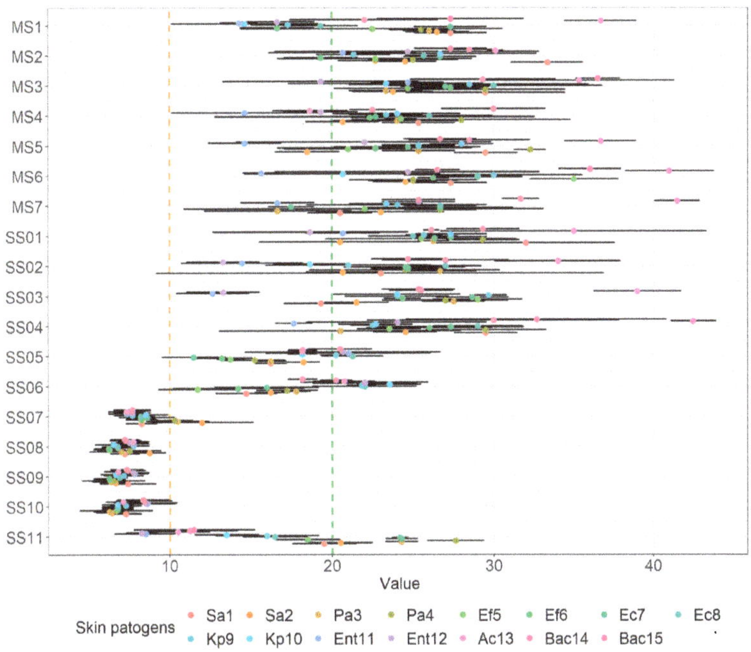

Figure 2. Means of inhibition zone together with 95% CI of various probiotics and other beneficial microbes against clinical skin pathogens using the agar spot assay.

As obvious in Figure 3, looking just at multi-strain ($M = 25.15$, $SD = 3.95$) and single-strain probiotics or beneficial microbes ($M = 16.74$, $SD = 1.74$), we showed that the latter has a statistically significant lower mean inhibition zone ($t = -7.553$, $p < 0.001$), which is also indicated in Figure 3. Grouping the data along the lines for species we observed the average means of the inhibition zone in descending order as follows: multi-strain probiotics that contained mainly lactobacilli strains and single-strain lactobacilli ($M = 25.20$, $SD = 4.20$), single strain bifidobacteria probiotics ($M = 18.05$, $SD = 3.09$), probiotic yeast strain and *Bacillus* strain ($M = 16.98$, $SD = 6.31$), and finally the *Propionibacterium* ($M = 7.48$, $SD = 0.42$). These means had a statistically significant difference ($F(3.56) = 47.38$, $p < 0.001$).

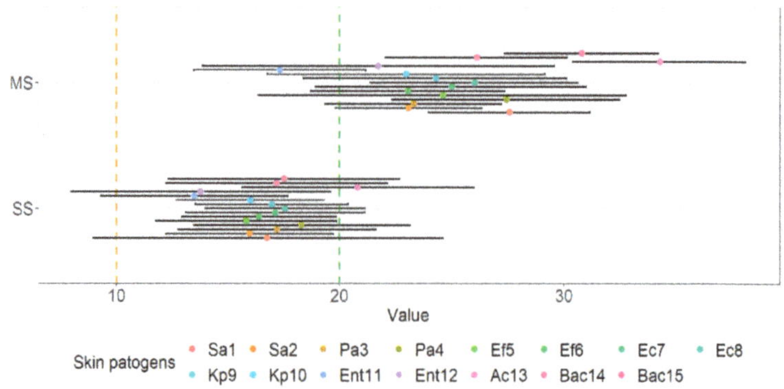

Figure 3. Means of inhibition zone together with 95% CI of multi-strain and single-strain probiotics against clinical skin pathogens using the agar spot assay.

A post-hoc HSD test comparing pairs showed that the mean zone of inhibition against wound pathogens for all lactobacilli-containing probiotics was higher than others and the mean zone of inhibition of the *Propionibacterium* strains was lower than the others, which can also be at least partially indicated in Figure 4.

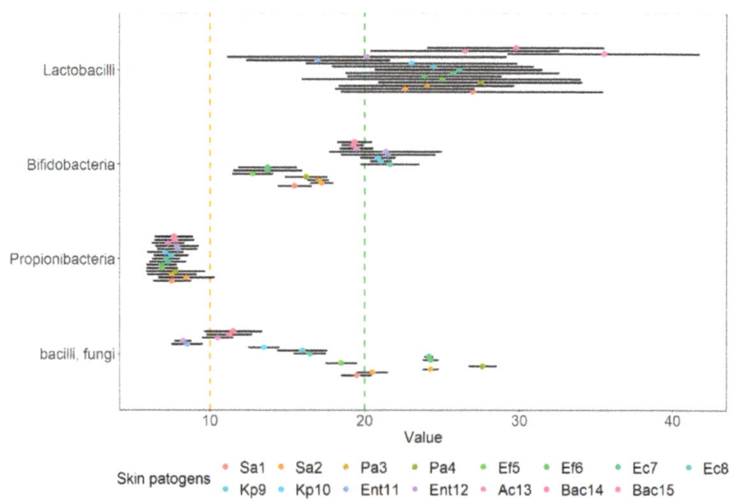

Figure 4. Means of inhibition zones together with 95% CI of probiotics and other beneficial microbes, divided into groups according to main species, against clinical skin pathogens using the agar spot assay.

3.3. Co-Culturing for Microbial Competition Assay

The evaluation of the average log step reduction for all investigated probiotics and other beneficial microbes against the clinical pathogens, isolated from various skin wounds using co-culturing is noted in Table 9. The scale of a log step reduction of more than 6 was considered strong inhibition, between 3 and 6 was considered intermediate inhibition and less than 3 was considered low inhibition. All results of the average log step reduction and percentage of reduction are noted in Supplementary Table S2.

Table 9. Evaluation of the antagonistic effect of various probiotics and other beneficial microbes against clinical skin pathogens using the co-culturing assay.

	Evaluation of Log Step Reduction Using the Co-Culturing Assay *														
	Sa1	**Sa2**	**Pa3**	**Pa4**	**Ef5**	**Ef6**	**Ec7**	**Ec8**	**Kp9**	**Kp10**	**Ent11**	**Ent12**	**Ac13**	**Bac14**	**Bac15**
MS1	2+	2+	3+	3+	2+	2+	3+	2+	+	2+	3+	3+	3+	3+	3+
MS2	2+	2+	3+	3+	2+	2+	2+	2+	+	2+	3+	3+	3+	3+	2+
MS3	2+	2+	2+	3+	2+	2+	3+	3+	+	3+	3+	3+	3+	3+	2+
MS4	2+	2+	3+	3+	2+	2+	3+	3+	3+	3+	2+	3+	3+	3+	3+
MS5	2+	2+	3+	3+	+	2+	3+	3+	3+	2+	3+	3+	3+	2+	3+
MS6	2+	2+	3+	2+	+	2+	2+	+	+	2+	2+	3+	3+	2+	+
MS7	2+	2+	3+	3+	2+	2+	3+	3+	3+	3+	3+	3+	3+	2+	2+
SS01	2+	+	3+	3+	+	+	2+	2+	2+	+	2+	3+	+	2+	+
SS02	2+	2+	3+	3+	2+	+	2+	2+	2+	+	3+	3+	2+	2+	+
SS03	2+	2+	3+	3+	2+	2+	2+	3+	+	2+	3+	3+	2+	+	+
SS04	3+	3+	3+	2+	2+	2+	3+	3+	3+	2+	3+	3+	2+	3+	2+
SS05	2+	2+	3+	2+	2+	2+	3+	3+	+	3+	3+	3+	2+	3+	2+
SS06	3+	3+	3+	3+	+	2+	2+	3+	2+	3+	3+	3+	2+	+	+
SS07	2+	+	3+	3+	2+	+	2+	2+	2+	2+	2+	2+	2+	+	+
SS08	+	+	+	+	+	+	+	+	+	+	+	+	2+	2+	+
SS09	+	+	+	+	+	+	+	+	+	+	+	+	+	+	+
SS10	2+	+	2+	+	+	+	+	2+	2+	2+	2+	2+	2+	+	3+
SS11	2+	2+	+	+	+	+	2+	+	+	+	+	2+	+	+	+

* a log step reduction of more than 6 was considered strong inhibition (3+), between 3 and 6 was considered intermediate inhibition (2+) and less than 3 was considered low inhibition (+).

As obvious from Table 9 strong reduction of pathogens (log step reduction of more than 6 log steps) was found for three multiple-strain probiotics (MS4, MS5, and MS7), whilst a low reduction of pathogens was found for three single-strain probiotics including *Propionibacterium freudenreichii* DSM 20271, *Propionibacterium propionici* DSM 20272, and *Saccharomyces boulardii* (SS8, SS9, and SS11 respectively). All other probiotics achieved an intermediate reduction of pathogens (log step reduction between 3 and 6 log steps). One clinical pathogen of the *Enterobacter* genus was less resistant as an average log step reduction above 6 log steps was achieved for all probiotics and one clinical pathogen of *Enterococcus faecalis* was most resistant as the average log step reduction under 3 log steps was achieved for all probiotics. Comparing the log step reduction of the wound pathogens after co-culturing with probiotics and other beneficial microbes we observed lower inhibition compared to agar spot assays for all probiotics and other beneficial microbes ($F(17.252) = 12.08, p < 0.001$).

The same was observed when comparing the inhibition effect of multi-strain and single-strain probiotics against wound pathogens ($t = -3.962$, $p < 0.001$), where multi-strain probiotics ($M = 5.62$, $SD = 129$) achieved a higher log step reduction of all challenge pathogens than single-strain probiotics ($M = 3.94$, $SD = 1.03$). When comparing the inhibition effect of probiotic species against all pathogens we found a statistical difference ($F(3.56) = 26.79, p < 0.001$). When comparing pairs with HSD post-hoc tests, we showed that there was no statistical difference between the inhibition effect of lactobacilli-containing probiotics ($M = 5.30$, $SD = 1.20$) and bifidobacteria-containing probiotics ($M = 5.51$, $SD = 1.38$) against the wound pathogens. However, there was a difference when comparing both lactobacilli and bifidobacteria containing probiotics to single-strain beneficial microbes that contained propionibacteria, the *Bacillus* species, and the probiotic yeast. Additionally, there was no difference between the probiotic yeast ($M = 2.25$, $SD = 1.46$) and the beneficial microbes that contained propionibacteria ($M = 2.78$, $SD = 0.94$).

3.4. Agar Well Diffusion Assay

Below are the results of the mean zone of inhibition for all investigated cell-free supernatants (S) (Table 10) and neutralized cell-free supernatants (NS) (Table 10) of probiotics and other beneficial microbes against the clinical pathogens, isolated from various skin wounds. Exact values of inhibition zones and standard deviation are noted in Supplementary Tables S3 and S4.

Table 10. Evaluation of the antagonistic effect of various cell-free supernatants of probiotics and other beneficial microbes against clinical skin pathogens using the agar well diffusion assay.

	Evaluation of Zone of Inhibition Using the Agar Well Diffusion Assay *														
	Sa1	**Sa2**	**Pa3**	**Pa4**	**Ef5**	**Ef6**	**Ec7**	**Ec8**	**Kp9**	**Kp10**	**Ent11**	**Ent12**	**Ac13**	**Bac14**	**Bac15**
MS1	2+	2+	2+	2+	2+	2+	2+	2+	2+	2+	2+	+	2+	2+	+
MS2	2+	2+	+	+	+	+	2+	2+	2+	2+	2+	+	+	–	+
MS3	2+	+	+	2+	2+	+	2+	2+	+	2+	2+	+	+	2+	+
MS4	2+	+	+	2+	2+	–	2+	2+	+	2+	2+	2+	+	+	2+
MS5	2+	+	–	2+	+	–	2+	2+	+	2+	+	2+	+	+	2+
MS6	+	+	2+	–	+	+	+	3+	+	+	2+	2+	+	+	2+
MS7	2+	2+		2+	–	2+	2+	+	2+	+	+	+	+	+	+
SS01	2+	2+	2+	+	–	–	2+	3+	+	+	+	–	+	–	+
SS02	2+	+	+	2+	+	+	2+	2+	+	2+	+	2+	+	–	–
SS03	–	2+	+	+	–	–	2+	2+	+	–	2+	2+	+	–	+
SS04	2+	+	–	+	+	–	+	2+	+	+	2+	+	+	–	–
SS05	2+	2+	2+	+	–	–	2+	2+	2+	2+	+		+	–	–
SS06	+	–		+	+	+	2+	2+	2+	2+	2+	2+	2+	–	–
SS07	–	+	2+	2+	–	–	–	–	–	–	+	–	+	2+	–
SS08	+	+	–	–	–	–	+	–	–	–	2+	–	–	–	–
SS09	+	2+	–	–	–	–	+	–	–	–	–	–	–	–	–
SS10	+	+	2+	+	–	2+	+	+	+	+	–	+	–	+	–
SS11	+	–	2+	2+	+	–	–	2+	–	–	+	–	+	+	–

* Cell-free supernatant after filtration; more than 20 mm was considered strong inhibition (3+), between 11 and 20 mm was considered intermediate inhibition (2+), and less than 10 mm was considered low inhibition (+). The diameter of the colony is included. If no zone of inhibition was detected the result was reported as <6 mm (–).

As obvious from Table 10, the cell-free supernatants of all multiple-strain probiotics except MS7 exhibited an intermediate average inhibition (zone of inhibition was between 10 and 20 mm). All cell-free supernatants of single-strain probiotics and MS7 exhibited only a low inhibition (the zone of inhibition was less than 10 mm). Only two probiotics (MS6 and SS01) exhibited high inhibition of cell-free supernatant, both for the same clinical strain of *Escherichia coli*. No pathogen stood out in its resistance against the cell-free supernatants. All results show a lower inhibition ability of the cell-free supernatant compared to probiotics.

As obvious from Table 11 (Supplementary Table S4) the neutralized cell-free supernatants of all probiotic strains exhibited only low average inhibition for all investigated clinical pathogens from wounds. Only one neutralized cell-free supernatant of *Limosilactobacillus reuteri* DSM 17938 exhibited a strong inhibition against one clinical strain from the *Bacteroides* genus. No pathogen stood out in its resistance against the neutralized cell-free supernatants. All results show a lower ability of the neutralized cell-free supernatant compared to cell-free supernatants.

Table 11. Evaluation of the antagonistic effect of various neutralized cell-free supernatants of probiotics and other beneficial microbes against clinical skin pathogens using the agar well diffusion assay.

	Evaluation of Zone of Inhibition Using the Agar Well Diffusion Assay *														
	Sa1	**Sa2**	**Pa3**	**Pa4**	**Ef5**	**Ef6**	**Ec7**	**Ec8**	**Kp9**	**Kp10**	**Ent11**	**Ent12**	**Ac13**	**Bac14**	**Bac15**
MS1	–	+	2+	–	–	–	–	2+	–	–	–	–	–	–	–
MS2	–	+	+	+	+	–	2+	2+	–	2+	–	–	+	2+	+
MS3	–	+	+	+	2+	+	2+	+	2+	2+	–	–	+	–	+
MS4	–	+	–	–	+	+	–	2+	–	2+	2+	2+	–	–	2+
MS5	–	–	–	–	–	–	–	–	–	2+	–	+	–	–	2+
MS6	+	–	–	–	–	–	–	–	–	–	–	–	+	–	2+
MS7	+	2+	–	–	2+	2+	2+	–	–	–	+	2+	–	–	–
SS01	–	2+	+	2+	–	–	2+	2+	–	–	2+	–	–	–	–
SS02	–	2+	–	–	–	–	2+	2+	–	–	2+	–	–	3+	–
SS03	–	2+	–	2+	–	–	2+	2+	–	–	2+	2+	–	–	–
SS04	+	2+	–	–	–	–	+	2+	–	2+	2+	–	+	–	–
SS05	+	+	–	–	–	–	–	2+	–	–	2+	–	+	2+	–
SS06	+	–	–	–	2+	–	2+	+	2+	2+	+	+	+	–	2+
SS07	+	+	–	–	–	–	–	–	–	–	–	–	+	–	–
SS08	2+	+	–	–	–	–	–	–	–	–	2+	–	–	–	–
SS09	+	2+	–	–	–	–	+	+	–	–	2+	–	–	–	–
SS10	–	+	–	–	–	–	+	+	2+	–	–	–	–	–	–
SS11	+	–	2+	–	–	–	–	–	–	–	–	–	+	–	–

* Neutralized cell-free supernatant after filtration with pH = 7 by addition of NaOH. More than 20 mm was considered strong inhibition (3+), between 11 and 20 mm was considered intermediate inhibition (2+), and less than 10 mm was considered low inhibition (+). The diameter of the colony is included. If no zone of inhibition was detected, the result was reported as <6 mm.

The visual comparison of the results of the inhibition zones of cell-free supernatants (S) and neutralized cell-free supernatants (NS) of all probiotics and beneficial microbes against clinical pathogens are noted in Figure 5. Figure 6 displays the comparison of the inhibition zones of S and NS of multi- and single- strain probiotics and microbes against clinical wound pathogens and Figure 6 displays the results of the inhibition zones of S and NS for all probiotics and beneficial microbes, divides into main species against the wound pathogens.

Statistically comparing results of the inhibition of cell-free supernatant and neutralized cell-free supernatants agar well diffusion, S and NS respectfully (Figure 6), with the aforementioned groups we observed, that there was no statistically significant interaction between agar well diffusion results and all probiotics of beneficial microbes (F(17.504) = 1.281, $p = 0.199$), but there was a simple main effect on various probiotics ($p < 0.001$) and agar well diffusion ($p < 0.001$) on the mean zone.

Looking at multi- and single- strain probiotics (Figure 7), we can observe a statistically significant interaction between the effect of both S and N supernatants using the agar well diffusion assay against wound pathogens (F(1.56) = 7.475, $p = 0.008$) as well as simple main effects, more precisely mean zones of supernatants of multi-stain probiotics were higher compared to single-stain supernatants ($p < 0.001$) and inhibition was higher for S compared to NS ($p < 0.001$). There was also no interaction between the inhibition of cell-free supernatants of probiotics and other beneficial microbes, divided into main species (F(3.112) = 2.740, $p = 0.610$) against all wound pathogens. However, both supernatants exhibited significant simple main effects, higher for S than NS ($p = 0.007$) and also higher for bifidobacteria-containing single strain probiotics ($p = 0.030$) and lactobacilli-containing probiotics ($p = 0.001$) than for propionibacteria-containing single strain beneficial microbes ($p = 0.005$).

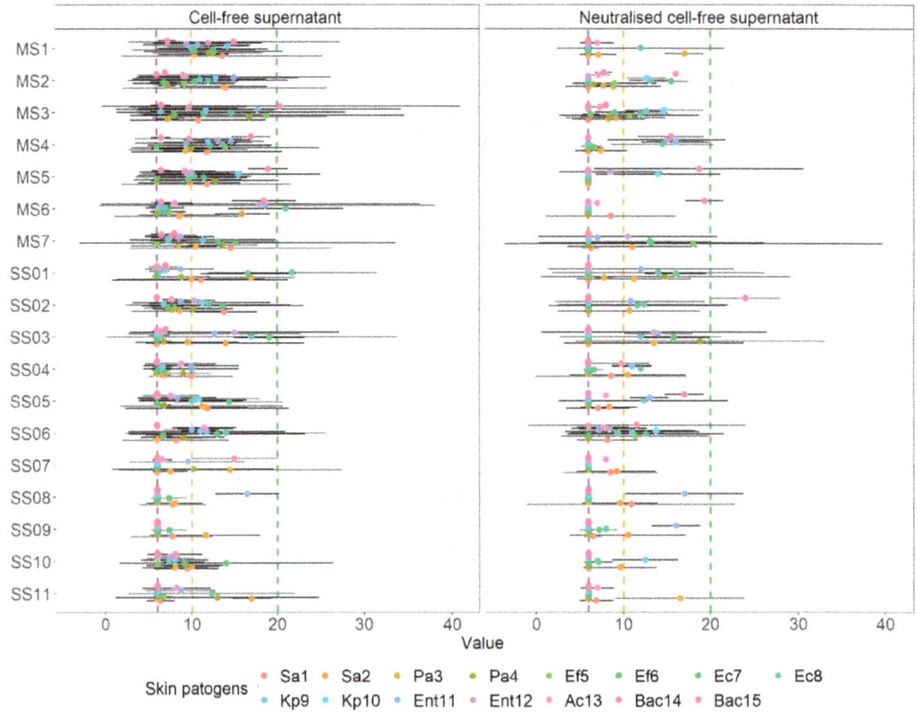

Figure 5. Means of inhibition zone together with 95% CI of various cell-free supernatants of probiotics and other beneficial microbes against clinical skin pathogens using the agar well diffusion assay.

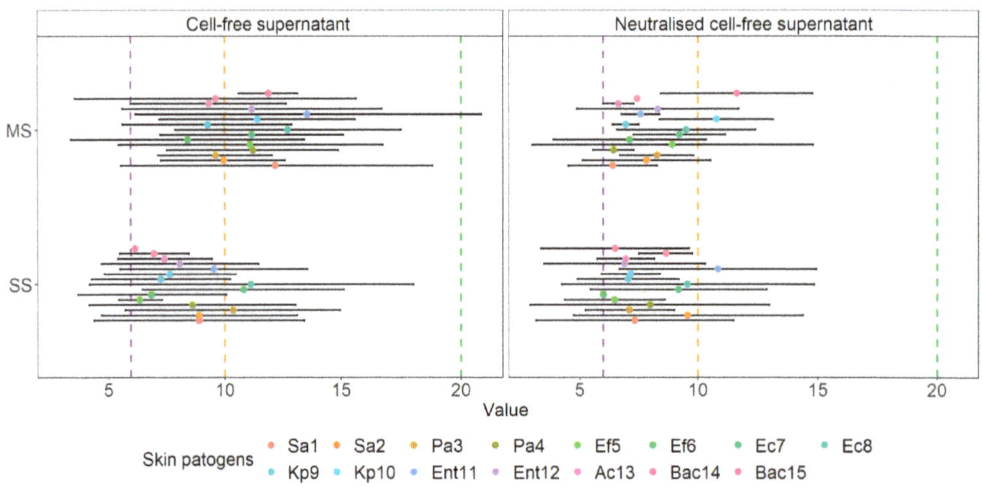

Figure 6. Means of the inhibition zone together with 95% CI of cell-free supernatants of multi-strain and single-strain probiotics against clinical skin pathogens using the agar well diffusion assay.

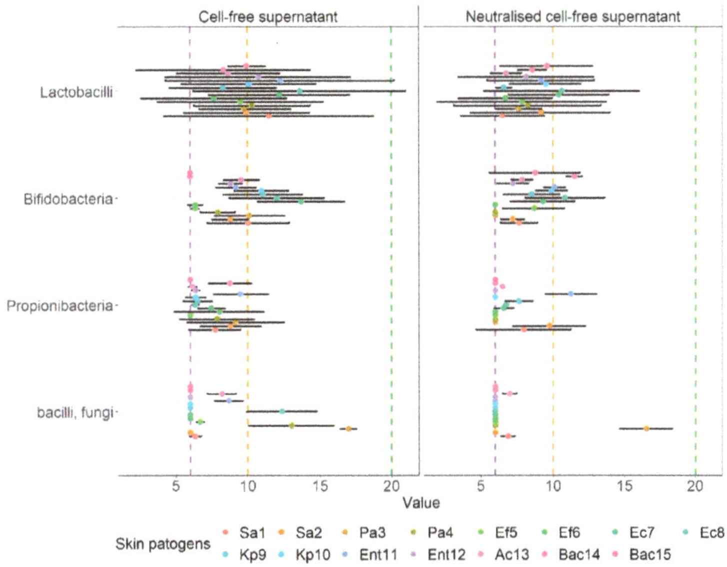

Figure 7. Means of inhibition zone together with 95% CI of cell-free supernatants of probiotics and other beneficial microbes, divided into groups according to main species, against clinical skin pathogens using the agar well diffusion assay.

4. Discussion

One important attribute of probiotics and probiotic candidates is their antimicrobial effect against pathogens. It is a well-known attribute of the lactobacilli and bifidobacteria genera [62]. The antimicrobial effect against pathogens is mostly attributed to the production of metabolites such as bacteriocins, organic acids, short-chain fatty acids, and hydrogen peroxide. Other important mechanisms of action of probiotics include competitive exclusion, immune modulation, stimulation of host defenses, and the production of signaling molecules that trigger changes in gene expression [55,63,64]. However, appropriate methodology is important in order to determine realistic and repeatable results. Our study used three different in vitro methods for determining the antimicrobial effect: the agar-spot assay, the co-culturing assay, and the agar-well diffusion assay. The first two methods utilised live microbes, whilst the last method utilised cell-free supernatant or postbiotics. The methods presented differences in the results. All dietary supplements achieved a certain level of inhibition of all pathogens, although there were variations between strains and multi-strain supplements, where the latter exhibited higher inhibition of the clinical pathogens than the single strain ($p < 0.05$), regardless of the method.

Our analysis of the collected data showed that the means of inhibition of probiotics and other beneficial microbes against all wound pathogens were statistically different ($F(17.252) = 40.5$, $p < 0.001$), where single strain beneficial microbes containing propionibacteria and *Bacillus* species (SS07-SS10) exhibited smaller inhibition zones against wound pathogens compared to all other probiotics and all multi-strain probiotics exhibited larger inhibition than single-strain probiotics. Looking at interactions via two-way ANOVA analysis, we observed a statistically significant interaction between multi- and single- strain probiotics or beneficial microbes and agar well diffusion ($F(1.56) = 7.475$, $p = 0.008$) as well as higher simple main effects for mean zones of multi-stain probiotics compared to single-stains ($p < 0.001$) and mean zones in agar well diffusion was lower in neutralized supernatant compared to the supernatant ($p < 0.001$) against wound pathogens.

The three methods to assess the antimicrobial effect of probiotics and other beneficial microbes or their metabolites against skin pathogens deployed in this study are based on phenotype characteristics that can be used for culturable microorganisms [65]. The most time-consuming is the co-culturing assay which requires the preparation of 10-fold dilutions for the enumeration of the pathogen after incubation with probiotics to determine the reduction effect. There are several modifications to this method, including incubation time, media type, and final detection method [55,66–68]. The advantage of this method is that both the probiotics and the pathogens are in a liquid environment enabling more simulation of the natural environment than existing in a colony on a solid surface, where immobilisation restricts growth. Such an environment causes different dynamics, less growth restriction, quorum sensing, and planktonic growth of both the probiotic and pathogen microbes [69–71]. Both the agar-well diffusion assay and the agar spot assay are conducted on solid media and require the measurement of the zone of inhibition against challenge pathogens. The agar spot assay investigates the inhibition effect of microbes, grown in a colony, whilst the agar-well diffusion assay investigates the inhibition effect of the cell-free supernatant, either in direct form or neutralized to eliminate the effect of organic acids. Both methods also exhibit several modifications with regard to solid media preparation, incubation conditions, initial concentration, and diffusion of metabolites [21,22,25,66,72–74]. Some authors measured either the whole diameter of the zone of inhibition which includes the diameter of the formed probiotic colony or the well with the supernatant [22,66,72,73], whilst other authors measured only the radius of the inhibition zone [25,55,74]. In our study, all cell-free supernatants of lactobacilli-containing multi-strain dietary supplements exhibited some inhibition as the average inhibition was intermediate for all except MS7. However, the cell-free supernatants of single-strain lactobacilli achieved average low inhibition. In the study by Lopes and co-authors [25], all investigated lactobacilli strains exhibited inhibition against examples of possible wound pathogens including *Escherichia coli*, *Pseudomonas aeruginosa*, and *Staphylococcus aureus*. However, when examining the results, it is obvious that in some cases, the radius of the zone of inhibition is only 1 mm, meaning the inhibition was also low for some strains, as found in our study. Similarly, in the study by Tejero-Sarinena and co-authors [74] the radii of the zones of inhibition of the non-adjusted cell-free supernatant of various lactobacilli and bifidobacteria strains were low, between 0.7 mm and up to 2 mm.

Neutralization of the culture supernatants with alkali vastly reduced the antagonistic effects of all our multiple-strain dietary supplements and our single strains thus indicating that the main mechanism of antagonism was the production of organic acids, such as lactic acid, propionic acid, butyric acid, and that bacteriocinogenic potentials were only partially used. On the other hand, the neutralized cell-free supernatant of *Lactiplantibacillus plantarum* DSM 2601 (SS04), *Propionibacterium freudenreichii* DSM 20271 (SS08), and *Propionibacterium acidipropionici* DSM 20272 (SS09) exhibited a somewhat higher average inhibition than the non-adapted cell-free supernatant, thus indicating that bacteriocins, such as perhaps plataricins, pediocins, or other neutral metabolites were produced [75–77]. The neutralized cell-free supernatant *Bifidobacterium animalis* subsp. *lactis* BB12 had a higher antagonistic effect than the non-neutralized for some pathogens (*Enterococcus faecium*, *Escherichia coli*, *Klebsiella pneumoniae*, and *Bacteroides* spp.) but not for *Pseudomonas aeruginosa*, *Enterobacter* spp. or *Acinetobacter* spp. Similarly, in the study by Fredua-Agyeman and co-authors [78] the neutralized cell-free supernatants of BB12 and *Lactobacillus acidophilus* La-5 did not show inhibition against *Pseudomonas aeruginosa*. Additionally, in the study by Lopes and co-authors [25], the antimicrobial activity of the cell-free supernatant was also attributed to organic acid production as the neutralized supernatant did not exhibit inhibition. The same conclusions were also found in the study by Tejero-Sarinena and co-authors [74].

In order to enable some comparison between all three methods, we created a scale of the co-culturing method based on the disinfection requirement for medical devices according to the Food and Drug Administration [79], where a log step reduction of 6 log steps or more is considered a strong reduction of the pathogen. This was based on the disinfection requirement of disinfectants where the log step reduction of 6 log steps corresponds to a 99.9999% reduction in the case of the initial pathogen concentration of 10^6 cfu. All chosen probiotics achieved a log step reduction for all challenge pathogens. According to the scale, three of our five chosen multi-strain dietary supplements (MS4, MS5, MS7) exhibited a strong average reduction of pathogens, whilst the other multi-strain dietary supplements achieved intermediate log step reduction of the pathogen. Eight of our chosen single-strain beneficial microbes achieved an average intermediate reduction of pathogens, whilst three achieved a low average reduction of the pathogen (*Propionibacterium freudenreichii* DSM 20271, *Propionibacterium acidipropionici* DSM 20272, and *Saccharomyces boulardii*). Other studies using the co-culturing method found that probiotics or probiotic candidates caused a reduction of pathogens, including *Escherichia coli* and *Staphylococcus aureus* co-cultured with *Lactobacillus acidophilus* La5 and *Bifidobacterium longum* ATCC 15707 [67], *Staphylococcus aureus*, and *Pseudomonas aeruginosa* co-cultured with *Limosilactobacillus fermentum* [68], *Escherichia coli*, *Salmonella* Enteritidis, *Salmonella* serotype (ser.) Typhimurium, *Staphylococcus intermedius*, *Klebsiella oxytoca*, and other pathogens co-cultured with lactobacilli isolated from piglet feces [66].

Molecular methods are much less time-consuming than classical phenotype methods that cannot easily distinguish between various species of the same genera and are also applicable for enumeration [80,81]. Although we did not conduct all species-specific PCR protocols for all species declared in all dietary supplements, we found positive results for all the PCR protocols that we conducted to detect genera or species thus proving that the reliability of the labelling system of probiotic supplements has improved compared to previous years [43,82]. Despite the recent division of lactobacilli into 23 novel genera [15] we found positive bands for all lactobacilli-containing dietary supplements using the primers pairs LbMA1-rev/R-16-1 [42], and it is obvious that these new genera share a common DNA section. Almost no dietary supplements used this new nomenclature. Another interesting finding was the positive band for *Lacticaseibacillus paracasei* Shirota (SS01, Yakult®), using the primer pairs Prl/CasII for the *casei* species, published by Walter and co-authors in the year 2000 [44]. As *Lacticaseibacillus paracasei* Shirota was reclassified from the *casei* species [83] after the publication of the primers, it is obvious that they are not species-specific and share a common DNA section as they belong to closely related species [84]. Recently the heterogenous genus of propionibacteria was divided into cutaneous (*Cutibacterium* spp.) and dairy propionic acid-producing bacteria (*Propionibacterium* and *Acidipropionibacterium* spp.) [41]; however, using the primer pairs PB1/PB2 [50] all propionibacteria: *Propionibacterium freudenreichii* subsp. *Freudenreichii* DSM 20271, *Propionibacterium freudenreichii* subsp. *Shermanii, and Acidipropionibacterium acidipropionici* DSM 20271 (SS08-SS10, respectively) were detected. Additionally, the primer pair CS1/SC2 [54] was used to detect *Saccharomyces cerevisiae* and we found a positive band for MS6 and SS11, which both contain *Saccharomyces boulardii* according to the manufacturers, confirming it is in fact a variant of *Saccharomyces cerevisiae* [54]. These findings also indicate that all manufacturers are not up to date with taxonomic changes.

The probiotic *Lacticaseibacillus rhamnosus* GG, also known as LGG, was the first lactobacilli strain to be patented in 1989 and has proven health benefits as shown by systematic reviews of several clinical studies, focused on antibiotic-associated diarrhoea [85], paediatric diarrhoea [86], gastroenteritis [87] and respiratory tract infections in children [88]. It is a biofilm-forming and immunomodulating probiotic that has shown antimicrobial effect against several pathogens [89] and is often used in in vitro studies as a reference strain for examining the antimicrobial effect of potential new probiotic strains [90,91]. In our study, this strain was SS03 and it also exhibited strong inhibition of most clinical pathogens using the agar spot method. However, using the co-culturing method, our results show that only an intermediate inhibition rate was achieved, thus implying that complex mechanisms of the probiotic are at work in different circumstances and that promising in vitro results using one method does not necessarily correlate with other methods [55] or correlate to statistically significant health benefits in clinical studies [92,93].

Lacticaseibacillus paracasei Shirota and *Limosilactobacillus reuteri* DSM 17938 are also well-researched probiotic strains (SS01 and SS02). Both strains exhibited the same results as SS03, namely strong average inhibition using the agar spot assay and an intermediate inhibition rate using the co-culturing assay. The same results were also found for the less-researched strain *Lactiplantibacillus plantarum* DSM 2601 (SS04). The latest clinical studies of *Lacticaseibacillus paracasei* Shirota (Yakult®), find consumption leads to improvement of depressive symptoms [83], lipid metabolism and intestinal microbiota [94], digestive disorders [95], and immunological function [96]. *Lacticaseibacillus paracasei* Shirota has also shown antifungal activity [97] and, similarly to our study, antibacterial activity against *Escherichia coli* and *Bacteroides* spp. [98]. *Limosilactobacillus reuteri* DSM 17938 (BioGaia®) has replaced the original strain *Limosilactobacillus reuteri* ATCC 55730 as it does not contain plasmid-borne antibiotic resistance and both strains exhibit success in the treatment of acute gastroenteritis, especially in children [99]. Although *Limosilactobacillus reuteri* DSM 17938 exhibits strong antimicrobial potential against major gastric and enteric bacterial pathogens and rotavirus [100], it did not prove effective as eradication therapy for infection with *Helicobacter pylori*, thus indicating that further studies are needed to establish the role of probiotics as adjuvant therapy, as the authors concluded [101].

Two well-known strains of the same species of bifidobacteria were used as single-strain probiotics: *Bifibacterium animalis* subs. *lactis* HN019 and BB-12 (SS05 and SS06, respectively). Both strains exhibited comparable results using the agar spot assay. On the other hand, there were differences in individual results for the inhibition of pathogens for co-culturing and the agar-well diffusion assay, although the average inhibitions were almost the value, proving that many probiotics traits are indeed strain-specific [11] and cannot be generalized to all representatives of the same species. The strain HN019 proved successful against periodontal pathogens in a recent clinical trial [102] and is a well-known probiotic with immune-enhancing properties [103]. In an in vitro study using the co-culturing method *Bifidobacterium animalis* subs. *Lactis*, BB-12 successfully inhibited the growth of *Clostridoides difficile* (previously known as *Clostridium difficile*) [104]. This strain reduced the risk of respiratory infections in infancy in a clinical study [105].

The challenge propionic acid-producing bacteria used in our study included three strains: *Propionibacterium freudenreichii* subsp. *freudenreichii* DSM 20271, *Acidipropionibacterium acidipropionici* DSM 2072, and *Propionibacterium freudenreichii* subsp. *Shermanii* (SS08, 09, and 10, respectively), and achieved only intermediate, low, or even no average inhibition of pathogens, depending on the method. The common feature of these three bacteria is the ability to produce propionic acid. Our results are similar to the study by Dyshlyuk and co-authors [106] where moderate antimicrobial activity using a version of the agar spot method was found for *Propionibacterium jensenii* B-6085 and *Propionibacterium thoenii* B-6082, but not for *Propionibacterium freudenreichii* B-11921 and *Propionibacterium acidipropionici* B-5723 against pathogens *Escherichia coli* ATCC 25922, *Salmonella enterica* ATCC 14028, *Staphylococcus aureus* ATCC 25923, *Pseudomonas aeruginosa* B6643, *Proteus vulgaris* ATCC 63, and *Listeria monocytogenes* ATCC 7644. *Propionibacterium freudenreichii* subsp. *freudenreichii* DSM 20271 is known to produce cobalamin or vitamin B12 [107] and *Propionibacterium freudenreichii* subsp. *shermanii* has shown probiotic effect as part of multi-strain dietary supplements in clinical studies against irritable bowel syndrome-related intestinal microbiota stabilization [108], intestinal microbiota changes during anti-*Helicobacter pylori* treatment [109].

Our challenge spore-forming representative *Bacillus coagulans* MTCC 5260 (Prolife®) also achieved only intermediate, low, or even no average inhibition of pathogens, depending on the method. Probiotic *Bacillus* strains used either in spore or vegetative forms have shown antimicrobial, anticancer, antioxidant, and vitamin production properties. However, they can also produce toxins and biogenic amines and transfer antibiotic resistance genes; therefore, their safety is a concern. Studies on the microbiome using probiotic *Bacillus* strains are limited in humans [110]. The strain MTCC 5260 is also known as Unique IS2 and ATCC PTA-11748 [111] and has documented clinical efficacy against constipation [112]. It also exhibits antimicrobial effectiveness as it was efficient as an adjuvant in the treatment of bacterial vaginosis [113].

Saccharomyces cerevisiae var. *boulardii* (SS11) is the only representative of probiotic fungi used in our study and it achieved average intermediate or low pathogen reduction, depending on the method used. It is a well-known probiotic that produces various bioactive compounds and is mostly known for its role in treating gastrointestinal diseases [114,115]. Together with *Lacticaseibacillus rhamnosus* GG, it is even one of the few probiotics recommended by the ESPGHAN (European Society for Paediatric Gastroenterology, Hepatology, and Nutrition) and ESPID (European Society for Paediatric Infectious Diseases) [116,117] for treating acute gastroenteritis in children. *Saccharomyces boulardii* has also been proposed as an alternative to treating bacterial infections [114], however, our results do not support this claim for our challenge wound pathogens.

In our study, the multi-strain dietary supplement MS7 (Bio-Kult®) was effective in strong average inhibition found against most clinical pathogens using the agar spot method. This dietary supplement was also the most effective mixture against *Enterococcus faecalis* in another study using the agar spot test [72]. In a clinical study, this multi-strain probiotic was also associated with significant improvement in symptoms in patents with diarrhoea-predominated irritable bowel syndrome [118].

Several commercial dietary supplements including OMNi-BiOTic® Hetox, OMNi-BiOTic® 6, OMNi-BiOTic® Stress repair, OMNi-BiOTic® Flora plus+, and OMNi-BiOTic® Activ (MS1 to MS5) achieved strong average inhibition against all pathogens in our study. MS1, MS4, and MS5 also achieved strong average inhibition using the co-culturing method, whilst the average inhibition of cell-free supernatant was intermediate or even low. The lower effect of cell-free probiotic supernatant indicates that bacterial response is important in cell-cell signaling and/or bacteria-host interaction. The multi-strain dietary supplement MS2 was also used in a clinical study that found that this multi-strain probiotic might be a well-tolerated tool to positively influence the gastrointestinal quality of life as well as mental and somatic health, cognition, and immune response and potentially have effects on psychiatric symptoms [119]. In another clinical study, this multi-strain probiotic positively influenced the gastrointestinal tract of patients with diarrhoea-predominated irritable bowel syndrome [120]. In another study, the multi-strain postbiotic supernatant of the dietary supplement OMNi-BiOTic AAD10 with similar composition exhibited positive antibacterial and antifungal effects in vitro [121].

Our results show that several dietary supplements were efficient in reducing the pathogen loads of the investigated clinical pathogens. The concept that certain bacteria can destroy other, even pathogenic bacteria, especially with respect to the skin, is not new and many historic researchers, such as Metchnikoff, Nissle, Cantini, and others have investigated and proven this concept [9,122,123]. More than a decade ago, Howard and co-authors concluded that probiotics could be beneficial in the prevention and treatment of wound infections [124]. Probiotics also give positive results for wound healing, wound-epithelization, and neovascularization [125]; however, as such treatment represents a shift in the doctrine of wound treatments where using bacteria to fight bacteria is not intuitive [126,127], many more studies are needed to establish a consensus on the efficacy of using probiotics against skin pathogens.

5. Conclusions

The scientific evidence of the health benefits of using probiotics and postbiotics for wounds is becoming more extensive and, therefore, an important possible application of probiotics in the future. In light of our results, it seems that each clinical pathogen was differently susceptible to different probiotic strains, although in general the multispecies probiotics were more efficient than the single-strain probiotics; however, the method deployed also impacted the results. Perhaps a new approach such as a "probiogram" or "postbiogram" as an analogue to antibiograms could be a possibility in the future in finding the most efficient targeted probiotic strains, cell-free supernatants, or neutralized cell-free supernatants that have the highest antagonistic effect against individual clinical wound pathogens. Additionally, more robust, well-designed clinical trials of probiotics targeting different clinical skin pathogens are needed to establish more knowledge on the exact efficacy and mechanisms of individual probiotics against pathogens to draw evidence-based conclusions for clinical recommendations.

Supplementary Materials: The following supporting information can be downloaded at: https://www.mdpi.com/article/10.3390/microorganisms10122518/s1, (Tables S1–S4). S1: Results of the antagonistic effect of various probiotics and other beneficial microbes against clinical skin pathogens using the agar spot assay. S2: Results of the antagonistic effect of various probiotics and other beneficial microbes against clinical skin pathogens using the co-culturing assay. S3: Results of the antagonistic effect of various cell-free supernatants of probiotics and other beneficial microbes against clinical skin pathogens using the agar well diffusion assay. S4: Results of the antagonistic effect of various neutralised cell-free supernatants of probiotics and other beneficial microbes against clinical skin pathogens using the agar well diffusion assay.

Author Contributions: Conceptualization, S.F. and A.S.; methodology, S.F and M.S.; formal analysis, S.F.; investigation, S.F.; resources, S.F.; data curation, S.F. and P.K.; writing—original draft preparation, S.F.; writing—review and editing, S.F., P.M.V., A.S., and P.K.; visualization, funding acquisition, S.F. All authors have read and agreed to the published version of the manuscript.

Funding: This research was partially funded by Institut Allergosan, Austria.

Data Availability Statement: No additional data is available.

Conflicts of Interest: All authors declare no conflict of interest.

References

1. McLoughlin, I.J.; Wright, E.M.; Tagg, J.R.; Jain, R.; Hale, J.D.F. Skin microbiome—The next frontier for probiotic intervention. *Probiotics Antimicrob. Proteins* **2022**, *14*, 630–647. [CrossRef] [PubMed]
2. Perez Perez, G.I.; Gao, Z.; Jourdain, R.; Ramirez, J.; Gany, F.; Clavaud, C.; Demaude, J.; Breton, L.; Blaser, M.J. Body site is a more determinant factor than human population diversity in the healthy skin microbiome. *PLoS ONE* **2016**, *11*, e0151990. [CrossRef] [PubMed]
3. Mihai, M.M.; Preda, M.; Lungu, I.; Gestal, M.C.; Popa, M.I.; Holban, A.M. Nanocoatings for chronic wound repair—Modulation of microbial colonization and biofilm formation. *Int. J. Mol. Sci.* **2018**, *19*, 1179. [CrossRef] [PubMed]
4. Scalise, A.; Bianchi, A.; Tartaglione, C.; Bolletta, E.; Pierangeli, M.; Torresetti, M.; Marazzi, M.; Di Benedetto, G. Microenvironment and microbiology of skin wounds: The role of bacterial biofilms and related factors. *Semin. Vasc. Surg.* **2015**, *28*, 151–159. [CrossRef] [PubMed]
5. Peral, M.C.; Huaman Martinez, M.A.; Valdez, J.C. Bacteriotherapy with *Lactobacillus plantarum* in burns. *Int. Wound J.* **2009**, *6*, 73–81. [CrossRef]
6. Daeschlein, G. Antimicrobial and antiseptic strategies in wound management. *Int. Wound J.* **2013**, *10*, 9–14. [CrossRef]
7. Bowler, P.G.; Duerden, B.I.; Armstrong, D.G. Wound microbiology and associated approaches to wound management. *Clin. Microbiol. Rev.* **2001**, *14*, 244–269. [CrossRef]
8. Mahnic, A.; Breznik, V.; Bombek Ihan, M.; Rupnik, M. Comparison between cultivation and sequencing based approaches for microbiota analysis in swabs and biopsies of chronic wounds. *Front. Med.* **2021**, *8*, 607255. [CrossRef]
9. Fijan, S.; Frauwallner, A.; Langerholc, T.; Krebs, B.; Ter Haar Née Younes, J.A.; Heschl, A.; Mičetić Turk, D.; Rogelj, I. Efficacy of using probiotics with antagonistic activity against pathogens of wound infections: An integrative review of literature. *Biomed Res. Int.* **2019**, *2019*, 7585486. [CrossRef]
10. Martinengo, L.; Olsson, M.; Bajpai, R.; Soljak, M.; Upton, Z.; Schmidtchen, A.; Car, J.; Järbrink, K. Prevalence of chronic wounds in the general population: Systematic review and meta-analysis of observational studies. *Ann. Epidemiol.* **2019**, *29*, 8–15. [CrossRef]
11. Hill, C.; Guarner, F.; Reid, G.; Gibson, G.R.; Merenstein, D.J.; Pot, B.; Morelli, L.; Canani, R.B.; Flint, H.J.; Salminen, S.; et al. The international scientific association for probiotics and prebiotics consensus statement on the scope and appropriate use of the term probiotic. *Nat. Rev. Gastroenterol. Hepatol.* **2014**, *11*, 506–514. [CrossRef]
12. Reid, G.; Younes, J.A.; Van der Mei, H.C.; Gloor, G.B.; Knight, R.; Busscher, H.J. Microbiota restoration: Natural and supplemented recovery of human microbial communities. *Nat. Rev. Microbiol.* **2011**, *9*, 27–38. [CrossRef] [PubMed]
13. Cecchini, M.; Langer, J.; Slawomirski, L. *Antimicrobial Resistance in G7 Countries and beyond," Economic Issues, Policies and Options for Action: Oecd Report*; OECD: Paris, France, 2015.
14. Salminen, S.; Collado, M.C.; Endo, A.; Hill, C.; Lebeer, S.; Quigley, E.M.M.; Sanders, M.E.; Shamir, R.; Swann, J.R.; Szajewska, H.; et al. The international scientific association of probiotics and prebiotics (ISAPP) consensus statement on the definition and scope of postbiotics. *Nat. Rev. Gastroenterol. Hepatol.* **2021**, *18*, 649–667. [CrossRef] [PubMed]

15. Zheng, J.; Wittouck, S.; Salvetti, E.; Franz, C.; Harris, H.M.B.; Mattarelli, P.; O'Toole, P.W.; Pot, B.; Vandamme, P.; Walter, J.; et al. A taxonomic note on the genus *Lactobacillus*: Description of 23 novel genera, emended description of the genus *Lactobacillus* beijerinck 1901, and union of Lactobacillaceae and Leuconostocaceae. *Int. J. Syst. Evol. Microbiol.* **2020**, *70*, 2782–2858. [CrossRef] [PubMed]
16. Ramos, A.N.; Sesto Cabral, M.E.; Noseda, D.; Bosch, A.; Yantorno, O.M.; Valdez, J.C. Antipathogenic properties of *Lactobacillus plantarum* on *Pseudomonas aeruginosa*: The potential use of its supernatants in the treatment of infected chronic wounds. *Wound Repair Regen.* **2012**, *20*, 552–562. [CrossRef]
17. Valdéz, J.C.; Peral, M.C.; Rachid, M.; Santana, M.; Perdigón, G. Interference of *Lactobacillus plantarum* with *Pseudomonas aeruginosa* in vitro and in infected burns: The potential use of probiotics in wound treatment. *Clin. Microbiol. Infect.* **2005**, *11*, 472–479. [CrossRef]
18. Jones, M.L.; Ganopolsky, J.G.; Labbé, A.; Prakash, S. A novel nitric oxide producing probiotic patch and its antimicrobial efficacy: Preparation and in vitro analysis. *Appl. Microbiol. Biotechnol.* **2010**, *87*, 509–516. [CrossRef]
19. Chan, A.P.; Choi, Y.; Brinkac, L.M.; Krishnakumar, R.; Depew, J.; Kim, M.; Hinkle, M.K.; Lesho, E.P.; Fouts, D.E. Multidrug resistant pathogens respond differently to the presence of co-pathogen, commensal, probiotic and host cells. *Sci. Rep.* **2018**, *8*, 8656. [CrossRef]
20. Mohammedsaeed, W.; McBain, A.J.; Cruickshank, S.M.; O'Neill, C.A.; Schaffner, D.W. *Lactobacillus rhamnosus* GG inhibits the toxic effects of *Staphylococcus aureus* on epidermal keratinocytes. *Appl. Environ. Microbiol.* **2014**, *80*, 5773–5781. [CrossRef]
21. Shu, M.; Wang, Y.; Yu, J.; Kuo, S.; Coda, A.; Jiang, Y.; Gallo, R.L.; Huang, C.-M. Fermentation of *Propionibacterium acnes*, a commensal bacterium in the human skin microbiome, as skin probiotics against methicillin-resistant *Staphylococcus aureus*. *PLoS ONE* **2013**, *8*, e55380. [CrossRef]
22. Al-Malkey, M.K.; Ismeeal, M.C.; Al-Hur, F.J.; Mohammed, S.W.; Nayyel, H.J. Antimicrobial effect of probiotic *Lactobacillus* spp. on *Pseudomonas aeruginosa*. *J. Contemp. Med. Sci.* **2017**, *3*, 218–223. [CrossRef]
23. Prince, T.; McBain, A.J.; O'Neill, C.A. *Lactobacillus reuteri* protects epidermal keratinocytes from *Staphylococcus aureus*-induced cell death by competitive exclusion. *Appl. Environ. Microbiol.* **2012**, *78*, 5119–5126. [CrossRef] [PubMed]
24. Thomas, J.G.; Motlagh, H.; Povey, S.B.; Percival, S.L. 2—The role of micro-organisms and biofilms in dysfunctional wound healing. In *Advanced Wound Repair Therapies*; Farrar, D., Ed.; Woodhead Publishing: Sawston, UK, 2011; pp. 39–76.
25. Lopes, E.G.; Moreira, D.A.; Gullón, P.; Gullón, B.; Cardelle-Cobas, A.; Tavaria, F.K. Topical application of probiotics in skin: Adhesion, antimicrobial and antibiofilm in vitro assays. *J. Appl. Microbiol.* **2017**, *122*, 450–461. [CrossRef]
26. Moghadam, S.S.; Khodaii, Z.; Zadeh, S.F.; Ghooshchian, M.; Aghmiyuni, Z.F.; Shabestari, T.M. Synergistic or antagonistic effects of probiotics and antibiotics- alone or in combination- on antimicrobial-resistant *Pseudomonas aeruginosa* isolated from burn wounds. *Arch. Clin. Infect. Dis.* **2018**, *13*, e63121. [CrossRef]
27. Argenta, A.; Satish, L.; Gallo, P.; Liu, F.; Kathju, S. Local application of probiotic bacteria prophylaxes against sepsis and death resulting from burn wound infection. *PLoS ONE* **2016**, *11*, e0165294. [CrossRef] [PubMed]
28. Satish, L.; Gallo, P.H.; Johnson, S.; Yates, C.C.; Kathju, S. Local probiotic therapy with *Lactobacillus plantarum* mitigates scar formation in rabbits after burn injury and infection. *Surg. Infect.* **2017**, *18*, 119–127. [CrossRef] [PubMed]
29. Sürmeli, M.; Maçin, S.; Akyön, Y.; Kayikçioğlu, A.U. The protective effect of *Lactobacillus plantarum* against meticillin-resistant *Staphylococcus aureus* infections: An experimental animal model. *J. Wound Care* **2019**, *28*, s29–s34. [CrossRef]
30. Brachkova, M.I.; Marques, P.; Rocha, J.; Sepodes, B.; Duarte, M.A.; Pinto, J.F. Alginate films containing *Lactobacillus plantarum* as wound dressing for prevention of burn infection. *J. Hosp. Infect.* **2011**, *79*, 375–377. [CrossRef]
31. Ong, J.S.; Taylor, T.D.; Yong, C.C.; Khoo, B.Y.; Sasidharan, S.; Choi, S.B.; Ohno, H.; Liong, M.T. *Lactobacillus plantarum* USM8613 aids in wound healing and suppresses *Staphylococcus aureus* infection at wound sites. *Probiotics Antimicrob. Proteins* **2020**, *12*, 125–137. [CrossRef]
32. Jones, M.; Ganopolsky, J.G.; Labbé, A.; Gilardino, M.; Wahl, C.; Martoni, C.; Prakash, S. Novel nitric oxide producing probiotic wound healing patch: Preparation and in vivo analysis in a new zealand white rabbit model of ischaemic and infected wounds. *Int. Wound J.* **2012**, *9*, 330–343. [CrossRef]
33. Binda, S.; Hill, C.; Johansen, E.; Obis, D.; Pot, B.; Sanders, M.E.; Tremblay, A.; Ouwehand, A.C. Criteria to qualify microorganisms as probiotic in foods and dietary supplements. *Front. Microbiol.* **2020**, *11*, 1662. [CrossRef]
34. Peral, M.C.; Rachid, M.M.; Gobbato, N.M.; Martinez, M.A.H.; Valdez, J.C. Interleukin-8 production by polymorphonuclear leukocytes from patients with chronic infected leg ulcers treated with *Lactobacillus plantarum*. *Clin. Microbiol. Infect.* **2010**, *16*, 281–286. [CrossRef] [PubMed]
35. Komatsu, S.; Sakamoto, E.; Norimizu, S.; Shingu, Y.; Asahara, T.; Nomoto, K.; Nagino, M. Efficacy of perioperative synbiotics treatment for the prevention of surgical site infection after laparoscopic colorectal surgery: A randomized controlled trial. *Surg. Today* **2016**, *46*, 479–490. [CrossRef] [PubMed]
36. El-Ghazely, M.H.; Mahmoud, W.H.; Atia, M.A.; Eldip, E.M. Effect of probiotic administration in the therapy of pediatric thermal burn. *Ann. Burn. Fire Disasters* **2016**, *29*, 268–272.
37. Mayes, T.; Gottschlich, M.M.; James, L.E.; Allgeier, C.; Weitz, J.; Kagan, R.J. Clinical safety and efficacy of probiotic administration following burn injury. *J. Burn Care Res.* **2015**, *36*, 92–99. [CrossRef] [PubMed]

38. Kotzampassi, K.; Stavrou, G.; Damoraki, G.; Georgitsi, M.; Basdanis, G.; Tsaousi, G.; Giamarellos-Bourboulis, E.J. A four-probiotics regimen reduces postoperative complications after colorectal surgery: A randomized, double-blind, placebo-controlled study. *World J. Surg.* **2015**, *39*, 2776–2783. [CrossRef]
39. Aisu, N.; Tanimura, S.; Yamashita, Y.; Yamashita, K.; Maki, K.; Yoshida, Y.; Sasaki, T.; Takeno, S.; Hoshino, S. Impact of perioperative probiotic treatment for surgical site infections in patients with colorectal cancer. *Exp. Ther. Med.* **2015**, *10*, 966–972. [CrossRef]
40. Moraffah, F.; Kiani, M.; Abdollahi, M.; Yoosefi, S.; Vatanara, A.; Samadi, N. In vitro-in vivo correlation for the antibacterial effect of *Lactiplantibacillus plantarum* as a topical healer for infected burn wound. *Probiotics Antimicrob. Proteins* **2022**, *14*, 675–689. [CrossRef]
41. Bücher, C.; Burtscher, J.; Domig, K.J. Propionic acid bacteria in the food industry: An update on essential traits and detection methods. *Compr. Rev. Food Sci. Food Saf.* **2021**, *20*, 4299–4323. [CrossRef]
42. Dubernet, S.; Desmasures, N.; Guéguen, M. A PCR-based method for identification of lactobacilli at the genus level. *FEMS Microbiol. Lett.* **2002**, *214*, 271–275. [CrossRef]
43. Sul, S.Y.; Kim, H.J.; Kim, T.W.; Kim, H.Y. Rapid identification of *Lactobacillus* and *Bifidobacterium* in probiotic products using multiplex PCR. *J. Microbiol. Biotechnol.* **2007**, *17*, 490–495. [PubMed]
44. Walter, J.; Tannock, G.W.; Tilsala-Timisjarvi, A.; Rodtong, S.; Loach, D.M.; Munro, K.; Alatossava, T. Detection and identification of gastrointestinal *Lactobacillus* species by using denaturing gradient gel electrophoresis and species-specific PCR primers. *Appl. Environ. Microbiol.* **2000**, *66*, 297–303. [CrossRef] [PubMed]
45. Takahashi, H.; Fujita, T.; Suzuki, Y.; Benno, Y. Monitoring and survival of *Lactobacillus gasseri* SBT2055 in the human intestinal tract. *Microbiol. Immunol.* **2006**, *50*, 867–870. [CrossRef] [PubMed]
46. Gaspar, C.; Palmeira-de-Oliveira, R.; Martinez-de-Oliveira, J.; Neves, J.D.; Pestana, P.G.; Rolo, J.; Donders, G.; Palmeira-de-Oliveira, A. Development and validation of a new one step multiplex-pcr assay for the detection of ten *Lactobacillus* species. *Anaerobe* **2019**, *59*, 192–200. [CrossRef]
47. Bernhard, A.E.; Field, K.G. Identification of nonpoint sources of fecal pollution in coastal waters by using host-specific 16S ribosomal DNA genetic markers from fecal anaerobes. *Appl. Environ. Microbiol.* **2000**, *66*, 1587–1594. [CrossRef]
48. Sheu, S.J.; Hwang, W.Z.; Chiang, Y.C.; Lin, W.H.; Chen, H.C.; Tsen, H.Y. Use of Tuf gene-based primers for the PCR detection of probiotic *Bifidobacterium* species and enumeration of bifidobacteria in fermented milk by cultural and quantitative real-time pcr methods. *J. Food Sci.* **2010**, *75*, M521–M527. [CrossRef]
49. Deasy, B.M.; Rea, M.C.; Fitzgerald, G.F.; Cogan, T.M.; Beresford, T.P. A rapid PCR based method to distinguish between *Lactococcus* and *Enterococcus*. *Syst. Appl. Microbiol.* **2000**, *23*, 510–522. [CrossRef]
50. Rossi, F.; Torriani, S.; Dellaglio, F. Genus- and species-specific PCR-based detection of dairy propionibacteria in environmental samples by using primers targeted to the genes encoding 16S rRNA. *Appl. Environ. Microbiol.* **1999**, *65*, 4241–4244. [CrossRef]
51. Majeed, M.; Nagabhushanam, K.; Arumugam, S.; Alli, F. Novel PCR Primers and Methods Thereof for the Identification of *Bacillus coagulans*. WO 2017/058741, 6 April 2017.
52. Wattiau, P.; Renard, M.E.; Ledent, P.; Debois, V.; Blackman, G.; Agathos, S.N. A PCR test to identify *Bacillus subtilis* and closely related species and its application to the monitoring of wastewater biotreatment. *Appl. Microbiol. Biotechnol.* **2001**, *56*, 816–819. [CrossRef]
53. Cheng, S.; McCleskey, F.K.; Gress, M.J.; Petroziello, J.M.; Liu, R.; Namdari, H.; Beninga, K.; Salmen, A.; DelVecchio, V.G. A PCR assay for identification of *Enterococcus faecium*. *J. Clin. Microbiol.* **1997**, *35*, 1248–1250. [CrossRef]
54. Mitterdorfer, G.; Mayer, H.K.; Kneifel, W.; Viernstein, H. Clustering of *Saccharomyces boulardii* strains within the species *S. cerevisiae* using molecular typing techniques. *J. Appl. Microbiol.* **2002**, *93*, 521–530. [CrossRef] [PubMed]
55. Fijan, S.; Šulc, D.; Steyer, A. Study of the in vitro antagonistic activity of various single-strain and multi-strain probiotics against *Escherichia coli*. *Int. J. Environ. Res. Public Health* **2018**, *15*, 1539. [CrossRef] [PubMed]
56. Igbafe, J.; Kilonzo-Nthenge, A.; Nahashon, S.N.; Mafiz, A.I.; Nzomo, M. Probiotics and antimicrobial effect of *Lactiplantibacillus plantarum, Saccharomyces cerevisiae,* and *Bifidobacterium longum* against common foodborne pathogens in poultry. *Agriculture* **2020**, *10*, 368. [CrossRef]
57. Shokryazdan, P.; Sieo, C.C.; Kalavathy, R.; Liang, J.B.; Alitheen, N.B.; Faseleh Jahromi, M.; Ho, Y.W. Probiotic potential of lactobacillus strains with antimicrobial activity against some human pathogenic strains. *Biomed Res. Int.* **2014**, *2014*, 927268. [CrossRef] [PubMed]
58. Davis, W.W.; Stout, T.R. Disc plate method of microbiological antibiotic assay. I. Factors influencing variability and error. *Appl. Microbiol.* **1971**, *22*, 659–665. [CrossRef] [PubMed]
59. Fijan, S. Antimicrobial effect of probiotics against common pathogens. In *Probiotics and Prebiotics in Human Nutrition and Health*; Rao, V., Rao, L., Eds.; InTech: Rijeka, Croatia, 2016; ISBN 978-953-51-2476-4. [CrossRef]
60. Tranberg, A.; Klarin, B.; Johansson, J.; Påhlman, L.I. Efficacy of *Lactiplantibacillus plantarum* 299 and *299v* against nosocomial oropharyngeal pathogens in vitro and as an oral prophylactic treatment in a randomized, controlled clinical trial. *MicrobiologyOpen* **2021**, *10*, e1151. [CrossRef] [PubMed]
61. Holder, I.A.; Boyce, S.T. Agar well diffusion assay testing of bacterial susceptibility to various antimicrobials in concentrations non-toxic for human cells in culture. *Burns* **1994**, *20*, 426–429. [CrossRef]

62. Varela-Trinidad, G.U.; Domínguez-Díaz, C.; Solórzano-Castanedo, K.; Íñiguez-Gutiérrez, L.; Hernández-Flores, T.D.J.; Fafutis-Morris, M. Probiotics: Protecting our health from the gut. *Microorganisms* **2022**, *10*, 1428. [CrossRef]
63. Millette, M.; Luquet, F.M.; Lacroix, M. In vitro growth control of selected pathogens by *Lactobacillus acidophilus*- and *Lactobacillus casei*-fermented milk. *Lett. Appl. Microbiol.* **2007**, *44*, 314–319. [CrossRef]
64. Saxelin, M.; Tynkkynen, S.; Mattila-Sandholm, T.; de Vos, W.M. Probiotic and other functional microbes: From markets to mechanisms. *Curr. Opin. Biotechnol.* **2005**, *16*, 204–211. [CrossRef]
65. Wang, H.; Jiang, X. Isolation and characterization of competitive exclusion microorganisms from animal wastes-based composts against *Listeria monocytogenes*. *J. Appl. Microbiol.* **2022**, *132*, 4531–4543. [CrossRef] [PubMed]
66. Dowarah, R.; Verma, A.K.; Agarwal, N.; Singh, P.; Singh, B.R. Selection and characterization of probiotic lactic acid bacteria and its impact on growth, nutrient digestibility, health and antioxidant status in weaned piglets. *PLoS ONE* **2018**, *13*, e0192978. [CrossRef] [PubMed]
67. El-Kholy, A.M.; El-Shinawy, S.H.; Meshref, A.M.S.; Korny, A.M. Screening of antagonistic activity of probiotic bacteria against some food-borne pathogens. *J. Appl. Environ. Microbiol.* **2014**, *2*, 53–60. [CrossRef]
68. Varma, P.; Nisha, N.; Dinesh, K.R.; Kumar, A.V.; Biswas, R. Anti-infective properties of *Lactobacillus fermentum* against *Staphylococcus aureus* and *Pseudomonas aeruginosa*. *J. Mol. Microbiol. Biotechnol.* **2011**, *20*, 137–143. [CrossRef]
69. Jeanson, S.; Floury, J.; Gagnaire, V.; Lortal, S.; Thierry, A. Bacterial colonies in solid media and foods: A review on their growth and interactions with the micro-environment. *Front. Microbiol.* **2015**, *6*, 1284. [CrossRef]
70. Fijan, S. Influence of the growth of *Pseudomonas aeruginosa* in milk fermented by multispecies probiotics and kefir microbiota. *J. Probiotics Health* **2015**, *4*, 1–6. [CrossRef]
71. Skandamis, P.N.; Jeanson, S. Colonial vs. Planktonic type of growth: Mathematical modeling of microbial dynamics on surfaces and in liquid, semi-liquid and solid foods. *Front. Microbiol.* **2015**, *6*, 1178. [CrossRef]
72. Chapman, C.M.; Gibson, G.R.; Todd, S.; Rowland, I. Comparative in vitro inhibition of urinary tract pathogens by single- and multi-strain probiotics. *Eur. J. Nutr.* **2013**, *52*, 1669–1677. [CrossRef]
73. Monteiro, C.; Do Carmo, M.; Melo, B.; Alves, M.; Dos Santos, C.; Monteiro, S.; Bomfim, M.; Fernandes, E.; Monteiro-Neto, V. In vitro antimicrobial activity and probiotic potential of *Bifidobacterium* and *Lactobacillus* against species of *Clostridium*. *Nutrients* **2019**, *11*, 448. [CrossRef]
74. Tejero-Sariñena, S.; Barlow, J.; Costabile, A.; Gibson, G.R.; Rowland, I. In vitro evaluation of the antimicrobial activity of a range of probiotics against pathogens: Evidence for the effects of organic acids. *Anaerobe* **2012**, *18*, 530–538. [CrossRef]
75. Favaro, L.; Barretto Penna, A.L.; Todorov, S.D. Bacteriocinogenic lab from cheeses—Application in biopreservation? *Trends Food Sci. Technol.* **2015**, *41*, 37–48. [CrossRef]
76. Piwowarek, K.; Lipińska, E.; Hać-Szymańczuk, E.; Kieliszek, M.; Ścibisz, I. *Propionibacterium* spp.—Source of propionic acid, vitamin b12, and other metabolites important for the industry. *Appl. Microbiol. Biotechnol.* **2018**, *102*, 515–538. [CrossRef] [PubMed]
77. El Issaoui, K.; Khay, E.O.; Abrini, J.; Zinebi, S.; Amajoud, N.; Senhaji, N.S.; Abriouel, H. Molecular identification and antibiotic resistance of bacteriocinogenic lactic acid bacteria isolated from table olives. *Arch. Microbiol.* **2021**, *203*, 597–607. [CrossRef] [PubMed]
78. Fredua-Agyeman, M.; Gaisford, S. Assessing inhibitory activity of probiotic culture supernatants against *Pseudomonas aeruginosa*: A comparative methodology between agar diffusion, broth culture and microcalorimetry. *World J. Microbiol. Biotechnol.* **2019**, *35*, 49. [CrossRef] [PubMed]
79. FDA. *Content and Format of Premarket Notification [510(k)] Submissions for Liquid Chemical Sterilants/ High Level Disinfectants*; U.S. Department of Health And Human Services, Food and Drug Administration: Washington, DC, USA, 2000.
80. Matijašić, B.B.; Obermajer, T.; Rogelj, I. Quantification of *Lactobacillus gasseri, Enterococcus faecium* and *Bifidobacterium infantis* in a probiotic otc drug by real-time pcr. *Food Control* **2010**, *21*, 419–425. [CrossRef]
81. Goldstein, E.J.C.; Tyrrell, K.L.; Citron, D.M. Lactobacillus species: Taxonomic complexity and controversial susceptibilities. *Clin. Infect. Dis.* **2015**, *60*, S98–S107. [CrossRef]
82. Patro, J.N.; Ramachandran, P.; Lewis, J.L.; Mammel, M.K.; Barnaba, T.; Pfeiler, E.A.; Elkins, C.A. Development and utility of the fda 'gutprobe' DNA microarray for identification, genotyping and metagenomic analysis of commercially available probiotics. *J. Appl. Microbiol.* **2015**, *118*, 1478–1488. [CrossRef]
83. Otaka, M.; Kikuchi-Hayakawa, H.; Ogura, J.; Ishikawa, H.; Yomogida, Y.; Ota, M.; Hidese, S.; Ishida, I.; Aida, M.; Matsuda, K.; et al. Effect of *Lacticaseibacillus paracasei* strain Shirota on improvement in depressive symptoms, and its association with abundance of actinobacteria in gut microbiota. *Microorganisms* **2021**, *9*, 1026. [CrossRef]
84. Kim, E.; Yang, S.M.; Cho, E.J.; Kim, H.Y. Novel real-time PCR assay for *Lactobacillus casei* group species using comparative genomics. *Food Microbiol.* **2020**, *90*, 103485. [CrossRef]
85. Szajewska, H.; Kołodziej, M. Systematic review with meta-analysis: *Lactobacillus rhamnosus* GG in the prevention of antibiotic-associated diarrhoea in children and adults. *Aliment. Pharmacol. Ther.* **2015**, *42*, 1149–1157. [CrossRef]
86. Li, Y.-T.; Xu, H.; Ye, J.-Z.; Wu, W.-R.; Shi, D.; Fang, D.-Q.; Liu, Y.; Li, L.-J. Efficacy of *Lactobacillus rhamnosus* GG in treatment of acute pediatric diarrhea: A systematic review with meta-analysis. *World J. Gastroenterol.* **2019**, *25*, 4999–5016. [CrossRef] [PubMed]
87. Szajewska, H.; Skórka, A.; Ruszczyński, M.; Gieruszczak-Białek, D. Meta-analysis: *Lactobacillus* GG for treating acute gastroenteritis in children—Updated analysis of randomised controlled trials. *Aliment. Pharmacol. Ther.* **2013**, *38*, 467–476. [CrossRef] [PubMed]

88. Laursen, R.P.; Hojsak, I. Probiotics for respiratory tract infections in children attending day care centers—A systematic review. *Eur. J. Pediatr.* **2018**, *177*, 979–994. [CrossRef] [PubMed]
89. Capurso, L. Thirty years of *Lactobacillus rhamnosus* GG: A review. *J. Clin. Gastroenterol.* **2019**, *53* (Suppl. 1), S1–S41. [CrossRef] [PubMed]
90. Jang, H.J.; Lee, N.-K.; Paik, H.-D. Probiotic characterization of *Lactobacillus brevis* Ku15153 showing antimicrobial and antioxidant effect isolated from kimchi. *Food Sci. Biotechnol.* **2019**, *28*, 1521–1528. [CrossRef]
91. Wang, Y.; Jia, D.; Wang, J.-H.; Li, H.-H.; Liu, J.-L.; Liu, A.-H.; Wang, J.-M.; Guan, G.-Q.; Luo, J.-X.; Yin, H.; et al. Assessment of probiotic adhesion and inhibitory effect on *Escherichia coli* and *Salmonella* adhesion. *Arch. Microbiol.* **2021**, *203*, 6267–6274. [CrossRef]
92. Papadimitriou, K.; Zoumpopoulou, G.; Foligné, B.; Alexandraki, V.; Kazou, M.; Pot, B.; Tsakalidou, E. Discovering probiotic microorganisms: In vitro, in vivo, genetic and omics approaches. *Front. Microbiol.* **2015**, *6*, 58. [CrossRef]
93. Vinderola, G.; Gueimonde, M.; Gomez-Gallego, C.; Delfederico, L.; Salminen, S. Correlation between in vitro and in vivo assays in selection of probiotics from traditional species of bacteria. *Trends Food Sci. Technol.* **2017**, *68*, 83–90. [CrossRef]
94. Li, X.; Liu, Y.; Guo, X.; Ma, Y.; Zhang, H.; Liang, H. Effect of *Lactobacillus casei* on lipid metabolism and intestinal microflora in patients with alcoholic liver injury. *Eur. J. Clin. Nutr.* **2021**, *75*, 1227–1236. [CrossRef]
95. Mai, T.T.; Thi Thu, P.; Thi Hang, H.; Trang, T.T.T.; Yui, S.; Shigehisa, A.; Tien, V.T.; Dung, T.V.; Nga, P.B.; Hung, N.T.; et al. Efficacy of probiotics on digestive disorders and acute respiratory infections: A controlled clinical trial in young vietnamese children. *Eur. J. Clin. Nutr.* **2021**, *75*, 513–520. [CrossRef]
96. Macnaughtan, J.; Figorilli, F.; García-López, E.; Lu, H.; Jones, H.; Sawhney, R.; Suzuki, K.; Fairclough, S.; Marsden, J.; Moratella, A.; et al. A double-blind, randomized placebo-controlled trial of probiotic *Lactobacillus casei* Shirota in stable cirrhotic patients. *Nutrients* **2020**, *12*, 1651. [CrossRef] [PubMed]
97. Paniágua, A.L.; Correia, A.F.; Pereira, L.C.; de Alencar, B.M.; Silva, F.B.A.; Almeida, R.M.; de Medeiros Nóbrega, Y.K. Inhibitory effects of *Lactobacillus casei* Shirota against both *Candida auris* and *Candida* spp. Isolates that cause vulvovaginal candidiasis and are resistant to antifungals. *BMC Complement. Med. Ther.* **2021**, *21*, 237. [CrossRef] [PubMed]
98. Li, X.; Liang, H. Effects of *Lactobacillus casei* on iron metabolism and intestinal microflora in rats exposed to alcohol and iron. *Turk. J. Gastroenterol.* **2022**, *33*, 470–476. [CrossRef] [PubMed]
99. Szajewska, H.; Urbańska, M.; Chmielewska, A.; Weizman, Z.; Shamir, R. Meta-analysis: *Lactobacillus reuteri* strain DSM 17938 (and the original strain ATCC 55730) for treating acute gastroenteritis in children. *Benef. Microbes* **2014**, *5*, 285–293. [CrossRef] [PubMed]
100. Liévin-Le Moal, V.; Servin, A.L. Anti-infective activities of lactobacillus strains in the human intestinal microbiota: From probiotics to gastrointestinal anti-infectious biotherapeutic agents. *Clin. Microbiol. Rev.* **2014**, *27*, 167–199. [CrossRef]
101. Moreno Márquez, C.; Fernández Álvarez, P.; Valdés Delgado, T.; Castro Laria, L.; Argüelles Arias, F.; Caunedo Álvarez, A.; Gómez Rodríguez, B.J. Randomized, double-blind, placebo-controlled clinical trial on the usefulness of probiotic *Lactobacillus reuteri* in bismuth-containing quadruple eradication therapy for infection with *Helicobacter pylori*. *Rev. Española Enferm. Dig.* **2021**. [CrossRef]
102. Invernici, M.M.; Furlaneto, F.A.C.; Salvador, S.L.; Ouwehand, A.C.; Salminen, S.; Mantziari, A.; Vinderola, G.; Ervolino, E.; Santana, S.I.; Silva, P.H.F.; et al. *Bifidobacterium animalis* subsp *lactis* HN019 presents antimicrobial potential against periodontopathogens and modulates the immunological response of oral mucosa in periodontitis patients. *PLoS ONE* **2020**, *15*, e0238425. [CrossRef]
103. Arunachalam, K.; Gill, H.; Chandra, R. Enhancement of natural immune function by dietary consumption of *Bifidobacterium lactis* (HN019). *Eur. J. Clin. Nutr.* **2000**, *54*, 263–267. [CrossRef]
104. Fredua-Agyeman, M.; Stapleton, P.; Basit, A.W.; Beezer, A.E.; Gaisford, S. In vitro inhibition of *Clostridium difficile* by commercial probiotics: A microcalorimetric study. *Int. J. Pharm.* **2017**, *517*, 96–103. [CrossRef]
105. Taipale, T.; Pienihäkkinen, K.; Isolauri, E.; Larsen, C.; Brockmann, E.; Alanen, P.; Jokela, J.; Söderling, E. *Bifidobacterium animalis* subsp. *lactis* BB-12 in reducing the risk of infections in infancy. *Br. J. Nutr.* **2011**, *105*, 409–416. [CrossRef]
106. Dyshlyuk, L.S.; Milentyeva, I.S.; Asyakina, L.K.; Ostroumov, L.A.; Osintsev, A.M.; Pozdnyakova, A.V. Using *Bifidobacterium* and *Propionibacterium* strains in probiotic consortia to normalize the gastrointestinal tract. *Braz. J. Biol.* **2022**, *84*, e256945. [CrossRef] [PubMed]
107. Koskinen, P.; Deptula, P.; Smolander, O.P.; Tamene, F.; Kammonen, J.; Savijoki, K.; Paulin, L.; Piironen, V.; Auvinen, P.; Varmanen, P. Complete genome sequence of *Propionibacterium freudenreichii* DSM 20271(T). *Stand. Genom. Sci.* **2015**, *10*, 83. [CrossRef] [PubMed]
108. Lyra, A.; Krogius-Kurikka, L.; Nikkilä, J.; Malinen, E.; Kajander, K.; Kurikka, K.; Korpela, R.; Palva, A. Effect of a multispecies probiotic supplement on quantity of irritable bowel syndrome-related intestinal microbial phylotypes. *BMC Gastroenterol.* **2010**, *10*, 110. [CrossRef]
109. Myllyluoma, E.; Ahlroos, T.; Veijola, L.; Rautelin, H.; Tynkkynen, S.; Korpela, R. Effects of anti-*Helicobacter pylori* treatment and probiotic supplementation on intestinal microbiota. *Int. J. Antimicrob. Agents* **2007**, *29*, 66–72. [CrossRef] [PubMed]
110. Lee, N.-K.; Kim, W.-S.; Paik, H.-D. Bacillus strains as human probiotics: Characterization, safety, microbiome, and probiotic carrier. *Food Sci. Biotechnol.* **2019**, *28*, 1297–1305. [CrossRef] [PubMed]
111. Ahire, J.J.; Neelamraju, J.; Madempudi, R.S. Behavior of *Bacillus coagulans* unique IS2 spores during passage through the simulator of human intestinal microbial ecosystem (shime) model. *LWT* **2020**, *124*, 109196. [CrossRef]

112. Madempudi, R.S.; Neelamraju, J.; Ahire, J.J.; Gupta, S.K.; Shukla, V.K. *Bacillus coagulans* unique IS2 in constipation: A double-blind, placebo-controlled study. *Probiotics Antimicrob. Proteins* **2020**, *12*, 335–342. [CrossRef] [PubMed]
113. Ratna Sudha, M.; Yelikar, K.A.; Deshpande, S. Clinical study of *Bacillus coagulans* unique IS-2 (ATCC PTA-11748) in the treatment of patients with bacterial vaginosis. *Indian J. Microbiol.* **2012**, *52*, 396–399. [CrossRef]
114. Abid, R.; Waseem, H.; Ali, J.; Ghazanfar, S.; Muhammad Ali, G.; Elasbali, A.M.; Alharethi, S.H. Probiotic yeast *Saccharomyces*: Back to nature to improve human health. *J. Fungi* **2022**, *8*, 444. [CrossRef]
115. Dinleyici, E.C.; Kara, A.; Ozen, M.; Vandenplas, Y. *Saccharomyces boulardii* CNCM I-745 in different clinical conditions. *Expert Opin. Biol. Ther.* **2014**, *14*, 1593–1609. [CrossRef]
116. Szajewska, H.; Guarino, A.; Hojsak, I.; Indrio, F.; Kolacek, S.; Shamir, R.; Vandenplas, Y.; Weizman, Z. Use of probiotics for management of acute gastroenteritis: A position paper by the espghan working group for probiotics and prebiotics. *J. Pediatr. Gastroenterol. Nutr.* **2014**, *58*, 531–539. [CrossRef] [PubMed]
117. Guarino, A.; Ashkenazi, S.; Gendrel, D.; Lo Vecchio, A.; Shamir, R.; Szajewska, H. European society for pediatric gastroenterology, hepatology, and nutrition/european society for pediatric infectious diseases evidence-based guidelines for the management of acute gastroenteritis in children in Europe: Update 2014. *J. Pediatr. Gastroenterol. Nutr.* **2014**, *59*, 132–152. [CrossRef] [PubMed]
118. Ishaque, S.M.; Khosruzzaman, S.M.; Ahmed, D.S.; Sah, M.P. A randomized placebo-controlled clinical trial of a multi-strain probiotic formulation (Bio-Kult®) in the management of diarrhea-predominant irritable bowel syndrome. *BMC Gastroenterol.* **2018**, *18*, 71. [CrossRef] [PubMed]
119. Reininghaus, E.Z.; Wetzlmair, L.C.; Fellendorf, F.T.; Platzer, M.; Queissner, R.; Birner, A.; Pilz, R.; Hamm, C.; Maget, A.; Rieger, A.; et al. Probiotic treatment in individuals with euthymic bipolar disorder: A pilot-study on clinical changes and compliance. *Neuropsychobiology* **2020**, *79*, 71–79. [CrossRef] [PubMed]
120. Moser, A.M.; Spindelboeck, W.; Halwachs, B.; Strohmaier, H.; Kump, P.; Gorkiewicz, G.; Högenauer, C. Effects of an oral synbiotic on the gastrointestinal immune system and microbiota in patients with diarrhea-predominant irritable bowel syndrome. *Eur. J. Nutr.* **2019**, *58*, 2767–2778. [CrossRef]
121. Kienesberger, B.; Obermüller, B.; Singer, G.; Arneitz, C.; Gasparella, P.; Klymiuk, I.; Horvath, A.; Stadlbauer, V.; Magnes, C.; Zügner, E.; et al. Insights into the composition of a co-culture of 10 probiotic strains (Omni Biotic® AAD10) and effects of its postbiotic culture supernatant. *Nutrients* **2022**, *14*, 1194. [CrossRef]
122. Sprunt, K.; Leidy, G. The use of bacterial interference to prevent infection. *Can. J. Microbiol.* **1988**, *34*, 332–338. [CrossRef]
123. Florey, H.W. Use of micro-organisms for therapeutic purposes. *Br. Med. J.* **1945**, *2*, 635–642. [CrossRef]
124. Howard, J.C.; Reid, G.; Gan, B.S. Probiotics in surgical wound infections: Current status. *Clin. Investig. Med.* **2004**, *27*, 274–281.
125. Bekiaridou, A.; Karlafti, E.; Oikonomou, I.M.; Ioannidis, A.; Papavramidis, T.S. Probiotics and their effect on surgical wound healing: A systematic review and new insights into the role of nanotechnology. *Nutrients* **2021**, *13*, 4265. [CrossRef]
126. Reid, G.; Jass, J.; Sebulsky, M.T.; McCormick, J.K. Potential uses of probiotics in clinical practice. *Clin. Microbiol. Rev.* **2003**, *16*, 658–672. [CrossRef] [PubMed]
127. Siddharthan, R.; Chapek, M.; Warren, M.; Martindale, R. Probiotics in prevention of surgical site infections. *Surg. Infect. (Larchmt)* **2018**, *19*, 781–784. [CrossRef] [PubMed]

microorganisms MDPI

Review

The Weissella Genus: Clinically Treatable Bacteria with Antimicrobial/Probiotic Effects on Inflammation and Cancer

Sadia Ahmed [1,2,3,4,†], Sargun Singh [1,2,†], Vaidhvi Singh [1,2], Kyle D. Roberts [1], Arsalan Zaidi [3,4] and Alexander Rodriguez-Palacios [1,2,5,6,*]

1 Division of Gastroenterology and Liver Disease, Case Western Reserve University School of Medicine, Cleveland, OH 44106, USA
2 Digestive Health Research Institute, Case Western Reserve University School of Medicine, Cleveland, OH 44106, USA
3 National Probiotic Laboratory, National Institute for Biotechnology and Genetic Engineering (NIBGE-C), Faisalabad 38000, Pakistan
4 Pakistan Institute of Engineering and Applied Sciences, Nilore, Islamabad 45650, Pakistan
5 University Hospitals Research and Education Institute, University Hospitals Cleveland Medical Center, Cleveland, OH 44106, USA
6 Department of Molecular Biology and Microbiology, Case Western Reserve University School of Medicine, Cleveland, OH 44106, USA
* Correspondence: axr503@case.edu
† These authors contributed equally to this work.

Abstract: *Weissella* is a genus earlier considered a member of the family *Leuconostocaceae,* which was reclassified into the family *Lactobacillaceae* in 1993. Recently, there have been studies emphasizing the probiotic and anti-inflammatory potential of various species of *Weissella*, of which *W. confusa* and *W. cibaria* are the most representative. Other species within this genus include: *W. paramesenteroides, W. viridescens, W. halotolerans, W. minor, W. kandleri, W. soli, W. ghanensis, W. hellenica, W. thailandensis, W. fabalis, W. cryptocerci, W. koreensis, W. beninensis, W. fabaria, W. oryzae, W. ceti, W. uvarum, W. bombi, W. sagaensis, W. kimchi, W. muntiaci, W. jogaejeotgali, W. coleopterorum, W. hanii, W. salipiscis,* and *W. diestrammenae. Weissella confusa, W. paramesenteroides, W. koreensis,* and *W. cibaria* are among the few species that have been isolated from human samples, although the identification of these and other species is possible using metagenomics, as we have shown for inflammatory bowel disease (IBD) and healthy controls. We were able to isolate *Weissella* in gut-associated bacteria (post 24 h food deprivation and laxatives). Other sources of isolation include fermented food, soil, and skin/gut/saliva of insects/animals. With the potential for hospital and industrial applications, there is a concern about possible infections. Herein, we present the current applications of *Weissella* on its antimicrobial and anti-inflammatory mechanistic effects, the predisposing factors (e.g., vancomycin) for pathogenicity in humans, and the antimicrobials used in patients. To address the medical concerns, we examined 28 case reports focused on *W. confusa* and found that 78.5% of infections were bacteremia (of which 7 were fatal; 1 for lack of treatment), 8 were associated with underlying malignancies, and 8 with gastrointestinal procedures/diseases of which 2 were Crohn's disease patients. In cases of a successful resolution, commonly administered antibiotics included: cephalosporin, ampicillin, piperacillin-tazobactam, and daptomycin. Despite reports of *Weissella*-related infections, the evolving mechanistic findings suggest that *Weissella* are clinically treatable bacteria with emerging antimicrobial and probiotic benefits ranging from oral health, skin care, obesity, and inflammatory diseases to cancer.

Keywords: probiotic; antimicrobial; anti-inflammatory; anticancer; GRAS; starter culture; food; gut

Citation: Ahmed, S.; Singh, S.; Singh, V.; Roberts, K.D.; Zaidi, A.; Rodriguez-Palacios, A. The Weissella Genus: Clinically Treatable Bacteria with Antimicrobial/Probiotic Effects on Inflammation and Cancer. *Microorganisms* **2022**, *10*, 2427. https://doi.org/10.3390/microorganisms10122427

Academic Editor: Sabina Fijan

Received: 24 October 2022
Accepted: 2 December 2022
Published: 7 December 2022

1. Introduction

The *Weissella* genus has begun to take center stage in the past few years owing to its probiotic potential and its many prospective applications, ranging from the healthcare industry to the skin care and food industries. Due to its ability to thrive in stomach acid and bile, adherence to intestinal cells, and its antimicrobial potential against other pathogenic microorganisms including but not limited to *Staphylococcus aureus*, *Listeria monocytogenes*, *Salmonella typhi*, and *Salmonella enterica*, most *Weissella* species meet the pre-requisites needed to be classified as a probiotic. The only limitation to its widespread use is the lack of a significant volume of research at the moment and a handful of reported cases of pathogenicity. However, a bulk of these cases are a result of some preexisting disposition or comorbidity associated with the host. Despite such pathogenic potential, we set to investigate to what extent this genus is clinically treatable with common antimicrobials in the event of its identification in human infections. Herein, we primarily describe the antimicrobial and anti-inflammatory potential of *Weissella* and summarize the commonly used antibiotics in clinical settings where humans were diagnosed and treated/cured of *Weissella* infections.

The *Weissella* genus was first considered a member of the family *Leuconostocaceae* due to their significant shared similarities [1] but later on differentiated into a distinguished genus, which was named '*Weissella*' after the German microbiologist Norbert Weiss [2]. It was reclassified based on the phenotypic and genotypic analysis by Collins in 1993 [3]. The bacteria in this genus are non-spore-forming, generally non-motile, Gram-positive and catalase-negative [4] in nature that exist as either rods or ovoid-shaped cocci [5] belonging to the phylum *Firmicutes* and the family *Lactobacillaceae.* These bacteria are found to thrive in various ecological environments such as soil [6], plants, freshwater lakes [7], spontaneously fermented vegetables, and animal foods [8,9]. They can also be present as commensals on the skin surface and in the saliva and gastrointestinal tract of humans and animals as regular residents. The gastrointestinal tract is particularly thought to be a reservoir for colonization by *Weissella* [10].

2. Taxonomy and Sources of Isolation

According to the taxonomy database at The National Center for Biotechnology Information (NCBI, txid46255), as of October 2022, *Weissella* consists of 22 species: *Weissella bombi*, *Weissella ceti*, *Weissella cibaria*, *Weissella coleopterorum*, *Weissella confusa*, *Weissella diestrammenae*, *Weissella halotolerans*, *Weissella hanii*, *Weissella hellenica*, *Weissella jogaejeotgali*, *Weissella kandleri*, *Weissella koreensis*, *Weissella minor*, *Weissella muntiaci*, *Weissella oryzae*, *Weissella paramesenteroides*, *Weissella sagaensis*, *Weissella salipiscis*, *Weissella soli*, *Weissella thailandensis*, *Weissella uvarum*, and *Weissella viridescens*. However, Teixeira et al. [11] (February 2021) reported that 25 species of *Weissella* have been validated, whereas Fanelli et al. [12] grouped 26 species of *Weissella* into 6 phylogenetic clusters. Outside of the NCBI database, six more species are found in the 'List of Prokaryotic names with Standing in Nomenclature' database (https://lpsn.dsmz.de/genus/weissella) accessed on 18 October 2022: *Weissella beninensis*, *Weissella cryptocerci*, *Weissella fabalis*, *Weissella fabaria*, *Weissella ghanensis*, and *Weissella kimchi.* Twenty-six of these species are validly published under the International Code of Nomenclature (ICNP), except for *Weissella salipiscis*. It is to be noted that when taking into consideration the NCBI and LPSN databases together, 28 species of *Weissella* have been reported (Table 1).

Of the known species of *Weissella*, two (*W. confusa* and *W. cibaria*) have been reported from human or animal clinical infections [13]. However, the metagenome analysis of human fecal samples obtained from IBD patients and controls in our laboratory revealed the presence of several *Weissella* species (Singh et al. unpublished data). All known species of *Weissella* and their varied sources of isolation include: meat (*W. viridescens*, *W. halotolerans*, *W. minor*), fermented animal and plant-based food items (*W. confusa*, *W. jogaejeotgali*, *W. kimchi*, *W. hellenica*, *W. thailandensis*, *W. koreensis*, *W. ghanensis*, *W. sagaensis*, *W. beninensis*, *W. fabaria*, *W. fabalis*, *W. oryzae*, *W. hanii*, and *W. salipiscis*), animal/insect sources (*W. ceti*,

W. diestrammenae, *W. cryptocerci*, *W. bombi*, *W. muntiaci*, and *W. coleopterorum*), wine/wine grapes (*W. paramesenteroides* and *W. uvarum*), soil (*W. soli* and *W. kandleri*), and human samples (*W. cibaria*).

Table 1. Summary of sources of most common *Weissella* species.

S. No.	Bacterial Name	Source	Ref.
1	*W. viridescens*	Cured meat	[14]
2	*W. paramesenteroides*	Wine	[15]
3	*W. confusa*	Fermented Greek sausage	[16]
4	*W. kandleri*	Namib desert	[17]
5	*W. halotolerans*	Meat products	[18]
6	*W. minor*	Meat products	[18]
7	*W. hellenica*	Fermented Greek sausage	[16]
8	*W. thailandensis*	Fermented fish	[19]
9	*W. soli*	Soil	[20]
10	*W. cibaria*	Malaysian food and human samples	[21]
11	*W. koreensis*	Kimchi	[22]
12	*W.ghanensis*	Ghanaian cocoa fermentation	[23]
13	*W. beninensis*	Submerged cassava fermentations	[24]
14	*W. fabaria*	Ghanaian cocoa fermentation	[25]
15	*W. ceti*	Beaked whales	[26]
16	*W. fabalis*	Cocoa bean fermentations	[27]
17	*W. oryzae*	Fermented rice grains	[28]
18	*W. diestrammenae*	Gut of a camel cricket	[29]
19	*W. uvarum*	Wine grapes	[30]
20	*W. cryptocerci*	Gut of the insect	[31]
21	*W. bombi*	Bumble bee gut	[32]
22	*W. jogaejeotgali*	Korean fermented seafood	[33]
23	*W. kimchi*	Kimchi	[34]
24	*W. muntiaci*	Feces of Formosan barking deer	[1]
25	*W. sagaensis*	Traditional Chinese yogurt	[35]
26	*W. hanii*	kimchi	[36]
27	*W. salipiscis*	fermented fish	[37]
28	*W. coleopterorum*	Intestine of the diving beetle	[38]

As published in the BV-BRC (Bacterial and Viral Bioinformatics Resource Center) database, as of 19 October 2022, the genome for the genus *Weissella* (Taxonomy Id: 46255) has been reported a total of 448 times, of which the genome for *Weissella cibaria* has been reported the most (n = 168), followed by *Weissella confusa* (n = 128) and *Weissella paramesentroides* (n = 44). The sources of isolation being: human (n = 95), insect (n = 20), avian (n = 9), nonhuman mammal (n = 21), plants, and fermented food sources. The genomes of five species have been reported as isolated from humans: *Weissella cibaria*, *Weissella paramesenteroides*, *Weissella koreensis*, and *Weissella confusa*. With respect to the genome size, *Weissella* has a smaller pool of genes compared to other fecal commensal bacteria belonging to the genera *Parabacteroides*, *Bacteroides*, *Lactobacillus*, and *Pediococcus*. As investigated in our laboratory (Singh et al., unpublished), the genome size and the coding sequence (CDS) of the *Weissella* genus are much smaller than the other fecal bacteria.

3. Safety and Virulence Genes

The safety of *W. confusa* has always been a controversial subject due to reports of its isolation from human clinical samples. Although not formally assigned to a risk group by the American Biological Safety Association (ABSA), it has been allocated to Risk Group 1 microorganisms by the German Committee for Biological Agents. The American Type Culture Collection (ATCC) recommends using the strain ATCC 10881^{TM} under biosafety level 1 [7], which makes it unlikely to cause disease in healthy individuals.

Some potential virulence determinants, such as hemolysin, collagen, and adhesin, have been discovered in some of the species of the genus *Weissella* through genome analysis [6], but their role and transferable potential across *Weissella* are still unknown. As in other lactic acid bacteria (LAB), hemolysin genes are universally present in the genus, but their role in pathogenicity remains unproven. The presence of some adhesins may be a desired characteristic in favor of the probiotic potential of *Weissella*. For example, a fibronectin-binding protein (*FbpA*) present in *W. cibaria* strains inhibits the biofilm formed by *S. aureus*, thus being protective against *S. aureus* infections. While there is some evidence to suggest the role of the gut-colonizing potential of *FbpA* in establishing infection in a host, one cannot ignore that the ability for gut-colonization is essential to the probiotic potential of *Weissella*, as demonstrated by Wang et al. [39]. Similarly, mucus-binding proteins play a crucial role in the adhesion of probiotic bacteria to the host gut [40].

It is important to establish the safety of bacteria designed for human consumption to ensure that the organisms are well tolerated and do not pose a health threat when properly administered. To evaluate the safety of these organisms, animal models are typically given higher doses than would be administered to a human. Lyophilized *W. confusa* orally administered to rats at a concentration of 92×10^8 CFU/kg body weight/day for 90 days did not show any evidence of mental or physical ailment when evaluated using a combination of behavioral tests as well as physical examination. Blood cell counts did not show a significant difference in erythrocyte, white blood cell, or lymphocyte concentrations in untreated versus treated rats when controlling for sex [41].

4. Opportunistic Infections That Respond to Antibiotics

There have been documented case reports of *Weissella* causing infections in immunocompromised patients, earning it the reputation of an opportunistic pathogen [3]. Reported cases include sepsis, endocarditis, post-operative osteomyelitis, abscess, and meningitis [3,42], among others, as listed in Table 2. Of the listed occurrences of *Weissella* infections, barring a select few, all were successfully resolved with antibiotics. Of the 28 reported cases, 22 were reported to have developed bacteremia, either alone or in conjunction with other infections, such as endocarditis. Fifteen of these people survived upon clinical intervention with antibiotics. At least one reported patient with Crohn's disease who developed bacteremia had a history of probiotic consumption, which was postulated to be the source of *Weissella's* introduction into the host body. The probiotic product composition was, however, never investigated.

Table 2. Antimicrobials and clinical infections with *Weissella confusa:* humans respond very well to antibiotics.

Age, Sex	Underlying Conditions	Clinical Infection	Treatment	Outcome	Ref.
12, F	Gastrostomy	Bacteremia	Cephalosporin	Cured	[43]
71, M	Cecal carcinoma	Bacteremia	Cephalosporin	Cured	[43]
-	-	Organ colonization	Ampicillin	Cured	[44]
49, M	None	Abscess infection	Cephalothin	Cured	[45]
-	-	Organ colonization	Ampicillin	Cured	[46]
46, M	Abdominal aortic dissection repair, coronary artery bypass grafting, parenteral nutrition	Bacteremia	Piperacillin-tazobactam	Cured	[47]
49, M	Alcohol abuse history, treatment with corticosteroids	Endocarditis, Bacteremia	None	Fatal	[48]
65, M	Aortic insufficiency	Infective endocarditis	Penicillin G, gentamicin, moxifloxacin, cefoperazone	Cured	[49]
56.6, 6F, and 4M	Malignancy (4), chronic steroid use (3), chemotherapy (3), abdominal surgery (4), polymicrobial infection (5), central catheter (6)	Bacteremia	Vancomycin, ceftazidime, ampicillin-sulbactam, amoxicillin-clavulanate, gentamicin, ciprofloxacin, and trimethoprim-sulfamethoxazole	Cured (4), Fatal (6)	[50]

Table 2. *Cont.*

Age, Sex	Underlying Conditions	Clinical Infection	Treatment	Outcome	Ref.
34, M	Hematopoietic stem cell transplant recipient	Bacteremia	Vancomycin, aztreonam, and daptomycin	Cured	[51]
58, M	Severe burns, polymicrobial infection, central catheter	Bacteremia	Vancomycin, imipenem, and daptomycin	Cured	[51]
54, M	Hepatocellular carcinoma, Liver transplant, hepatic artery thrombosis, diabetes	Bacteremia	Metronidazole and levofloxacin	Cured	[52]
48, M	Gastroesophageal adenocarcinoma	Bacteremia	Cefoperazone-sulbactam Metronidazole	Cured	[53]
60, F	Hypertension, aortic intramural hematoma	Bacteremia	Teicoplanin and piperacillin-tazobactam	Cured	[54]
94, F	Osteoarthritis, total knee arthroplasty	Prosthetic joint	Levofloxacin	Cured	[55]
63, F	Crohn's disease with gastrointestinal strictures, central venous catheter	Bacteremia	Piperacillin/tazobactam	Cured	[56]
14, M	Medulloblastoma, surgery, chemo and radiotherapy, polymicrobial infection	Bacteremia	Clindamycin, amikacin	Cured	[57]
78, M	Immunodeficiency	Meningitis	Ampicillin		[3]
25, M	Crohn's disease, short bowel syndrome, intestinal failure	Bacteremia	Meropenem, metronidazole, and cefuroxime	Cured	[10]

It is to be noted that the prevalence of *Weissella* in human samples is often under- and over-reported owing to its shared characteristics and features with members of the *Leuconostocaceae* family. Any recently reported increases in isolation can be attributed to associated comorbidities, such as immunodeficiency, history of invasive procedures, organ transplantation, as well as the use of vancomycin, to which *Weissella* is resistant. For most of the healthy population, the source of exposure to *Weissella* is food culture. In available reports, wherever *W. confusa* has been isolated from human tissues, there have been predisposing factors, such as immunocompromised state, prior vancomycin exposure, central venous catheter insertion, history of gastrointestinal procedures/pathologies, and so forth. Due to *Weissella's* gastrointestinal inhabitance, surgical procedures may lead to the translocation of *Weissella* into the bloodstream. History of vancomycin usage should not be discounted in cases of bacteremia where there has been a possible disruption of the gut microbiome secondary to vancomycin use, allowing for naturally present/ingested *Lactobacilli*/*Weissella* to thrive and overgrow [10].

In the case of a potential infection, it is to be noted that all *Weissella* isolates have been found to be susceptible to antibiotics such as ampicillin, penicillin G, chloramphenicol, erythromycin, doxycycline, minocycline, quinupristin/dalfopristin, gentamicin, and streptomycin [58]. However, some strains exhibit resistance to penicillin [59] and ampicillin [60], and all are resistant to vancomycin [61]. The resistance of *Weissella* is reported to be intrinsic, which may be due to the lack of D-Ala-D-lactate, a target site in their cell wall for vancomycin [42]. Although vancomycin has justifiably been suggested to be a risk factor for *Weissella* infections, their use has been reported in *Weissella* infections in humans that responded to therapy in which other antibiotics were also administered (Table 2). For clinicians, therefore, it is important to be mindful of using vancomycin as the antibiotic of choice in cases of bacteremia that show growth of Gram-positive cocci, not disregarding a possible *Weissella* infection and the in vitro resistance mentioned against vancomycin. There remains, in general, a scarcity of information on the pathogenic pathways of *Weissella* [62].

5. Probiotic/Postbiotic Potential and Health Benefits of *Weissella*

There has been a growing interest in studying the different strains of *Weissella* being isolated from diverse ecological environments due to the multitude of prospective pharmacological functions associated with them. Newly isolated *Weissella* strains must meet several criteria to be considered potential probiotic organisms. As per the consensus statement issued by a panel of experts convened by the International Scientific Association for Probiotics and Prebiotics (ISAPP) (October 2013), probiotics are defined as "live microorganisms that, when administered in adequate amounts, confer a health benefit on the host" [63]. In addition to demonstrating a positive effect on the host, a probiotic organism intended for oral ingestion should be able to tolerate conditions in the digestive tract and adhere to the intestinal lining [64]. Lakra et al. [65] sought to evaluate the ability of two newly isolated *Weissella* species to colonize the digestive tract. *W. confusa* MD1 and *W. cibaria* MD2 were both found to be capable of adhering to HT-29 intestinal epithelial cells and mucous surfaces in vitro. Scanning electron microscopy images provided visual confirmation of both strains' ability to adhere to the intestinal cells in the presence of mucin without damaging the underlying cells [65]. The *W. confusa* strain, Lb. Con, was also evaluated to determine its ability to break down the pesticide chlorpyrifos. The strain showed excellent ability to grow on glucose-free MRS medium supplemented with varying concentrations of chlorpyrifos while being able to degrade 25% of the pesticide. This finding may find its implementation in cases of in vivo and food pesticide toxicity [66].

Recently, much attention has been drawn to the application of postbiotics due to their safety and beneficial advantage in the health and industrial sectors over live bacteria [67, 68]. As per the consensus statement issued by a panel of experts convened by ISAPP (2019), postbiotics are defined as a "preparation of inanimate microorganisms and/or their components that confers a health benefit on the host" [69]. These are microbial cells that have been intentionally rendered inactive and may or may not produce metabolites or have cell components that confer the established health benefits. Several research studies have characterized the diversity of metabolites produced by the varied species in the *Weissella* genus [70]. Postbiotics such as bioactive peptides, exopolysaccharides (EPS), enzymes, organic acids, short-chain fatty acids (SCFA), and similar by-products play an important role in the 'biopreservation' of food and possess antimicrobial, immunomodulatory, and anti-inflammatory properties [71,72].

5.1. Antimicrobial Potential

Exopolysaccharides. Of the many metabolites produced by *Weissella*, most previous avenues of research have primarily focused on the diversity of exopolysaccharides (EPSs) secreted by the *Weissella* species. These are a diverse class of macromolecules that help these bacteria execute a variety of functions while protecting them and aiding in general survival [8,73]. The many properties of EPSs include antibacterial, antifungal, antioxidant, and anti-inflammatory functions [9,74], as well as their growth-promoting potential [75,76]. In addition, *Weissella* seems to play an important role in the reduction of a depression-like state [77] and in the strengthening of the gut epithelial barrier [78].

Among *Weissella* species, *W. confusa* is one of the most important EPS producers [11], and different *W. confusa* strains, such as *W. confusa* VP30, XG-3, and KR780676, produce several EPSs with distinct functions. The *W. confusa* strain KR780676 has been shown to produce a galactan EPS that can resist enzymatic degradation in the gastrointestinal tract, as demonstrated in in vitro studies [79]. Interestingly, this EPS also promoted the growth of several probiotic species, including *L. plantarum* and *L. fermentum*, using in vitro screening. When the strain was orally administered to mice, similar effects were observed, as illustrated by an increase in the relative abundance of the probiotic *Lactobacillus* and *Bifidobacterium* species in the stools of these mice [79].

Weissella cibaria strains have shown an extensive ability to thwart the population growth of pathogenic microorganisms. Park et al. [75] showed that the EPS produced by *W. cibaria* promoted the growth and the antibacterial activity of a well-established probiotic bacteria, *L. rhamnosus*. The concentration ranges for EPS in the growth media of *L. rhamnosus* defined its antibacterial activity against a range of pathogens. A lower concentration of EPS was sufficient to inhibit the growth of *L. monocytogenes* and *S. aureus*, thereby demonstrating a higher antibacterial activity against these bacteria, whereas the growth of *B. cereus* and *E. coli* was inhibited at higher concentrations of EPS. In another study, a higher yield of EPS was observed in the *W. cibaria* strain W27 (isolated from kimchi) when supplemented with sucrose. This also improved the bacteriocinogenic potential of the bacteria by inducing its surface property [80]. The assimilation of sucrose helped enable the hydrophobic nature of W27 that ultimately contributed to the enhanced antibacterial activity as observed against *S. aureus*, *B. cereus*, and *E. coli*. Another study by Yeu et al. [81] illustrated the inhibitory effect that the EPS derived from *W. cibaria EIR/P2* has on biofilm formation by the pathogens responsible for upper respiratory tract infections: *S. aureus*, *Streptococcus pyogenes*, *Streptococcus pneumoniae*, and *Moraxella catarrhalis*. With respect to removal of preformed biofilms, the best effect was elucidated against *S. aureus*, while no effect against *S. pneumoniae* was observed. In a separate study, the strain *W. cibaria* JW 15 was shown to exhibit an anti-biofilm effect against *S. aureus*, *L. monocytogenes*, *S. enterica*, and *S. typhi* [82,83], which are all known to possess pathogenic potential capable of causing serious infections.

Bacteriocins. Probiotic bacteria also show bactericidal or bacteriostatic activity though the synthesis of small ribosomal peptides called bacteriocins [84]. Only six purified bacteriocins have been reported to be produced by *W. cibaria*, *W. paramesenteroides*, and *W. hellenica* [85]. Srionnual et al. [86] were the first to report a unique bacteriocin called *'weissellicin 110'* produced by the *W. cibaria* strain, 110 (isolated from a Thai fermented fish product). However, the full amino acid sequence information of *w110* was not calculated by the author at the time. The bacteriocin *w110* showed narrow-spectrum antibacterial activity against some Gram-positive bacteria but did not inhibit the growth of food-borne pathogen *L. monocytogenes*. In 2017, Li et al. [87] deduced the full amino acid sequence of *w110* through whole-genome sequencing and classified it into class IId due to the presence of unique genes and for having a 21-amino-acid N-terminal leader peptide. Teixeira et al. reported similar narrow-spectrum antibacterial activity against LAB in another strain of *W. cibaria, W25* [88]. Based on AntiSMASH analysis, the authors proposed the synthesis of two different types of bacteriocins in *W. cibaria* W25 being produced by the translation of lassopeptide (MicJ25) and RiPP-like bacteriocin_IIc genes, but the identification of these bacteriocin-producing genes was not confirmed by the Bagel 4 software (http://bagel4.molgenrug.nl/, accessed on 18 October 2022).

Another novel bacteriocin named *'weissellicin D'* was reported to be synthesized by *W. hellenica* strain D1501 (isolated from fermented meat) [89]. The thermostable *'weissellicin D'* exhibited a broad range of antibacterial activity against many food-borne pathogens, such as *E. coli*, *S. aureus*, and *L. monocytogenes*. It was also found to inhibit the growth of yeasts and molds that included *Candida albicans*, *Debaryomyces*, *Mucor*, *Saccharomyces cerevisiae*, and *Kluyveromyces marxianus*. However, possibly due to its autoimmunity, *weissellicin D* did not affect the growth activity of its own producer strain. Later, Chen et al. [90] successfully exploited the antagonistic activity of *W. hellenica* D1501 to improve the shelf life of tofu by simultaneously co-culturing it with spoilage bacteria such as *E. coli*, *S. aureus*, and *K. gibsonii* in soymilk.

Bacteriocin *'weissellicin A'* was identified and characterized from the strain *W. paramesenteroides* DX. It was heat-resistant and showed activity against a range of Gram-positive bacteria [91]. The bacteriocin was classified into class II and predicted to affect the integrity of plasma membranes of pathogens, causing an efflux of required nutritious cellular metabolites, thereby resulting in cell death. The thermostable and acid-resistant potential of this bacteriocin can be used in the preservation of acidic foods at an industrial scale. Another industrially important bacteriocin, *'weissellicin L'*, produced by *W. hellenica* was reported by Leong et al. [92], and the nucleotide characterization performed by Chen et al. [93] a year later declared it to be unique. The bacteriocin *weissellicin L* strongly inhibits

L. monocytogenes and, therefore, can be used in the biopreservation of chilled food, which is mostly contaminated by this spoilage pathogen.

The production of bacteriocins by any given *Weissella* strain depends on the nutrient composition and availability of vitamins in the culture medium where the bacterium grows. Isolated from Japanese pickles, *W. hellenica*, QU 13, as observed by Masuda et al. [94], was interestingly found to produce multiple bacteriocins. The authors discovered that QU13 produced two different bacteriocins, named *'weissellicin Y'* and *'weissellicin M'*, based on their nutritional preferences. While *weissellicin Y* is produced in the MRS media, *weissellicin M* is produced in the thiamine-rich media, APT. The vitamin, while accelerating the growth of *W. hellenica* QU 13 on the one hand, reduced the synthesis of *weissellicin Y* on the other. However, the production of *weissellicin M* was not affected by this transition at all. The mechanism of production of these bacteriocins is still not fully elucidated. Both *weissellicin Y* and *weissellicin M* possess broad antimicrobial spectra specifically targeted against *B. coagulans*. Between the two, *weissellicin M* showed comparatively higher antibacterial activity, as well as greater acid and thermal stability when compared to *weissellicin Y*.

Among other strains of *Weissella*, *W. confusa* Cys2-2 (isolated from ginger) was observed to produce a bacteriocin that showed bactericidal activity against the Gram-negative enterics, *E. coli*, *Salmonella*, and *Shigella* [95]. The Cys2-2 bacteriocin exerted this effect by altering the membrane integrity of target cells. Similar broad-spectrum antibacterial activity was reported by Goh and Philip [96] in another *W. confusa* strain, A3 (isolated from a dairy source), against bacteria such as *B. cereus*, *E. coli*, *P. aeruginosa*, and *M. luteus*. No inhibitory effect on the growth of *S. aureus* was noted. Yet another strain, *W. confusa* GCC_19R1, was found to have antibacterial activity against the Gram-negative bacteria *Stenotrophomonas maltophilia*, *Acinetobacter johnsonii*, *Achromobacter spanius*, and *Cedecea davisae* [85]. The bacteriocin synthesized by the *W. confusa* strain, PL9001, exhibits antagonistic activity against the pathogen *Helicobacter pylori* and can be used to treat H. pylori-induced gastritis and gastric ulcers [97,98]. *Weissella paramesenteroides* DFR-8 (isolated from cucumber) is reported to produce a thermostable bacteriocin that shows a broad-spectrum antimicrobial effect against both Gram-positive and -negative organisms [99]. In another study, the authors Pal and Ramana [99] reported the production of non-bacteriocin antimicrobial components from the same strain of *Weissella* that proved effective against Gram-negative bacteria.

BLIS: Apart from bacteriocins, some species of *Weissella* have also been found to produce 'bacteriocin-like inhibitory substance' (BLIS), a bacteriocin that has neither been fully characterized nor purified. For instance, *W. confusa MBF8-1* isolated from fermented soybean showed BLIS activity against *Leuc. mesenteroide* and other closely related species [100]. Encoded by a large plasmid, pWcMBF8-1, the strain *MBF8-1* produced a BLIS called *'weissellicin MBF'*.

Other metabolites/mechanisms: Some of the other miscellaneous mechanisms through which the *Weissella* spp. exhibit antimicrobial effects include: the production of hydrogen peroxide, organic acids (lactic, acetic, and citric acids), fatty acids, and specific proteins, e.g., N-acetylmuramidase. Lim et al. [101] successfully explored the antimicrobial activity of *W. cibaria*, CMU, against oral pathogens, possibly due to acid and hydrogen peroxide. Another strain of *W. cibaria*, KY10, isolated from shrimp gut, was shown to have bactericidal activity against *Vibrio parahaemolyticus* T.11 through the mechanism of organic acid release [102]. In another study, Dey et al. [103] examined the antibacterial activity of *W. confusa* DD_A7 isolated from kimchi and found that the DD-A7 strains trigger the oxidative stress to inhibit the growth of extended-spectrum β-lactamase (ESBL)-positive *E. coli*, which are emerging pathogens. *Weissella* also acts as an anti-mycobacterial, possibly due to its obligate heterofermentative nature, which makes it unique and prominent among other LAB. The authors emphasized that *Weissella* generates lactate and ethanol as by-products at equivalent concentrations that could have a stronger anti-mycobacterial effect than lactate alone [104].

5.2. Immunomodulatory and Anticancer Potential

Different species within the *Weissella* genus have been investigated for their role as potential probiotic organisms. There are several mechanisms through which such species may provide a positive health effect for the host. Probiotic organisms can modulate the immune system, reducing excess or unwarranted inflammation while simultaneously priming the host's defenses and immune system against pathogenic organisms. The multifaceted nature through which probiotic organisms interact with the host's immune system is still incompletely understood, although it is known that these organisms can change the way immune cells respond to identical stimuli.

One mechanism through which some species in the *Weissella* genus exert anti-inflammatory effects is through the modulation of the NF-κB-mediated signaling pathway. *W. confusa* down-regulates the expression of the iNOS gene responsible for regulating the production of nitric oxide (NO), a proinflammatory mediator produced from L-arginine [103]. *W. confusa* also metabolizes and uses arginine for its own growth, consequently further decreasing the NO levels in the body. Using a larval zebrafish model system, Dey and Kang [105] demonstrated that the EPS produced by *W. confusa* can reduce inflammation caused by *E. coli*-derived Shiga toxin. This finding is supported by previous research using a murine model system that found evidence that the EPS produced by *W. confusa* can modulate the immune system by increasing the level of Immunoglobulin A, the absence of which leads to intestinal inflammation [106].

Within the intestine, the intactness of the intestinal epithelial barrier is an important marker for gut health. Reduced adhesion between adjacent intestinal epithelial cells, termed "leaky gut", causes excessive permeability of the intestine and disruption of intestinal homeostasis, resulting in aberrant immune activation [107]. Restoration and maintenance of barrier function is an important trait to be assessed for in orally administered probiotics. One strain of *W. cibaria* named MW01, isolated from Chinese sauerkraut, was found to restore the barrier integrity of Caco-2 intestinal epithelial cells (IECs) after barrier dysfunction was induced by LPS [106]. The restoration of barrier integrity was accomplished through the inhibition of the nuclear translation of NF-κB and the subsequent blocking of the MLCK (mitogen-activated protein kinases) pathway. This led to the upregulation of the genes encoding for tight junction proteins (TJP), as well as increased TJP protein levels, and reduced release of proinflammatory cytokines such as IL-6, IL-8, and TNF-α [106]. This finding was corroborated by Silva et al. [108] when they analyzed the cell-free supernatant after administering *W. cibaria* CIATEJ BI–48.1 to confluent HT-29 monolayers. Several short-chain fatty acids (SFCAs) were detected, including acetic and butyric acids. SFCAs are important metabolites in the human digestive tract, where upward of 90% of butyrate in the intestinal lumen is metabolized to meet the energy needs of the colonocytes [109]. SCFAs such as acetic and butyric acid are also considered important modulators of inflammation in the colon and promote the upregulation of TJPs in the small intestine [110]

Probiotics in general can help reverse the gut dysbiosis implicated in the pathogenesis of several inflammatory and auto-immune conditions as well as cancers, and thereby help prevent these conditions or attenuate their severity. Certain strains of *Weissella* are being investigated as an adjuvant to conventional treatment in IBD. Various in vitro and animal models have explored this. For example, *W. paramesenteroides* WpK4 was able to reduce the disease activity index (DAI) as well as repair some of the mucosal damage in mice models with DSS-induced colitis. The bacteria also helped reduce the production of proinflammatory markers such as TNF- α, NO, IL-1β, and IL-6. As a consequence of this immunomodulation, the colitis mice also demonstrated reduced anxiety and depression-associated behavior [77]. The *W. confusa* strain F213 was shown to maintain the transepidermal resistance in an in vitro intestinal cell model employing Caco-2 cells where hydrogen peroxide was used to induce IBD. It also decreased intestinal permeability as well as helped maintain tight junctions, the disruption of which is seen in IBD [111]. The EPS purified from milk fermented with *W. confusa* VP30 was found to have a remarkable laxative effect on the constipated rat model and can be used to relieve constipation issues in humans [112].

Some strains of *Weissella*, such as *W. cibaria*, demonstrate anticancer effects against colorectal cancer by suppressing cell growth. Cha et al. [113] have a patent establishing the anticancer effect of *W. cibaria* against colorectal cancer cells by selectively suppressing the growth of cancer cells as opposed to normal cells. A review by Kwak et al. [114] investigated the benefits of kimchi LAB (*W. cibaria* and *L. plantarum*) against cancer and derived the same conclusion about *W. cibaria* [114]. The anti-proliferative activity of *W. cibaria* against cancer cells such as HeLa and Caco-2 has been confirmed in other studies as well [112,115,116]. A unique single-center study by Kwon et al. investigated the role of probiotic usage in post-operative cancer patients and noticed an increase in the populations of several beneficial bacteria that also included a very significant increase in the *Weissella* bacteria (a rise from 0.096% to 0.361%, $p < 0.004$) after 8 weeks of probiotic supplementation. This increase correlated with the improved quality of life in these patients [117]. Recently, Amer et al. [118] explored the anticancer effect of exopolysaccharides nanoparticles (EPS-NPs) produced by *W. paramesenteroides* MN2C2 against breast MCF-7, colon Caco-2, and liver HepG-2 malignant cells due to its potent antioxidant potential. These research findings could be an innovative alternative approach in cancer treatment, minimizing the use of prolonged conventional therapy and its adverse side effects.

Weissella cibaria has also been acknowledged for its ability to prime the immune system in immunocompromised hosts. Park and Lee [75] investigated the role of *W. cibaria* JW15 as a complementary treatment for counteracting the weakened immune system as a consequence of chemotherapy. Oral administration of the bacteria to immunosuppressed (secondary to cyclophosphamide) BALB/c mice resulted in increased splenocyte proliferation and elevated white blood cell counts (WBC). Additionally, *W. cibaria* JW15 was able to increase the production of TNFα, IL-1β, and IL-6 when challenged with purified lipopolysaccharide (LPS). Together, these results demonstrate the capacity of the *W. cibaria* JW15 strain to enhance the innate immune response in immunocompromised mice. Furthermore, no evidence of toxicity or bacteriemia because of the administered bacteria was observed in the treated mice despite their weakened immune system. Other studies by the same set of researchers in aged mice found similarly increased populations of red blood cells, WBS, and splenocytes along with elevated levels of cytokines IL-6 and IFN-γ [119]. When compared to the commercial strain of *L. rhamnosus* GG, the Weissella strain JW15 had higher immune-stimulating activity. In a nutshell, JW15′s use as a probiotic can help augment the immune response of the host [120]. In contrast to the aforementioned studies, the study by Seok et al. observed the suppression of the immune system in an in vitro murine model and illustrated the reduced expression of proinflammatory cytokines IL-6, IL-1β, and TNFα through inhibited activation of the NF-κB pathway in murine RAW 264.7 macrophages exposed to LPS. In addition, the administration of heat-killed *W. cibaria* JW15 resulted in the decreased production of nitric oxide and prostaglandin E2 [121].

5.3. Dental and Skin Health

Many of the traits that allow organisms such as *W. cibaria* and *W. confusa* to play a beneficial role in the gut also allow these organisms to be advantageous to other sites in the body that are susceptible to the same environmental challenges as the intestinal lumen. Particularly, *W. cibaria* has been shown to display probiotic effects in other body sites in animal studies. Using an induced-periodontitis mouse model, Kim et al. demonstrated that *W. cibaria* CMU could reduce the severity of periodontitis in a dose-dependent manner [106]. Mice treated with this strain displayed reduced alveolar bone loss and lower levels of the pathogenic *Porphyromonas gingivalis*. The authors posit that *W. cibaria* CMU has the capacity to adhere to saliva-coated surfaces in the mouth, as well as produce antimicrobial compounds that reduce the growth of pathogenic species that are responsible for the development of periodontitis. In a similar set of experiments, *W. cibaria* CMU was found to inhibit the co-aggregation of another oral pathogen, *Fusobacterium nucleatum*, which is also known to contribute to the development of periodontitis [122]. Additionally, this strain of *W. cibaria* reduced IL-6 and IL-8 secretion by oral epithelial cells previously challenged with *F. nucleatum*. A study by Kibar et al. into *W. cibaria* EIR/P-2 (isolated from bee pollen) displayed its antibacterial activity against *Streptococcus mutans*, a bacterium notorious for causing tooth decay and caries. The dextran derived from the EPS also had proliferative activity, which, coupled with the antimicrobial/antibiofilm functions, may find its application in the field of dentistry as an agent of 'periodontal healing and regeneration' [82].

Weissella has been found to have applications in the skin industry as a prospective treatment for many skin conditions. The WIKIM28 strain of *W. cibaria* has been shown to benefit patients with atopic dermatitis [79]. The authors used an induced-dermatitis model made by exposing mice to 2,4-dinitrochlorobenzene, which created skin lesions and thickened the epidermal layer mimicking dermatitis in humans. Oral administration of *W. cibaria* WIKIM28 resulted in a reduction of symptoms in treated mice, as observed through improved histological scoring of skin sections [79]. Additionally, the treated mice displayed an increased ratio of differentiated $CD4^{+}CD25^{+}Foxp3^{+}$ regulatory T cells (Tregs), along with a corresponding increase in IL-10 in polyclonal mesenteric lymphocytes, which is associated with a suppressed immune response. Taken together, these results indicate that *W. cibaria* WIKIM28 is capable of reducing inflammation in the skin through interactions between the spleen, lymphatic system, and the intestine.

5.4. Anti-Obesity

Probiotic bacteria help maintain metabolic homeostasis by producing active metabolites [123]. *Weissella* spp. such as *W. koreensis* (isolated from kimchi) have been reported to exhibit anti-obesity effects by regulating lipid metabolism. A study by Moon et al. [124] found that *W. koreensis* OK1-6 metabolized arginine into L-ornithine, a non-protein amino acid that down-regulated the expression of adipocyte-specific genes C/EBPα, aP2, SREBP1, and fatty acid synthase (FAS) in 3T3-L1 cells lines, thus preventing the accumulation of intracellular lipid and triglyceride inside the cells [125].

Choi et al. [126] reported the anti-obesity effect of *W. cibaria* MG5285 in mice with high-fat diet (HFD)-induced obesity. The study showed a significantly reduced expression of lipogenic proteins, including peroxisome proliferator-activated receptor γ (PPAR-γ), CCAAT/enhancer-binding protein α, FAS, and adipocyte-protein 2. In addition, sterol regulatory element-binding protein 1-c and its downstream protein FAS in the liver tissue were significantly decreased. These strains attenuated fat accumulation in the liver by upregulating the phosphorylation of AMP-activated protein kinase and acetyl-CoA carboxylase in the HFD-fed mice. Another study evaluated the anti-adipogenic capacity of whey that had been 'biotransformed' by *Weissella cibaria* in 3T3-L1 adipocytes and found a disruption in the intracellular signaling pathways responsible for the expression of obesity-associated genes and transcription factors, such as PPAR-γ, resulting in reduced accumulation of triglycerides and lipids inside the 3T3-L1 cells [127].

6. Starter Culture

Weissella spp. have tremendous functional and technological potential, which can improve the safety and nutritional and sensory characteristics of food. They enhance the flavor and improve the texture of food by producing beneficial metabolites such as organic acids, short-chain fatty acids, and esters during food fermentation [128–130]. They have been used as a starter culture in various ethnic foods in Europe and Asia [58]. For instance, *W. cibaria* and *W. confusa* have been studied in the fermentation of bread. These starter strains have been observed to enhance the production of functional components and rheological attributes of bread by eliminating the need for baker's yeast [131] and also improving the softness [132] and texture of gluten-free bread [133]. In addition, *W. cibaria* has been studied in the manufacturing of functional bread fortified with riboflavin [134]. The starter culture *Weissella viridescens* F2 was found to reduce the accumulation of biogenic amines and to improve the quality of Roucha during fermentation [135]. An ornithine-producing strain, *W. koreensis DB1*, isolated from kimchi, was studied in the fermentation of rice bran and found to have a significant effect on the organoleptic properties of rice. Such starter cultures offer a dual role by contributing to food functionality through its fermentation.

Despite having probiotic and outstanding starter culture potential, the genus has not yet been permitted for commercial use in the United States or the European Union nor acknowledged as a part of the International Dairy Federation Inventory [7,136] due to the lack of GRAS (Generally Recognized as Safe) status. The vast diversity of this genus and the deficiency of scientific literature could be possible reasons for not being granted the GRAS status by the FDA (Food and Drug Administration) or by the EFSA (European Food Safety Authority). The genus is also often classified as opportunistic pathogens, which seems to limit its application as potentially beneficial probiotics in the food industry despite being potentially powerful starters for food fermentation [85,137].

7. Conclusions

The *Weissella* genus comprises environmentally omnipresent bacteria that also happen to exist as commensals in the gastrointestinal tract of healthy vertebrates. In particular, the species *W. confusa* and *W. cibaria* have become subjects of extensive research in recent times, most of which involve an in-depth evaluation of their myriad health benefits as probiotics. The other end of the research looks into the cases of proposed opportunistic pathogenicity in people with underlying medical conditions. To date, the scales tip in favor of the many health benefits attributable to the use of these species, whereas cases of bacteremia, endocarditis, and meningitis remain low, with infections being seen in already-compromised hosts with predisposing factors, such as immunocompromised status, hospital procedures, and medical comorbidities, to name a few. Members of *Weissella* species are not generally known to infect healthy populations. All reported cases have been successfully treated with a variety of antimicrobials that the *Weissella* species remain sensitive to. The use of vancomycin is discouraged, however, owing to extensive resistance seen in this genus against the antibiotic. *Weissella's* remarkable antimicrobial and anti-inflammatory nature speak to its probiotic potential. With significant research backing the use of *Weissella,* the genus may come to be recognized as an important probiotic in the near future, with applications spanning across industries.

Author Contributions: All authors contributed to the review, discussion of the findings, and writing of the manuscript. All authors have read and agreed to the published version of the manuscript.

Funding: This review was primarily supported by NIH grant R21 DK118373 to A.R.-P., entitled "Identification of pathogenic bacteria in Crohn's disease". Partial support originated from the Litwin IBD Pioneers Award to A.R.-P. from the Crohn's and Colitis foundation (award number: 882584). Sadia Ahmed was sponsored by the International Research Support Initiative Program (IRSIP), Higher Education Commission of Pakistan (PIN: IRSIP 48 BMS 30).

Acknowledgments: Sadia is grateful to Arsalan Zaidi, Principal Scientist at Pakistan Atomic Energy Commission (PAEC, Pakistan), whose constant support helped Sadia Ahmed pursue her research aspirations at Case Western Reserve University in Rodriguez-Palacios' laboratory as part of her Ph.D program.

Conflicts of Interest: No conflict of interest.

References

1. Lin, S.-T.; Wang, L.-T.; Wu, Y.-C.; Guu, J.-R.J.; Tamura, T.; Mori, K.; Huang, L.; Watanabe, K. *Weissella muntiaci* sp. nov., isolated from faeces of Formosan barking deer (Muntiacus reevesi). *Int. J. Syst. Evol. Microbiol.* **2020**, *70*, 1578–1584. [CrossRef] [PubMed]
2. Fusco, V.; Quero, G.M.; Cho, G.S.; Kabisch, J.; Meske, D.; Neve, H.; Bockelmann, W.; Franz, C.M.A.P. The genus Weissella: Taxonomy, ecology and biotechnological potential. *Front. Microbiol.* **2015**, *6*, 155. [CrossRef] [PubMed]
3. Cheaito, R.A.; Awar, G.; Alkozah, M.; Cheaito, M.A.; El Majzoub, I. Meningitis due to Weissella Confusa. *Am. J. Emerg. Med.* **2020**, *38*, 1298.e1–1298.e3. [CrossRef] [PubMed]
4. Baugh, A.V.; Howarth, T.M.; West, K.L.; Kerr, L.E.J.; Love, J.; Parker, D.A.; Fedenko, J.R.; Tennant, R.K. Draft Genome Sequence of Weissella paramesenteroides STCH-BD1, Isolated from Ensiled Sorghum bicolor. *Microbiol. Resour. Announc.* **2021**, *10*, e01328-20. [CrossRef] [PubMed]
5. Björkroth, J.; Holzapfel, W. Genera Leuconostoc, Oenococcus and Weissella. *Prokaryotes* **2006**, *4*, 267–319. [CrossRef]

6. Rossi, F.; Amadoro, C.; Colavita, G. Members of the Lactobacillus Genus Complex (LGC) as Opportunistic Pathogens: A Review. *Microorganisms* **2019**, *7*, 126. [CrossRef] [PubMed]
7. Sturino, J.M. Literature-based safety assessment of an agriculture- and animal-associated microorganism: Weissella confusa. *Regul. Toxicol. Pharmacol.* **2018**, *95*, 142–152. [CrossRef] [PubMed]
8. Öz, E.; Kaban, G.; Barış, O.; Kaya, M. Isolation and identification of lactic acid bacteria from pastırma. *Food Control* **2017**, *77*, 158–162. [CrossRef]
9. Kang, B.K.; Cho, M.S.; Park, D.S. Red pepper powder is a crucial factor that influences the ontogeny of Weissella cibaria during kimchi fermentation. *Sci. Rep.* **2016**, *6*, 28232. [CrossRef]
10. Spiegelhauer, M.R.; Yusibova, M.; Rasmussen, I.K.B.; Fuglsang, K.A.; Thomsen, K.; Andersen, L.P. A case report of polymicrobial bacteremia with Weissella confusa and comparison of previous treatment for successful recovery with a review of the literature. *Access Microbiol.* **2020**, *2*, e000119. [CrossRef]
11. Teixeira, C.G.; Fusieger, A.; Milião, G.L.; Martins, E.; Drider, D.; Nero, L.A.; de Carvalho, A.F. Weissella: An Emerging Bacterium with Promising Health Benefits. *Probiotics Antimicrob. Proteins* **2021**, *13*, 915–925. [CrossRef]
12. Fanelli, F.; Montemurro, M.; Chieffi, D.; Cho, G.-S.; Franz, C.M.A.P.; Dell'Aquila, A.; Rizzello, C.G.; Fusco, V. Novel Insights Into the Phylogeny and Biotechnological Potential of Weissella Species. *Front. Microbiol.* **2022**, *13*, 914036. [CrossRef]
13. Vela, A.I.; Porrero, C.; Goyache, J.; Nieto, A.; Sánchez, B.; Briones, V.; Moreno, M.; Domínguez, L.; Fernandez-Garayzabal, J.F. Weissella confusa Infection in Primate (Cercopithecus mona). *Emerg. Infect. Dis.* **2003**, *9*, 1307–1309. [CrossRef]
14. Niven, C., Jr.; Evans, J.B. Lactobacillus viridescens nov. spec., a heterofermentative species that produces a green discoloration of cured meat pigments. *J. Bacteriol.* **1957**, *73*, 758–759. [CrossRef]
15. Garvie, E.I. The Growth Factor and Amino Acid Requirements of Species of the Genus Leuconostoc, including *Leuconostoc paramesenteroides* (sp. nov.) and Leuconostoc oenos. *J. Gen. Microbiol.* **1967**, *48*, 439–447. [CrossRef]
16. Collins, M.; Samelis, J.; Metaxopoulos, J.; Wallbanks, S. Taxonomic studies on some Leuconostoc-like organisms from fermented sausages: Description of a new genus Weissella for the Leuconostoc paramesenteroides group of species. *J. Appl. Bacteriol.* **1993**, *75*, 595–603. [CrossRef]
17. Holzapfel, W.H.; Van Wyk, E.P. *Lactobacillus kandleri* sp. nov., a new species of the subgenus betabacterium, with glycine in the peptidoglycan. *Zent. Für Bakteriol. Mikrobiol. Hyg.* **1982**, *3*, 495–502. [CrossRef]
18. Kandler, O.; Schillinger, U.; Weiss, N. *Lactobacillus halotolerans* sp. nov., nom. rev. and *Lactobacillus minor* sp. nov., nom. rev. *Syst. Appl. Microbiol.* **1983**, *4*, 280–285. [CrossRef]
19. Tanasupawat, S.; Shida, O.; Okada, S.; Komagata, K. *Lactobacillus acidipiscis* sp. nov. and *Weissella thailandensis* sp. nov., isolated from fermented fish in Thailand. *Int. J. Syst. Evol. Microbiol.* **2000**, *50*, 1479–1485. [CrossRef]
20. Magnusson, J.; Jonsson, H.; Schnürer, J.; Roos, S. *Weissella soli* sp. nov., a lactic acid bacterium isolated from soil. *Int. J. Syst. Evol. Microbiol.* **2002**, *52*, 831–834. [CrossRef]
21. Bjorkroth, J.; Schillinger, U.; Geisen, R.; Weiss, N.; Hoste, B.; Holzapfel, W.H.; Korkeala, H.; Vandamme, P. Taxonomic study of Weissella confusa and description of *Weissella cibaria* sp. nov., detected in food and clinical samples. *Int. J. Syst. Evol. Microbiol.* **2002**, *52*, 141–148. [CrossRef] [PubMed]
22. Lee, J.-S.; Lee, K.C.; Ahn, J.-S.; Mheen, T.-I.; Pyun, Y.-R.; Park, Y.-H. *Weissella koreensis* sp. nov., isolated from kimchi. *Int. J. Syst. Evol. Microbiol.* **2002**, *52*, 1257–1261. [CrossRef] [PubMed]
23. De Bruyne, K.; Camu, N.; Lefebvre, K.; De Vuyst, L.; Vandamme, P. *Weissella ghanensis* sp. nov., isolated from a Ghanaian cocoa fermentation. *Int. J. Syst. Evol. Microbiol.* **2008**, *58*, 2721–2725. [CrossRef] [PubMed]
24. Padonou, S.W.; Schillinger, U.; Nielsen, D.S.; Franz, C.M.A.P.; Hansen, M.; Hounhouigan, J.D.; Nago, M.C.; Jakobsen, M. *Weissella beninensis* sp. nov., a motile lactic acid bacterium from submerged cassava fermentations, and emended description of the genus Weissella. *Int. J. Syst. Evol. Microbiol.* **2010**, *60*, 2193–2198. [CrossRef] [PubMed]
25. De Bruyne, K.; Camu, N.; De Vuyst, L.; Vandamme, P. *Weissella fabaria* sp. nov., from a Ghanaian cocoa fermentation. *Int. J. Syst. Evol. Microbiol.* **2010**, *60*, 1999–2005. [CrossRef]
26. Vela, A.I.; Fernández, A.; de Quirós, Y.B.; Herráez, P.; Domínguez, L.; Fernández-Garayzábal, J.F. *Weissella ceti* sp. nov., isolated from beaked whales (Mesoplodon bidens). *Int. J. Syst. Evol. Microbiol.* **2011**, *61*, 2758–2762. [CrossRef]
27. Snauwaert, I.; Papalexandratou, Z.; De Vuyst, L.; Vandamme, P. Characterization of strains of *Weissella fabalis* sp. nov. and Fructobacillus tropaeoli from spontaneous cocoa bean fermentations. *Int. J. Syst. Evol. Microbiol.* **2013**, *63*, 1709–1716. [CrossRef]
28. Tohno, M.; Kitahara, M.; Inoue, H.; Uegaki, R.; Irisawa, T.; Ohkuma, M.; Tajima, K. *Weissella oryzae* sp. nov., isolated from fermented rice grains. *Int. J. Syst. Evol. Microbiol.* **2013**, *63*, 1417–1420. [CrossRef]
29. Oh, S.J.; Shin, N.-R.; Hyun, D.-W.; Kim, P.S.; Kim, J.Y.; Kim, M.-S.; Yun, J.-H.; Bae, J.-W. *Weisselladiestrammenae* sp. nov., isolated from the gut of a camel cricket (Diestrammena coreana). *Int. J. Syst. Evol. Microbiol.* **2013**, *63*, 2951–2956. [CrossRef]
30. Nisiotou, A.; Dourou, D.; Filippousi, M.-E.; Banilas, G.; Tassou, C. *Weissella uvarum* sp. nov., isolated from wine grapes. *Int. J. Syst. Evol. Microbiol.* **2014**, *64*, 3885–3890. [CrossRef]
31. Heo, J.; Hamada, M.; Cho, H.; Weon, H.-Y.; Kim, J.-S.; Hong, S.-B.; Kim, S.-J.; Kwon, S.-W. *Weissella cryptocerci* sp. nov., isolated from gut of the insect Cryptocercus kyebangensis. *Int. J. Syst. Evol. Microbiol.* **2019**, *69*, 2801–2806. [CrossRef] [PubMed]
32. Praet, J.; Meeus, I.; Cnockaert, M.; Houf, K.; Smagghe, G.; Vandamme, P. Novel lactic acid bacteria isolated from the bumble bee gut: *Convivina intestini* gen. nov., sp. nov., *Lactobacillus bombicola* sp. nov., and *Weissella bombi* sp. nov. *Antonie Van Leeuwenhoek* **2015**, *107*, 1337–1349. [CrossRef] [PubMed]

33. Lee, S.-H.; Ku, H.-J.; Ahn, M.-J.; Hong, J.-S.; Shin, H.; Lee, K.C.; Lee, J.-S.; Ryu, S.; Jeon, C.O.; Lee, J.-H. *Weissella jogaejeotgali* sp. nov., isolated from jogae jeotgal, a traditional Korean fermented seafood. *Int. J. Syst. Evol. Microbiol.* **2015**, *65*, 4674–4681. [CrossRef] [PubMed]
34. Choi, H.-J.; Cheigh, C.-I.; Kim, S.-B.; Lee, J.-C.; Lee, D.-W.; Choi, S.-W.; Park, J.-M.; Pyun, Y.-R. *Weissella kimchii* sp. nov., a novel lactic acid bacterium from kimchi. *Int. J. Syst. Evol. Microbiol.* **2002**, *52*, 507–511. [CrossRef] [PubMed]
35. Li, Y.Q.; Tian, W.L.; Gu, C.T. *Weissella sagaensis* sp. nov., isolated from traditional Chinese yogurt. *Int. J. Syst. Evol. Microbiol.* **2020**, *70*, 2485–2492. [CrossRef]
36. Jeong, S.-H.; Jung, J.-H.; Jang, J.; Lee, J.; Chun, J. Weissella hanii sp. nov., from kimchi. *Int. J. Syst. Evol. Microbiol.* **2011**. Available online: https://www.ncbi.nlm.nih.gov/Taxonomy/Browser/wwwtax.cgi?mode=Info&id=167774&lvl=3&lin=f (accessed on 18 October 2022).
37. Tanasupawat, S.; Sukontasing, S.; Okada, S.; Komagata, K. Weissella salipiscis sp. nov., from fermented fish (pla-ra) in Thailand. 2006. Available online: https://www.ncbi.nlm.nih.gov/Taxonomy/Browser/wwwtax.cgi?mode=Info&id=381421&lvl=3&lin=f (accessed on 18 October 2022).
38. Hyun, D.-W.; Lee, J.-Y.; Sung, H.; Kim, P.S.; Jeong, Y.-S.; Lee, J.-Y.; Yun, J.-H.; Choi, J.-W.; Han, J.E.; Lee, S.-Y.; et al. *Brevilactibacter coleopterorum* sp. nov., isolated from the intestine of the dark diving beetle Hydrophilus acuminatus, and *Weissella coleopterorum* sp. nov., isolated from the intestine of the diving beetle Cybister lewisianus. *Int. J. Syst. Evol. Microbiol.* **2021**, *71*, 004779. [CrossRef]
39. Wang, L.; Si, W.; Xue, H.; Zhao, X. A fibronectin-binding protein (FbpA) of *Weissella cibaria* inhibits colonization and infection of *Staphylococcus aureus* in mammary glands. *Cell. Microbiol.* **2017**, *19*, e12731. [CrossRef]
40. Quattrini, M.; Korcari, D.; Ricci, G.; Fortina, M. A polyphasic approach to characterize *Weissella cibaria* and *Weissella confusa* strains. *J. Appl. Microbiol.* **2019**, *128*, 500–512. [CrossRef]
41. Cupi, D.; Elvig-Jørgensen, S.G. Safety assessment of Weissella confusa—A direct-fed microbial candidate. *Regul. Toxicol. Pharmacol.* **2019**, *107*, 104414. [CrossRef]
42. Abriouel, H.; Lerma, L.L.; Muñoz, M.D.C.C.; Montoro, B.P.; Kabisch, J.; Pichner, R.; Cho, G.-S.; Neve, H.; Fusco, V.; Franz, C.M.A.P.; et al. The controversial nature of the Weissella genus: Technological and functional aspects versus whole genome analysis-based pathogenic potential for their application in food and health. *Front. Microbiol.* **2015**, *6*, 1197. [CrossRef]
43. Riebel, W.J.; Washington, J.A. Clinical and microbiologic characteristics of pediococci. *J. Clin. Microbiol.* **1990**, *28*, 1348–1355. [CrossRef]
44. Green, M.; Wadowsky, R.M.; Barbadora, K. Recovery of vancomycin-resistant gram-positive cocci from children. *J. Clin. Microbiol.* **1990**, *28*, 484–488. [CrossRef]
45. Bantar, C.E.; Relloso, S.; Castell, F.R.; Smayevsky, J.; Bianchini, H.M. Abscess caused by vancomycin-resistant Lactobacillus confusus. *J. Clin. Microbiol.* **1991**, *29*, 2063–2064. [CrossRef]
46. Green, M.; Barbadora, K.; Michaels, M. Recovery of vancomycin-resistant gram-positive cocci from pediatric liver transplant recipients. *J. Clin. Microbiol.* **1991**, *29*, 2503–2506. [CrossRef]
47. Olano, A.; Chua, J.; Schroeder, S.; Minari, A.; La Salvia, M.; Hall, G. *Weissella confusa* (Basonym: *Lactobacillus confusus*) Bacteremia: A Case Report. *J. Clin. Microbiol.* **2001**, *39*, 1604–1607. [CrossRef]
48. Flaherty, J.D.; Levett, P.N.; Dewhirst, F.E.; Troe, T.E.; Warren, J.R.; Johnson, S. Fatal Case of Endocarditis Due to *Weissella confusa*. *J. Clin. Microbiol.* **2003**, *41*, 2237–2239. [CrossRef]
49. Shin, J.H.; Kim, D.I.; Kim, H.R.; Kim, D.S.; Kook, J.-K.; Lee, J.N. Severe infective endocarditis of native valves caused by Weissella confusa detected incidentally on echocardiography. *J. Infect.* **2007**, *54*, e149–e151. [CrossRef]
50. Lee, M.-R.; Huang, Y.-T.; Liao, C.-H.; Lai, C.-C.; Lee, P.-I.; Hsueh, P.-R. Bacteraemia caused by Weissella confusa at a university hospital in Taiwan, 1997–2007. *Clin. Microbiol. Infect.* **2011**, *17*, 1226–1231. [CrossRef]
51. Salimnia, H.; Alangaden, G.; Bharadwaj, R.; Painter, T.; Chandrasekar, P.; Fairfax, M. Weissella confusa: An unexpected cause of vancomycin-resistant gram-positive bacteremia in immunocompromised hosts. *Transpl. Infect. Dis.* **2010**, *13*, 294–298. [CrossRef]
52. Harlan, N.; Kempker, R.; Parekh, S.; Burd, E.; Kuhar, D. Weissella confusa bacteremia in a liver transplant patient with hepatic artery thrombosis. *Transpl. Infect. Dis.* **2010**, *13*, 290–293. [CrossRef]
53. Kumar, A.; Augustine, D.; Sudhindran, S.; Kurian, A.M.; Dinesh, K.R.; Karim, S.; Philip, R. Weissella confusa: A rare cause of vancomycin-resistant Gram-positive bacteraemia. *J. Med. Microbiol.* **2011**, *60*, 1539–1541. [CrossRef] [PubMed]
54. Lee, W.; Cho, S.-M.; Kim, M.; Ko, Y.-G.; Yong, D.; Lee, K. *Weissella confusa* Bacteremia in an Immune-Competent Patient with Underlying Intramural Hematomas of the Aorta. *Ann. Lab. Med.* **2013**, *33*, 459–462. [CrossRef] [PubMed]
55. Medford, R.; Patel, S.N.; Evans, G.A. A confusing Case–*Weissella confusa* Prosthetic Joint Infection: A Case Report and Review of the Literature. *Can. J. Infect. Dis. Med. Microbiol.* **2014**, *25*, 173–175. [CrossRef] [PubMed]
56. Vasquez, A.; Pancholi, P.; Balada-Llasat, J.-M. Photo Quiz: Confusing Bacteremia in a Crohn's Disease Patient. *J. Clin. Microbiol.* **2015**, *53*, 759. [CrossRef] [PubMed]
57. Aberkane, S.; Didelot, M.-N.; Carrière, C.; Laurens, C.; Sanou, S.; Godreuil, S.; Jean-Pierre, H. Bactériémie à Weissella confusa: Un pathogène opportuniste sous-estimé. *Médecine et Maladies Infectieuses* **2017**, *47*, 297–299. [CrossRef]
58. Kankaya, D.A.; Tuncer, Y. Antibiotic resistance in vancomycin-resistant lactic acid bacteria (VRLAB) isolated from foods of animal origin. *J. Food Process. Preserv.* **2020**, *44*, e14468. [CrossRef]
59. Suwannaphan, S. Isolation, identification and potential probiotic characterization of lactic acid bacteria from Thai traditional fermented food. *AIMS Microbiol.* **2021**, *7*, 431–446. [CrossRef]

60. Oh, A.; Daliri, E.B.-M.; Oh, D.H. Screening for potential probiotic bacteria from Korean fermented soybean paste: In vitro and Caenorhabditis elegans model testing. *LWT* **2018**, *88*, 132–138. [CrossRef]
61. Mahmood, A. *Mechanistic Insights into Vancomycin Resistance in Clostridioides Difficile*; The University of Texas MD Anderson Cancer Center UT: Houston, TX, USA, 2020.
62. Panthee, S.; Paudel, A.; Blom, J.; Hamamoto, H.; Sekimizu, K. Complete Genome Sequence of Weissella hellenica 0916-4-2 and Its Comparative Genomic Analysis. *Front. Microbiol.* **2019**, *10*, 1619. [CrossRef]
63. Hill, C.; Guarner, F.; Reid, G.; Gibson, G.R.; Merenstein, D.J.; Pot, B.; Morelli, L.; Canani, R.B.; Flint, H.J.; Salminen, S.; et al. Expert consensus document: The International Scientific Association for Probiotics and Prebiotics consensus statement on the scope and appropriate use of the term probiotic. *Nat. Rev. Gastroenterol. Hepatol.* **2014**, *11*, 506–514. [CrossRef]
64. Huang, L.; Cui, K.; Mao, W.; Du, Y.; Yao, N.; Li, Z.; Zhao, H.; Ma, W. Weissella cibaria Attenuated LPS-Induced Dysfunction of Intestinal Epithelial Barrier in a Caco-2 Cell Monolayer Model. *Front. Microbiol.* **2020**, *11*, 2039. [CrossRef]
65. Lakra, A.K.; Domdi, L.; Hanjon, G.; Tilwani, Y.M.; Arul, V. Some probiotic potential of Weissella confusa MD1 and Weissella cibaria MD2 isolated from fermented batter. *LWT* **2020**, *125*, 109261. [CrossRef]
66. Hamoud, N.-E.; Sifour, M. Biodegradation of chlorpyrifos by a Weissella confusa strain and evaluation of some probiotic traits. *Arch. Microbiol.* **2021**, *203*, 3615–3621. [CrossRef]
67. Aggarwal, S.; Sabharwal, V.; Kaushik, P.; Joshi, A.; Aayushi, A.; Suri, M. Postbiotics: From emerging concept to application. *Front. Sustain. Food Syst.* **2022**, *6*, 887642. [CrossRef]
68. Moraes, R.M.; Schlagenhauf, U.; Anbinder, A.L. Outside the limits of bacterial viability: Postbiotics in the management of periodontitis. *Biochem. Pharmacol.* **2022**, *201*, 115072. [CrossRef]
69. Salminen, S.; Collado, M.C.; Endo, A.; Hill, C.; Lebeer, S.; Quigley, E.M.M.; Sanders, M.E.; Shamir, R.; Swann, J.R.; Szajewska, H.; et al. The International Scientific Association of Probiotics and Prebiotics (ISAPP) consensus statement on the definition and scope of postbiotics. *Nat. Rev. Gastroenterol. Hepatol.* **2021**, *18*, 649–667. [CrossRef]
70. Jin, H.; Jeong, Y.; Yoo, S.-H.; Johnston, T.V.; Ku, S.; Ji, G.E. Isolation and characterization of high exopolysaccharide-producing Weissella confusa VP30 from young children's feces. *Microb. Cell Factories* **2019**, *18*, 1–13. [CrossRef]
71. Zakuan, N.M.; Ling, F.H.; Yazan, L.S. Anti-microbial, anti-cancer and immunomodulatory properties of proteinaceous postbiotic metabolite produced by Lactobacillus plantarum I-UL4. *Mal. J. Med. Health Sci.* **2019**, *15*, 81–84.
72. Thorakkattu, P.; Khanashyam, A.C.; Shah, K.; Babu, K.S.; Mundanat, A.S.; Deliephan, A.; Deokar, G.S.; Santivarangkna, C.; Nirmal, N.P. Postbiotics: Current Trends in Food and Pharmaceutical Industry. *Foods* **2022**, *11*, 3094. [CrossRef]
73. Adesulu-Dahunsi, A.; Sanni, A.; Jeyaram, K. Production, characterization and In vitro antioxidant activities of exopolysaccharide from Weissella cibaria GA44. *LWT* **2018**, *87*, 432–442. [CrossRef]
74. Yuan, S.; Wang, Y.; Zhao, F.; Kang, L. Complete Genome Sequence of Weissella confusa LM1 and Comparative Genomic Analysis. *Front. Microbiol.* **2021**, *12*, 2830. [CrossRef] [PubMed]
75. Park, S.; Saravanakumar, K.; Sathiyaseelan, A.; Han, K.-S.; Lee, J.; Wang, M.-H. Polysaccharides of Weissella cibaria Act as a Prebiotic to Enhance the Probiotic Potential of Lactobacillus rhamnosus. *Appl. Biochem. Biotechnol.* **2022**, 1–13. [CrossRef] [PubMed]
76. Wang, W.; Li, S.; Heng, X.; Chu, W. Weissella confusa CGMCC 19,308 Strain Protects Against Oxidative Stress, Increases Lifespan, and Bacterial Disease Resistance in Caenorhabditis elegans. *Probiotics Antimicrob. Proteins* **2021**, *14*, 121–129. [CrossRef] [PubMed]
77. Sandes, S.; Figueiredo, N.; Pedroso, S.; Sant'Anna, F.; Acurcio, L.; Junior, M.A.; Barros, P.; Oliveira, F.; Cardoso, V.; Generoso, S.; et al. Weissella paramesenteroides WpK4 plays an immunobiotic role in gut-brain axis, reducing gut permeability, anxiety-like and depressive-like behaviors in murine models of colitis and chronic stress. *Food Res. Int.* **2020**, *137*, 109741. [CrossRef] [PubMed]
78. Prado, G.; Torrinha, K.; Cruz, R.; Gonçalves, A.; Silva, C.; Oliveira, F.; Nunes, A.; Gomes, M.; Caliari, M. *Weissella paramesenteroides* WpK4 ameliorate the experimental amoebic colitis by increasing the expression of MUC-2 and the intestinal epithelial regeneration. *J. Appl. Microbiol.* **2020**, *129*, 1706–1719. [CrossRef]
79. Lim, S.K.; Kwon, M.-S.; Lee, J.; Oh, Y.J.; Jang, J.-Y.; Lee, J.-H.; Park, H.W.; Nam, Y.-D.; Seo, M.-J.; Roh, S.W.; et al. Weissella cibaria WIKIM28 ameliorates atopic dermatitis-like skin lesions by inducing tolerogenic dendritic cells and regulatory T cells in BALB/c mice. *Sci. Rep.* **2017**, *7*, 40040. [CrossRef]
80. Yu, Y.-J.; Chen, Z.; Chen, P.T.; Ng, I.-S. Production, characterization and antibacterial activity of exopolysaccharide from a newly isolated Weissella cibaria under sucrose effect. *J. Biosci. Bioeng.* **2018**, *126*, 769–777. [CrossRef]
81. Yeu, J.-E.; Lee, H.-G.; Park, G.-Y.; Lee, J.; Kang, M.-S. Antimicrobial and Antibiofilm Activities of *Weissella cibaria* against Pathogens of Upper Respiratory Tract Infections. *Microorganisms* **2021**, *9*, 1181. [CrossRef]
82. Kibar, H.; Arslan, Y.E.; Ceylan, A.; Karaca, B.; Haliscelik, O.; Kiran, F. Weissella cibaria EIR/P2-derived exopolysaccharide: A novel alternative to conventional biomaterials targeting periodontal regeneration. *Int. J. Biol. Macromol.* **2020**, *165*, 2900–2908. [CrossRef]
83. Xiong, L.; Ni, X.; Niu, L.; Zhou, Y.; Wang, Q.; Khalique, A.; Liu, Q.; Zeng, Y.; Shu, G.; Pan, K.; et al. Isolation and Preliminary Screening of a Weissella confusa Strain from Giant Panda (Ailuropoda melanoleuca). *Probiotics Antimicrob. Proteins* **2018**, *11*, 535–544. [CrossRef]
84. Klaenhammer, T.R. Genetics of bacteriocins produced by lactic acid bacteria. *FEMS Microbiol. Rev.* **1993**, *12*, 39–85. [CrossRef]
85. Nath, S.; Roy, M.; Sikidar, J.; Deb, B.; Sharma, I.; Guha, A. Characterization and in-vitro screening of probiotic potential of novel Weissella confusa strain GCC_19R1 isolated from fermented sour rice. *Curr. Res. Biotechnol.* **2021**, *3*, 99–108. [CrossRef]

86. Srionnual, S.; Yanagida, F.; Lin, L.-H.; Hsiao, K.-N.; Chen, Y.-S. Weissellicin 110, a Newly Discovered Bacteriocin from *Weissella cibaria* 110, Isolated from Plaa-Som, a Fermented Fish Product from Thailand. *Appl. Environ. Microbiol.* **2007**, *73*, 2247–2250. [CrossRef]
87. Li, S.-W.; Chen, Y.-S.; Lee, Y.-S.; Yang, C.-H.; Srionnual, S.; Wu, H.-C.; Chang, C.-H. Comparative genomic analysis of bacteriocin-producing Weissella cibaria 110. *Appl. Microbiol. Biotechnol.* **2017**, *101*, 1227–1237. [CrossRef]
88. Teixeira, C.G.; Rodrigues, R.D.S.; Yamatogi, R.S.; Lucau-Danila, A.; Drider, D.; Nero, L.A.; de Carvalho, A.F. Genomic Analyses of *Weissella cibaria* W25, a Potential Bacteriocin-Producing Strain Isolated from Pasture in Campos das Vertentes, Minas Gerais, Brazil. *Microorg.* **2022**, *10*, 314. [CrossRef]
89. Chen, C.; Chen, X.; Jiang, M.; Rui, X.; Li, W.; Dong, M. A newly discovered bacteriocin from Weissella hellenica D1501 associated with Chinese Dong fermented meat (Nanx Wudl). *Food Control* **2014**, *42*, 116–124. [CrossRef]
90. Chen, C.; Rui, X.; Lu, Z.; Li, W.; Dong, M. Enhanced shelf-life of tofu by using bacteriocinogenic Weissella hellenica D1501 as bioprotective cultures. *Food Control* **2014**, *46*, 203–209. [CrossRef]
91. Papagianni, M.; Papamichael, E.M. Purification, amino acid sequence and characterization of the class IIa bacteriocin weissellin A, produced by Weissella paramesenteroides DX. *Bioresour. Technol.* **2011**, *102*, 6730–6734. [CrossRef]
92. Leong, K.-H.; Chen, Y.-S.; Lin, Y.-H.; Pan, S.-F.; Yu, B.; Wu, H.-C.; Yanagida, F. Weissellicin L, a novel bacteriocin from *sian-sianzih* -isolated *Weissella hellenica* 4-7. *J. Appl. Microbiol.* **2013**, *115*, 70–76. [CrossRef]
93. Chen, Y.-S.; Lee, Y.-S.; Wu, H.-C.; Chiang, C.-M.; Pan, S.-F.; Leong, K.-H. Identification of nucleotide sequence involved in Weissellicin L production. *SpringerPlus* **2014**, *3*, 617. [CrossRef]
94. Masuda, Y.; Perez, R.; Zendo, T.; Sonomoto, K. Nutrition-adaptive control of multiple-bacteriocin production by *Weissella hellenica* QU 13. *J. Appl. Microbiol.* **2015**, *120*, 70–79. [CrossRef] [PubMed]
95. Tenea, G.N.; Lara, M.I. Antimicrobial compounds produced by *Weissella confusa* Cys2-2 strain inhibit Gram-negative bacteria growth. *CyTA J. Food* **2019**, *17*, 105–111. [CrossRef]
96. Goh, H.F.; Philip, K. Purification and Characterization of Bacteriocin Produced by Weissella confusa A3 of Dairy Origin. *PLoS ONE* **2015**, *10*, e0140434. [CrossRef] [PubMed]
97. Techo, S.; Visessanguan, W.; Vilaichone, R.-K.; Tanasupawat, S. Characterization and Antibacterial Activity Against Helicobacter pylori of Lactic Acid Bacteria Isolated from Thai Fermented Rice Noodle. *Probiotics Antimicrob. Proteins* **2018**, *11*, 92–102. [CrossRef] [PubMed]
98. Nam, H.; Ha, M.; Bae, O.; Lee, Y. Effect of *Weissella confusa* Strain PL9001 on the Adherence and Growth of *Helicobacter pylori*. *Appl. Environ. Microbiol.* **2002**, *68*, 4642–4645. [CrossRef]
99. Pal, A.; Ramana, K. Isolation and preliminary characterization of a nonbacteriocin antimicrobial compound from Weissella paramesenteroides DFR-8 isolated from cucumber (Cucumis sativus). *Process Biochem.* **2009**, *44*, 499–503. [CrossRef]
100. Malik, A.; Sumayyah, S.; Yeh, C.-W.; Heng, N.C. Identification and sequence analysis of pWcMBF8-1, a bacteriocin-encoding plasmid from the lactic acid bacterium Weissella confusa. *FEMS Microbiol. Lett.* **2016**, *363*, fnw059. [CrossRef]
101. Lim, H.-S.; Yeu, J.-E.; Hong, S.-P.; Kang, M.-S. Characterization of Antibacterial Cell-Free Supernatant from Oral Care Probiotic Weissella cibaria, CMU. *Molecules* **2018**, *23*, 1984. [CrossRef]
102. Kanjan, P.; Kimtun, A.; Chaimongkol, S.; Sakpetch, P. Probiotic *Weissella cibaria* KY10 derived from digestive tract of healthy shrimp exhibits strong antibacterial effects against *Vibrio parahaemolyticus* causing AHPND in shrimp. *Aquac. Res.* **2022**, *53*, 2597–2607. [CrossRef]
103. Dey, D.K.; Khan, I.; Kang, S.C. Anti-bacterial susceptibility profiling of Weissella confusa DD_A7 against the multidrug-resistant ESBL-positive E. coli. *Microb. Pathog.* **2018**, *128*, 119–130. [CrossRef]
104. Stedman, A.; van Vliet, A.H.M.; Chambers, M.A.; Gutierrez-Merino, J. Gut commensal bacteria show beneficial properties as wildlife probiotics. *Ann. N. Y. Acad. Sci.* **2020**, *1467*, 112–132. [CrossRef]
105. Dey, D.K.; Kang, S.C. Weissella confusa DD_A7 pre-treatment to zebrafish larvae ameliorates the inflammation response against Escherichia coli O157:H7. *Microbiol. Res.* **2020**, *237*, 126489. [CrossRef]
106. Kim, M.J.; You, Y.O.; Kang, J.Y.; Kim, H.J.; Kang, M.S. Weissella cibaria CMU exerts an anti-inflammatory effect by inhibiting Aggregatibacter actinomycetemcomitans-induced NF-κB activation in macrophages. *Mol. Med. Rep.* **2020**, *22*, 4143–4150. [CrossRef]
107. Christovich, A.; Luo, X.M. Gut Microbiota, Leaky Gut, and Autoimmune Diseases. *Front. Immunol.* **2022**, *13*, 946248. [CrossRef]
108. Silva, M.S.; Ramos, C.L.; González-Avila, M.; Gschaedler, A.; Arrizon, J.; Schwan, R.F.; Dias, D.R. Probiotic properties of Weissella cibaria and Leuconostoc citreum isolated from tejuino—A typical Mexican beverage. *LWT* **2017**, *86*, 227–232. [CrossRef]
109. Devi, P.B.; Kavitake, D.; Jayamanohar, J.; Shetty, P.H. Preferential growth stimulation of probiotic bacteria by galactan exopolysaccharide from Weissella confusa KR780676. *Food Res. Int.* **2021**, *143*, 110333. [CrossRef]
110. Xu, H.-M.; Huang, H.-L.; Xu, J.; He, J.; Zhao, C.; Peng, Y.; Zhao, H.-L.; Huang, W.-Q.; Cao, C.-Y.; Zhou, Y.-J.; et al. Cross-Talk Between Butyric Acid and Gut Microbiota in Ulcerative Colitis Following Fecal Microbiota Transplantation. *Front. Microbiol.* **2021**, *12*, 658292. [CrossRef]
111. Fatmawati, N.N.D.; Gotoh, K.; Mayura, I.P.B.; Nocianitri, K.A.; Suwardana, G.N.R.; Komalasari, N.L.G.Y.; Ramona, Y.; Sakaguchi, M.; Matsushita, O.; Sujaya, I.N. Enhancement of intestinal epithelial barrier function by Weissella confusa F213 and Lactobacillus rhamnosus FBB81 probiotic candidates in an in vitro model of hydrogen peroxide-induced inflammatory bowel disease. *BMC Res. Notes* **2020**, *13*, 1–7. [CrossRef]

112. Park, S.-H.; Lee, M.-R.; Yang, S.Y.; Lee, J.Y.; Lee, H.H.; Seong, Y.-J.; Kim, B.; Kim, H.-J.; Jin, H.; Johnston, T.V.; et al. In vivo functional effects of Weissella confusa VP30 exopolysaccharides on loperamide-induced constipation in rats. *Food Sci. Biotechnol.* **2022**, *31*, 1703–1715. [CrossRef]
113. Cha, S.; Ahn, B.; Kim, J. Korea Food Research Institute, Assignee Weissella cibaria 148-2 Lactic Bacteria for Functional Healthy Effect and Makgeolli Containing the Same. Korea Patent KR 1020080133488, 2008.
114. Kwak, S.-H.; Cho, Y.-M.; Noh, G.-M.; Om, A.-S. Cancer Preventive Potential of Kimchi Lactic Acid Bacteria (Weissella cibaria, Lactobacillus plantarum). *J. Cancer Prev.* **2014**, *19*, 253–258. [CrossRef]
115. Ankaiah, D.; Mitra, S.; Srivastava, D.; Sivagnanavelmurugan, M.; Ayyanna, R.; Jha, N.; Venkatesan, A. Probiotic characterization of bacterial strains from fermented South Indian tomato pickle and country chicken intestine having antioxidative and antiproliferative activities. *J. Appl. Microbiol.* **2021**, *131*, 949–963. [CrossRef] [PubMed]
116. Alvim, L.; Sandes, S.; Silva, B.; Steinberg, R.; Campos, M.; Acurcio, L.; Arantes, R.; Nicoli, J.; Neumann, E.; Nunes, A. Weissella paramesenteroides WpK4 reduces gene expression of intestinal cytokines, and hepatic and splenic injuries in a murine model of typhoid fever. *Benef. Microbes* **2016**, *7*, 61–73. [CrossRef] [PubMed]
117. Kwon, H.; Chae, S.H.; Jung, H.J.; Shin, H.M.; Ban, O.-H.; Yang, J.; Kim, J.H.; Jeong, J.E.; Jeon, H.M.; Kang, Y.W.; et al. The effect of probiotics supplementation in postoperative cancer patients: A prospective pilot study. *Ann. Surg. Treat. Res.* **2021**, *101*, 281–290. [CrossRef] [PubMed]
118. Amer, M.; Eldiwany, A.; Elgammal, E.; Atwa, N.A.; Dawoud, I.; Rashad, F.M. Nano-exopolysaccharide from the probiotic Weissella paramesenteroides MN2C2: Production, characterization and anticancer activity. *Egypt. J. Chem.* **2021**. [CrossRef]
119. Park, H.-E.; Kang, K.-W.; Kim, B.-S.; Lee, S.-M.; Lee, W.-K. Immunomodulatory Potential of Weissella cibaria in Aged C57BL/6J Mice. *J. Microbiol. Biotechnol.* **2017**, *27*, 2094–2103. [CrossRef]
120. Park, H.-E.; Do, K.-H.; Lee, W.-K. The immune-modulating effects of viable Weissella cibaria JW15 on RAW 264.7 macrophage cells. *J. Biomed. Res.* **2019**, *34*, 36–43. [CrossRef]
121. Hitch, T.C.A.; Hall, L.J.; Walsh, S.K.; Leventhal, G.E.; Slack, E.; de Wouters, T.; Walter, J.; Clavel, T. Microbiome-based interventions to modulate gut ecology and the immune system. *Mucosal Immunol.* **2022**, *15*, 1095–1113. [CrossRef]
122. Adebayo-Tayo, B.; Ishola, R.; Oyewunmi, T. Characterization, antioxidant and immunomodulatory potential on exopolysaccharide produced by wild type and mutant Weissella confusa strains. *Biotechnol. Rep.* **2018**, *19*, e00271. [CrossRef]
123. Reid, G.; Younes, J.A.; Van Der Mei, H.C.; Gloor, G.B.; Knight, R.; Busscher, H.J. Microbiota restoration: Natural and supplemented recovery of human microbial communities. *Nat. Rev. Genet.* **2011**, *9*, 27–38. [CrossRef]
124. Moon, Y.; Soh, J.; Yu, J.; Sohn, H.; Cha, Y.; Oh, S. Intracellular lipid accumulation inhibitory effect of W eissella koreensis OK 1-6 isolated from K imchi on differentiating adipocyte. *J. Appl. Microbiol.* **2012**, *113*, 652–658. [CrossRef]
125. Jeong, Y.; Kim, H.; Lee, J.; Won, G.; Choi, S.-I.; Kim, G.-H.; Kang, C.-H. The Antioxidant, Anti-Diabetic, and Anti-Adipogenesis Potential and Probiotic Properties of Lactic Acid Bacteria Isolated from Human and Fermented Foods. *Fermentation* **2021**, *7*, 123. [CrossRef]
126. Choi, S.-I.; You, S.; Kim, S.; Won, G.; Kang, C.-H.; Kim, G.-H. Weissella cibaria MG5285 and Lactobacillus reuteri MG5149 attenuated fat accumulation in adipose and hepatic steatosis in high-fat diet-induced C57BL/6J obese mice. *Food Nutr. Res.* **2021**, *65*. [CrossRef]
127. Lee, J.S.; Hyun, I.K.; Seo, H.-J.; Song, D.; Kim, M.Y.; Kang, S.-S. Biotransformation of whey by Weissella cibaria suppresses 3T3-L1 adipocyte differentiation. *J. Dairy Sci.* **2021**, *104*, 3876–3887. [CrossRef]
128. Zhao, C.; Su, W.; Mu, Y.; Jiang, L.; Mu, Y. Correlations between microbiota with physicochemical properties and volatile flavor compo-nents in black glutinous rice wine fermentation. *Food Res. Int.* **2020**, *138*, 109800. [CrossRef]
129. Byakika, S.; Mukisa, I.M.; Byaruhanga, Y.B.; Muyanja, C. Probiotic Potential of Lactic Acid Starter Cultures Isolated from a Traditional Fermented Sorghum-Millet Beverage. *Int. J. Microbiol.* **2020**, *2020*, 7825943. [CrossRef]
130. Chen, C.; Liu, Y.; Tian, H.; Ai, L.; Yu, H. Metagenomic analysis reveals the impact of JIUYAO microbial diversity on fermentation and the volatile profile of Shaoxing-jiu. *Food Microbiol.* **2019**, *86*, 103326. [CrossRef]
131. Lopez, C.M.; Rocchetti, G.; Fontana, A.; Lucini, L.; Rebecchi, A. Metabolomics and gene-metabolite networks reveal the potential of Leuconostoc and Weissella strains as starter cultures in the manufacturing of bread without baker's yeast. *Food Res. Int.* **2022**, *162*, 112023. [CrossRef]
132. Galli, V.; Venturi, M.; Coda, R.; Maina, N.H.; Granchi, L. Isolation and characterization of indigenous Weissella confusa for in situ bacterial exopolysaccharides (EPS) production in chickpea sourdough. *Food Res. Int.* **2020**, *138*, 109785. [CrossRef]
133. Montemurro, M.; Pontonio, E.; Rizzello, C. Design of a "Clean-Label" Gluten-Free Bread to Meet Consumers Demand. *Foods* **2021**, *10*, 462. [CrossRef]
134. Hernández-Alcántara, A.M.; Chiva, R.; Mohedano, M.L.; Russo, P.; Ruiz-Masó, J.; del Solar, G.; Spano, G.; Tamame, M.; López, P. Weissella cibaria riboflavin-overproducing and dextran-producing strains useful for the development of functional bread. *Front. Nutr.* **2022**, *9*, 978831. [CrossRef]
135. Han, J.; Lin, X.; Liang, H.; Zhang, S.; Zhu, B.; Ji, C. Improving the safety and quality of Roucha using amine-degrading lactic acid bacteria starters. *Food Res. Int.* **2022**, *161*, 111918. [CrossRef] [PubMed]

136. Bourdichon, F.; Patrone, V.; Fontana, A.; Milani, G.; Morelli, L. Safety demonstration of a microbial species for use in the food chain: Weissella confusa. *Int. J. Food Microbiol.* **2020**, *339*, 109028. [CrossRef] [PubMed]
137. Fessard, A.; Remize, F. Why Are *Weissella* spp. Not Used as Commercial Starter Cultures for Food Fermentation? *Fermentation* **2017**, *3*, 38. [CrossRef]

microorganisms

Review

The Efficacy of Probiotics as Antiviral Agents for the Treatment of Rotavirus Gastrointestinal Infections in Children: An Updated Overview of Literature

Andrej Steyer [1], Dušanka Mičetić-Turk [2,3] and Sabina Fijan [3,*]

1 National Laboratory of Health, Environment and Food, Division of Public Health Microbiology, Grablovičeva 44, 1000 Ljubljana, Slovenia
2 Department of Paediatrics, Faculty of Medicine, University of Maribor, Taborska ulica 8, 2000 Maribor, Slovenia
3 Institute for Health and Nutrition, Faculty of Health Sciences, University of Maribor, Žitna ulica 15, 2000 Maribor, Slovenia
* Correspondence: sabina.fijan@um.si

Abstract: Enteric viruses, including the rotavirus, norovirus, and adenoviruses, are the most common cause of acute gastroenteritis. The rotavirus disease is especially prevalent among children, and studies over the past decade have revealed complex interactions between rotaviruses and the gut microbiota. One way to treat and prevent dysbiosis is the use of probiotics as an antiviral agent. This review focuses on the latest scientific evidence on the antiviral properties of probiotics against rotavirus gastroenteric infections in children. A total of 19 studies exhibited a statistically significant antiviral effect of probiotics. The main probiotics that were effective were *Saccharomyces cerevisiae* var. *boulardii, Lacticaseibacillus rhamnosus* GG, and various multi-strain probiotics. The underlying mechanism of the probiotics against rotavirus gastroenteric infections in children included immune enhancement and modulation of intestinal microbiota leading to shortening of diarrhoea. However, several clinical studies also found no significant difference in the probiotic group compared to the placebo group even though well-known strains were used, thus showing the importance of correct dosage, duration of treatment, quality of probiotics and the possible influence of other factors, such as the production process of probiotics and the influence of immunisation on the effect of probiotics. Therefore, more robust, well-designed clinical studies addressing all factors are warranted.

Keywords: probiotics; microbiota; rotaviruses

Citation: Steyer, A.; Mičetić-Turk, D.; Fijan, S. The Efficacy of Probiotics as Antiviral Agents for the Treatment of Rotavirus Gastrointestinal Infections in Children: An Updated Overview of Literature. *Microorganisms* **2022**, *10*, 2392. https://doi.org/10.3390/microorganisms10122392

Academic Editor: Denis Roy

Received: 27 October 2022
Accepted: 29 November 2022
Published: 2 December 2022

1. Introduction

Acute gastroenteritis is one of the most frequently reported infectious diseases in the world. The most common cause of acute gastroenteritis (AGE) is various enteric viruses, including rotaviruses, noroviruses, astroviruses, adenoviruses, and other less presentable viruses [1]. Most are icosahedral nonenveloped viruses, known to present stability in the environment, resistant to many physio-chemical conditions. Their stability in the environment and on various fomites is also crucial for indirect transmission via contaminated surfaces, food, and water [2]. As the infectious dose, particularly for noroviruses, is very low [3], indirect infections are possible, and each year we can follow reports on food and/or waterborne infections, mostly with noroviruses.

Rotaviruses are members of the *Reoviridae* family and are characterized by their non-enveloped, segmented, double-stranded RNA genome (11 segments). Each of the 11 genes code for a single gene product. Six of the proteins are found in the virus particle (vp1, vp2, vp3, vp4, vp6 and vp7), whereas the remaining five proteins are non-structural (NDP1–NSP5). The *Rotavirus* is classified into serogroups A to E based on antigenic properties. Only groups A to C have been shown to infect humans, and the most human *Rotavirus* disease is caused by the group A *Rotavirus*. The group A *Rotavirus* is further classified into G (serotypes) and P types based on identification of antigens on the outer capsid proteins. Group A rotavirus genotypes are classified by a nucleotide-sequence-based, complete genome classification system [4,5].

Rotavirus gastroenteritis is still an important public health concern. In particular, low-income countries are fighting against the rotavirus disease, especially affecting small children [6]. Rotavirus gastroenteritis is the leading global pathogen of diarrhoea-associated mortality with the highest death rate among children under 5 years worldwide. Since 2006, efficient vaccines have been available to protect children from severe rotavirus gastroenteritis [7]. However, there are still high numbers of acute rotavirus gastroenteritis in those countries.

During the post-marketing phase of rotavirus vaccines, one of the most exceptional findings was the difference in vaccine effectiveness, being much lower in low- and middle-income countries [8,9]. One of the possible explanations was the effect of histo-blood groups, which may contribute to the virus binding on these antigens [10,11]. In parallel, a new research area of the virus–bacteria interactions opened, showing that enteric viruses may bind to bacteria surface antigens, which may influence the early phases of virus pathogenesis [12–14]. Consequently, it is clear now that the pathogenesis of enteric viruses is dependent not only on virus pathogenetic factors or host determinants, but also on the environment. The microbiota is therefore of high importance and can influence the effectiveness of the rotavirus or other enteric virus infections. In addition, studies on probiotics are also promising in the prevention phase, and to some extent, also in the curative phase of AGE [15–17].

The gastrointestinal tract is one of the most microbiologically active ecosystems with a high density of bacteria and other microbes formulating the intestinal microbiota. This microbiota has several beneficial roles for its human host, including antimicrobial activity, competitive exclusion, immunomodulation, strengthening of the epithelial barrier function, as well as influencing the immune system, central nervous system, and endocrine system [18–21]. Recent evidence-based research shows that the gut microbiota is an ally for the interaction with most human cells via the microbiota-gut-brain axis, microbiota-gut-skin axis, microbiota-gut-lung axis, microbiota-gut-liver axis, microbiota-gut-vagina axis, and many more axes. The microbiota thus aids in achieving homeostasis of skin health, respiratory health, organ health, mental health, and so forth of its host [19–28]. The intestinal microbiota coexists with microbes that reach the intestine through food intake and influences the immune cells associated with the lamina propria through the production of metabolites, crucial for the maturation of immune cells in the mucosal immune system [19,21,29]. Disruption of the homeostasis between the intestinal microbiome and the host immune system can adversely impact viral immunity [30].

Rotaviruses infect the small intestine, an important site of colonization by the microbiota, and studies over the past decade have begun to reveal a complex set of interactions between rotaviruses and the gut microbiota, as rotavirus infection can temporarily alter the composition of the gut microbiota [13]. One way to treat and prevent dysbiosis is the use of probiotics. Probiotics are, by definition, "live microorganisms that, when administered in adequate amounts, confer a health effect on the host" [31]. Scientific evidence shows enough evidence to justify the use of probiotics for the treatment of several disorders, including gastrointestinal dysbiosis, antibiotic-associated diarrhoea, irritable bowel syndrome, and inflammatory bowel disease, as well as anxiety, depression, and wound healing [19,20,32–35]. In a review on the management of acute gastroenteritis in Jordanian children [36], it was emphasised that prevention of diarrhoea diseases should focus on the improvement of nutrition, hygiene, and sanitation. In the case of rotavirus gastroenteritis, the authors proposed the introduction of routine vaccination against the rotavirus, as well as the use of adjuvant therapies. One of these possible therapies is probiotics. Other reviews addressing gastrointestinal infections also conclude that probiotics are one of the possible adjuvant strategies for diarrhoea in children by resuming a healthy microbiota status following infection [37–39] A recently published review even suggested the potential of a combined lactic-acid bacteria vaccine as an alternative recombinant vaccine against the rotavirus [40]. Two reviews have already addressed the efficacy of using probiotics for rotavirus infection in children, one published in 2015 [41] and another published in 2020 [42]. The review from 2015 focussed on the duration of rotavirus diarrhoea in children, whilst the 2020 review found that probiotics could reduce the occurrence of acute rotavirus diarrhoea in children. Our review investigated the underlying antiviral mechanisms of probiotics against rotavirus infections in children, includes updated information, and focused on the effective mechanisms of probiotics.

2. Rotavirus Infection and the Gut

The *Rotavirus* infects the mature enterocytes in the middle and upper parts of the villi and in the enteroendocrine cells in the small intestine, which ultimately leads to diarrhea [43]. Rotavirus infection can temporarily alter the composition of the gut microbiota, thus leading to dysbiosis [13]. According to one study, dysbios is caused by a decrease in the amount of bifidobacteria, normal *Escherichia coli,* and an increase in the amount of lactose-negative *Escherichia.* In cases of pronounced dysbiosis in young children, the clinical course of rotavirus infection is aggravated and the period of rotavirus excretion is prolonged [44]. Other studies found that patients with diarrheal stools with rotavirus had more bacterial communites at the genus level containing specific diarrheal causative bacteria than those of healthy subjects, suggesting that co-infection with the virus and bacteria could have occurred in some diarrhea cases [43]. Gut dysbiosis due to viral infection could be associated with a reduction in the populations of common and beneficial bacterial species and the resulting loss of diversity, as well as the gain of harmful bacteria. It may also be due to variations in crosstalk via direct interaction between rotaviruses and bacteria in the gut [43,45].

A symptomatic infection with rotaviruses stimulates a strong humoral IgG immune response which lasts for a lifetime. While the IgG responses are easily recorded, it is generally thought that protection from rotavirus disease is mediated by local IgA antibodies [4,46].

3. Probiotics and the Antiviral Mechanisms

Probiotic administration stimulates the immune system by inducing a network of signals mediated by various metabolites. Some probiotic strains stimulate the immune response and are therefore beneficial for patients suffering from immune deficiency, whilst other strains inhibit the immune response and are therefore beneficial for patients with conditions with immune activation. Additionally, the effects of probiotic modulation on the immune cells can be observed in lymphocytes, hematopoietic stem cells, T cells, macrophages, natural killer cells, and dendritic cells. Additionally, molecules usually associated with pathogens, such as lipopolysaccharide of gram-negative bacteria or lipoteichoic acids of gram-positive bacteria, can be produced by probiotics and interact with different toll-like receptors, and incite NF-κB-mediated antiviral gene expression [19,34,35,47,48].

It is also known that respiratory viruses can cause changes in the gut microbiome, therefore probiotics are a possible medication to treat respiratory viral infections via gut-microbiota modulation and production of immunomodulatory agents. Interactions between probiotics, macrophages, and dendritic cells are seen in the lamina propria, resulting in natural killer (NK) cell activation, which triggers interferon gamma (IFN-γ) production to defend against viruses, and efficient immune cells go to infection sites via circulatory and lymphatic systems to protect against respiratory viruses [35,47,48].

Bacteriocins produced by probiotics have also proven effective against viral infections as they exhibit antimicrobial potential against viral pathogens by prevention of viral particle aggregation and blocking the sites of host cell receptors or inhibition of viral penetration into human cells [49–52].

All above-mentioned mechanisms collectively lead to the indirect consequence of a shorter infectious period and overall reduction in the risk of viral infection [53–55].

On the other hand, previous bacterial infections in children may increase the risk of rotavirus infections by disrupting the balance of the intestinal microbiota, leading to dysbiosis and increasing the ratio of pathogenic bacteria [56,57]. Co-infection with bacterial diarrhoea-related bacterial pathogens, such as *Escherichia, Shigella, Klebsiella,* and *Campylobacter* spp., can cause a more severe course of the rotavirus disease [43]. Although it is well-established that probiotics display antibacterial activities against common pathogenic bacteria, including competitive exclusion, bacteriocin production, enhancing intestinal barrier function, and stimulation of host antimicrobial defences [46], these bacterial infections can antagonise the antivirus effects of probiotics while they are fighting off bacterial pathogens. Therefore, more research into the complex mechanisms of actions of probiotics and pathogens is warranted.

4. Clinical Trials with Established Antiviral Effect of Probiotics against Rotaviruses

We used the search strategy: "probiotics" AND rotavirus in various databases (PubMed, Web of Science, Scopus) and included clinical trials, which found a statistically significant antiviral effect of probiotics in the treatment of rotavirus gastroenteritis. Clinical trials without the full text available and in languages other than English were excluded. Clinical trials where rotaviruses were not determined or detected were also excluded. A total of 19 clinical studies with a statistically significant antiviral effect of probiotics against rotaviruses were found. These studies were conducted in Argentina, Bangladesh, Bolivia, Brazil, Croatia, Denmark, Egypt, Greece, India, Israel, Italy, the Republic of Korea, the Netherlands, Poland, Portugal, Slovenia, Taiwan, Turkey, and the United Kingdom. The characteristics of the clinical trials are described in Table 1.

Table 1. Characteristics of 19 clinical studies with statistically significant antiviral effects of probiotics against rotaviruses.

Reference (First Author, Year) [1]	Study Design	Population	Intervention			Main Findings
			Active	Control	Duration	
Shin et al. (2020), Rep. Korea [58]	RCT	50 hospitalized children with rotavirus gastroenteritis, aged up to 6 years. 15 in novel probiotic group. 8 in control group (group II). 27 in group III (retrospectively analysed through medical records).	Group I: • *Lactiplantibacillus* [3] *plantarum* LRCC5310 Group III: • Probiotic *Saccharomyces cerevisiae* species [2] according to hospital treatment policy Dosage: not specified.	Group II: standard treatment	Up to 8 days	Group I (*Lactiplantibacillus plantarum* LRCC5310) showed a statistically significant improvement in the number of patients with persistent diarrhoea, number of defecation events per day, and total diarrhoea period compared to group II (control). Group I showed slight improvement in the number of patients with loose stools, number of defecation events, and diarrhoea duration compared to group III (*Saccharomyces cerevisiae*-containing probiotic formulation).
Park, Kwon, Ku, and Ji (2017), Korea [59]	Double-blind RCT	57 hospitalized infants with rotavirus disease, aged between 9 and 16 months. 28 in probiotic group. 29 in control group.	• *Bifidobacterium longum* BORI • *Lactobacillus acidophilus* AD031 Dosage: 2.2×10^9 cfu/g twice daily	Placebo	3 days	A significantly shorter duration of patients' diarrhoea was observed in the probiotic group compared to the placebo group. Symptoms such as duration of fever, frequency of diarrhoea, and frequency of vomiting tended to be ameliorated by the probiotic treatment; however, differences were not statistically significant between the two groups.
Das, Gupta, and Das (2016), India [60]	Double-blind RCT	60 children, aged between 3 months to 5 years, with watery diarrhoea and stool rotavirus positive. 30 in probiotic group. 30 in control group.	• *Saccharomyces cerevisiae* var. *boulardii* Dosage: 250 g sachets twice daily	Placebo	5 days	A significantly shorter duration of diarrhoea and hospitalization was observed in the intervention group. No significant difference was seen for fever and vomiting. There was also no difference between the two groups in the proportion of children requiring parenteral rehydration and persistence of diarrhoea lasting beyond day 7.

Table 1. *Cont.*

Reference (First Author, Year) [1]	Study Design	Population	Intervention			Main Findings
			Active	Control	Duration	
Lee et al. (2015), Rep. Korea [55]	Double-blind RCT	29 children between 3 months and 7 years with viral gastroenteritis (9 rotavirus infection). 13 in probiotic group. 16 in control group.	• *Bifidobacterium longum* IBG, • *Bifidobacterium lactis* BL, • *Lactobacillus acidophilus* LA, • *Lacticaseibacillus* [3] *rhamnosus* LRH, • *Lactiplantibacillus* [3] *plantarum* [2], • *Pediococcus pentosaceus* [2] Dosage: 10^9 cfu/g twice daily.	Placebo	1 week	The multi-strain probiotic significantly shortened the duration of diarrhoea and fever compared to the placebo. The mean duration of vomiting was shorter in the probiotic group, but the difference in the study groups was not statistically significant.
Aggarwal et al. (2014), India [61]	Open Label RCT	200 children with watery diarrhoea (41 positive for rotavirus in stool), aged between 6 months and 5 years. 100 in probiotic group. 100 control group.	Culturelle probiotic GG contains: • *Lacticaseibacillus* [3] *rhamnosus* GG (LGG) Dosage: 10^{10} cfu/g once daily	Standard treatment	5 days	A statistically significant decrease in the duration of diarrhoea, faster improvement in stool consistency, and reduction in average number of stools per day was observed in the probiotic group compared to standard treatment.
Huang et al. (2014), Taiwan [62]	Open Label RCT	159 hospitalized children with infectious gastroenteritis (42 rotavirus), aged between 3 months to 14 years. 82 in probiotic group. 77 in control group.	Bio-three contains [4]: • *Enterococcus faecalis* T-110, • *Clostridium butyricum* TO-A, • *Bacillus mesentericus* TO-A Dosage: 3.48×10^8 cfu/g 3 times daily	standard treatment	7 days	A statistically significant decrease in the duration of severe diarrhoea was observed in the probiotic group compared to standard treatment. In the patients with rotavirus, a statistically significant decrease in gastroenteritis (Vesikari score) and diarrhoea frequency was also observed in the probiotic group.

Table 1. *Cont.*

Reference (First Author, Year) [1]	Study Design	Population	Intervention			Main Findings
			Active	Control	Duration	
Sindhu et al. (2014), India [63]	Double-blind RCT	124 children with gastroenteritis, aged between 6 months to 5 years, infected either with rotavirus (82) or *Cryptosporidium* species (42). 65 in probiotic group. 59 in control group.	• *Lacticaseibacillus* [3] *rhamnosus* GG Dosage: 10^{10} cfu/g once per daily	Placebo	4 weeks	A statistically significant increase in the IgG levels post-intervention was observed in children with rotavirus diarrhoea receiving LGG after 4 weeks. Fewer children with rotavirus diarrhoea on LGG had repeated diarrhoeal episodes. No differences were found in duration of diarrhoea.
Corrêa, Penna, Lima, Nicoli, and Filho (2011), Brazil [64]	Double-blind RCT	186 hospitalized children (57.4% with rotavirus), aged between 6 to 48 months, with acute diarrhoea. 90 in probiotic group. 86 in control group.	• *Saccharomyces cerevisiae* var. *boulardii* Dosage: 4×10^9 cfu/g twice daily	Placebo	5 days	A statistically significant reduction in the duration of diarrhoea was observed in probiotic group compared to placebo.
Dalgic, Sancar, Bayraktar, Pullu, and Hasim (2011), Turkey [65]	Single Blind RCT	240 children with rotavirus diarrhoea, divided into eight groups. 60 in each group.	Group 1: • *Saccharomyces cerevisiae* var. *boulardii*, Group 2: zinc (Zn), Group 3: lactose-free formula (LF), Group 4: Zn and • *Saccharomyces boulardii* (SB), Group 5: LF and • *Saccharomyces boulardii* (SB), Group 6: Zn and LF, Group 7: Zn and LF and • *Saccharomyces boulardii* (SB), Group 8: control Dosage: 250 mg once daily	Standard treatment	5 days	A statistically significant reduction in diarrhoea duration and hospital stay was observed in groups 2 and 4 compared to standard treatment. A significant difference in the duration of hospitalization between groups 1 and 4; groups 2 and 7; groups 3 and 4; groups 4 and 5, and groups 4 and 7 was also found.

Table 1. *Cont.*

Reference (First Author, Year) [1]	Study Design	Population	Intervention			Main Findings
			Active	Control	Duration	
Grandy, Medina, Soria, Terán, and Araya (2010), Bolivia [66]	Double-blind RCT	64 hospitalized children with rotavirus infection, aged 1 to 23 months, divided in 3 groups. Group 1: single strain probiotic (20). Group 2: multi-strain probiotic (23). Group 3: control (21)	Group 1: • *Saccharomyces cerevisiae* var. *boulardii* Dosage: 4×10^{10} cfu/g twice daily Group 2: • *Lactobacillus acidophilus* [2], • *Lacticaseibacillus* [3] *rhamnosus* [2], • *Bifidobacterium longum* [2] • *Saccharomyces cerevisiae* var. *boulardii* Dosage: 1.25×10^{8} cfu/g twice daily	Placebo	5 days	Statistically significant decrease in duration of diarrhoea shorter duration of fever was observed in children who received the single-species probiotic compared to the placebo. Statistically significant fewer episodes of vomiting were observed with the multi-species probiotic compared to the placebo. When probiotic groups were merged, the statistical significance of changes increased (total duration of diarrhoea, fever, and vomiting).
Basu, Paul, Ganguly, Chatterjee, and Chandra (2009), India [67]	Double-blind RCT	559 hospitalized children (319 with rotavirus), aged up to 2 years, divided into 3 groups. 185 in group A. 188 in group B. 186 in group C.	Group A: control Group B: LGG, Dosage: 10^{10} cfu/g twice daily. Group C: LGG, Dosage: 10^{12} cfu/g twice daily	Standard treatment	7 days	A statistically significant lower frequency and the duration of diarrhoea, requirement for intravenous therapy, and hospital was observed in both the intervention groups compared with the control. There was no significant difference between the 2 intervention groups.
(Teran, Teran-Escalera, and Villarroel (2009), Bolivia [68]	Single Blind RCT	75 hospitalized children, aged from 28 days to 24 months with rotavirus diarrhoea, divided into three groups. Group 1: nitazoxanide (25). Group 2: probiotic group (25). Group 3: control (25).	Group 1: nitazoxanide Group 2: • *Lactobacillus acidophilus* [2], • *Lacticaseibacillus* [3] *rhamnosus* [2], • *Bifidobacterium longum* [2] and • *Saccharomyces cerevisiae* var. *boulardii* Dosage: 1.25×10^{8} cfu/g twice daily	Standard treatment	5 days	A statistically significant reduction in the duration of diarrhoea and hospital stay was observed in the probiotic group compared to standard treatment.

Table 1. *Cont.*

Reference (First Author, Year) [1]	Study Design	Population	Intervention			Main Findings
			Active	Control	Duration	
Dubey, Rajeshwari, Chakravarty, and Famularo (2008), 2008, India [69]	Double-blind RCT	230 hospitalized children with rotavirus diarrhoea, aged between 6 months and 2 years. 113 in probiotic group. 111 in control group.	VSL#3 contains [4]: 4 strains of lactobacilli species: • *Lactobacillus acidophilus*, • *Lacticaseibacillus* [3] *paracasei*, • *Lactobacillus bulgaricus*, • *Lactiplantibacillus* [2] *plantarum*, 3 strains of Bifidobacteria species: • *Bifidobacterium breve*, • *Bifidobacterium infantis*, • *Bifidobacterium longum*, • 1 strain of Streptococcus thermophilus Dosage: 9×10^{10} cfu/g twice daily	Placebo	4 days	A statistically significant lower mean stool frequency and improved stool consistency was observed after day 2 up to day 4. After day 4, the control group also showed spontaneous improvement. The overall recovery rates were significantly better in the probiotic group compared with the placebo. A statistically significant lower overall requirement for oral rehydration salts was found.
Narayanappa (2008), India [70]	Double-blind RCT	80 hospitalized children with rotavirus diarrhoea, aged between 3 months and 3 years. 40 in probiotic group. 40 in control group.	BIFILAC contains [4]: • *Enterococcus faecalis* [2], • *Clostridium butyricum* [2], • *Bacillus mesentericus* [2], • *Bacillus coagulans* [2] Dosage: 1 sachet three times daily	Placebo	Up to 14 days	A statistically significant reduction in the number of episodes (frequency) of diarrhoea, mean duration of diarrhoea, degree of dehydration, duration and volume of oral rehydration salt therapy, duration and volume of intravenous fluid therapy and duration of rotavirus shedding was observed in the probiotic group compared to the control group.

Table 1. *Cont.*

Reference (First Author, Year) [1]	Study Design	Population	Intervention			Main Findings
			Active	Control	Duration	
Szymański, Pejcz, et al. (2006), Poland [71]	Double-blind RCT	87 children with infectious diarrhoea (39 with rotavirus), aged between 2 months and 6 years. 49 in probiotic group. 44 in control group.	Lakcid L contains: • *Lacticaseibacillus* [3] *rhamnosus* 573L/1, • *Lacticaseibacillus* [3] *rhamnosus* 573L/2 • *Lacticaseibacillus* [3] *rhamnosus* 573L/3 Dosage: 1.2×10^9 cfu/g twice daily	Placebo	5 days	A statistically significant reduction in the duration of rotavirus diarrhoea, but not of diarrhoea of any aetiology, in children was observed in the probiotic group compared to the control group. Intervention shortened the time of intravenous rehydration.
Gaón et al. (2003), Argentina [72]	Double-blind RCT	89 hospitalized children with infectious diarrhoea (27% with rotavirus), aged between 6 and 24 months. 29 in control group (group 1). 30 in group 2. 30 in lactobacilli group 3.	Group 1: placebo Group 2: • *Saccharomyces cerevisiae* var. *boulardii*, Group 3: • *Lacticaseibacillus* [3] *casei* and • *Lactobacillus acidophilus* CERELA Dosage: 10^{10}–10^{12} cfu/g twice daily	Placebo	5 days	A statistically significant reduction in the duration of diarrhoea and number of stools in children was observed in all probiotic groups compared to the control group.
Rosenfeldt et al. (2002), Denmark [73]	Double-blind RCT	69 hospitalized children with infectious diarrhoea (66% with rotavirus), aged between 6 and 36 months. 30 in probiotic group. 39 in control group.	• *Lacticaseibacillus* [3] *rhamnosus* 19070-2, • *Limosilactobacillus reuteri* DSM 12246, Dosage: 2.2×10^{10} cfu twice daily.	Placebo	5 days	A statistically significant reduction in the duration of hospital stay was observed in the probiotic group compared to the placebo. The beneficial effects (duration of diarrhoea, loose stool, length of hospital stay) were most prominent in children treated early in the diarrhoeal phase.

Table 1. *Cont.*

Reference (First Author, Year) [1]	Study Design	Population	Intervention			Main Findings
			Active	Control	Duration	
Guandalini et al., (2000), European study [74]	Double-blind RCT	287 children with liquid or semiliquid stools (101 with rotavirus), aged between 1 month and 3 years. 147 in probiotic group. 140 in placebo group.	• *Lacticaseibacillus* [3] *rhamnosus* GG, Dosage: 10^{10} cfu in 250 mL of standard treatment solution. Solution added to patient according to need.	Standard treatment	Up to 7 days	A statistically significant reduction in duration of diarrhoea and duration of hospital stay in rotavirus-positive and rotavirus-negative children was observed in the probiotic group compared to the control group. In rotavirus-positive children, a significant reduction in number of average stools was also found in the probiotic group compared to the control group.
Guarino, Canani, Spagnuolo, Albano, and Di Benedetto (1997), Italy [75]	Double-blind RCT	100 children with diarrhoea (61 positive for rotavirus), aged between 3 and 36 months. 52 in probiotic group. 48 in control group.	• *Lacticaseibacillus* [3] *rhamnosus* GG [5], Dosage: 10^{10} cfu in 200 m twice daily	Standard treatment	Up to 5 days	A statistically significant reduction in the duration of diarrhoea rotavirus-positive and rotavirus-negative ambulatory children with diarrhoea was observed in the probiotic group compared to the control group. Furthermore, the duration of rotavirus excretion was reduced.

[1] Clinical studies in descending chronological order, arranged alphabetically. RCT: Randomised, controlled trial. [2] Strain not specified. [3] Nomenclature of species has been updated according to Zheng et al., 2020 [76], [4] information of strains not reported in published clinical trial but retrieved from public website. [5] The clinical trial incorrectly notes the probiotic as *Lactobacillus casei* GG.

Our review included two additional clinical trials [58,59] compared to the 2020 review [42] and several more compared to the 2015 review [41] which selected clinical trials published until the year 2013. Strain-specific antiviral activity of probiotic strains, as well as the concentration of probiotic supplements and duration of supplementation, seem to be the most important factors that influence the efficiency of probiotics on rotavirus disease in children [55].

Rotaviruses can cause significant diarrhoeal disease in infants and young ones of various mammalian and avian species [15]. According to European Society for Paediatric Gastroenterology, Hepatology and Nutrition/ESPGHAN/ESMAD, the standard recommended treatment for acute diarrhoea in children, whether due to the rotavirus, norovirus, bacterial or other infection, includes oral rehydration solutions (ORS) and continuance of feeding. Adjuvant therapy with micronutrients, probiotics, or anti-diarrhoea agents are also rendered useful. The recommended probiotics are *Lacticaseibacillus rhamnosus* GG (ATCC 53103), also known as LGG, and the yeast *Saccharomyces cerevisiae* var. *boulardii* [77–80].

The underlying mechanism against rotavirus infections is immune enhancement, as certain strains of lactobacilli promote immunological responses. This includes increasing concentrations of anti-rotavirus-specific IgA [55,81], reducing intestinal microbiota imbalance, enhancing the colonization of probiotics [82,83], and reducing the incidence of diarrhoea [84]. One important activity of probiotics is also increasing the clearance of stool rotavirus by reducing faecal rotavirus shedding, and thus aiding the epidemiological importance in the transmission of rotaviruses [85,86].

The beneficial effects of probiotics in the 19 studies noted in Table 1 have confirmed an antiviral effect of certain probiotics, leading to shortening of diarrhoea in children due to rotavirus enteritis after supplementation. Some studies divided the intervention groups of children into more than one group to ascertain the effect of different combinations of probiotics or different concentrations on rotavirus diarrhoea. Five of these studies investigated the single-strain probiotic *Lactobacillus rhamnosus* GG [61,63,67,74,75]. *Saccharomyces boulardii* was investigated in six studies [58,60,64–66,72]. One aforementioned study [58] also investigated the effectiveness of *Lactiplantibacillus plantarum* LRCC5310 on rotavirus infection. Two other aforementioned studies [66,72] also investigated multi-strain probiotics. All the remaining studies investigated various multi-strain probiotics [55,59,62,68–71,73].

Two abstracts of additional studies in the English language were found [87,88] that noted a beneficial effect of the probiotics in the abstract, but a full text with all relevant data was not available despite contacting the authors; therefore, they were also not included in Table 1. Two studies [83,89] in the Chinese language also found a beneficial effect of probiotics for the prevention of diarrhoea in children, some of which tested positive for the rotavirus in stool samples; however, they were not included in Table 1 as only the abstract was in English. According to the abstracts, both studies found that probiotic supplementation with lactobacilli and/or bifidobacteria (species not specified in abstract) significantly decreased the incidence and duration of diarrhoea. Another study in the French language [90] also found an antiviral effect of *Saccharomyces cerevisiae* var. *boulardii* supplementation in children with acute diarrhoea (15 with rotavirus infection) and found a significant decrease in the duration of diarrhoea. The latter three mentioned studies were not included in Table 1 due to language barriers.

The effect of different multi-strain probiotics on rotavirus diarrhoea was significant after supplementation with *Bifidobacterium longum* BORI and *Lactobacillus acidophilus* AD031 [59], *Bifidobacterium longum* IBG, *Bifidobacterium lactis* BL, *Lactobacillus acidophilus* LA, *Lacticaseibacillus rhamnosus* LRH, *Lactiplantibacillus plantarum*, and *Pediococcus pentosaceus* [55], *Enterococcus faecalis* T-110, *Clostridium butyricum* TO-A and *Bacillus mesentericus* TO-A [62], unspecified strains of *Lactobacillus acidophilus*, *Lacticaseibacillus rhamnosus*, and *Saccharomyces boulardii* [66,68], VSL#3, containing four lactobacilli strains, three bifidobacteria strains, and one strain of *Streptococcus thermophilus* [69], BIFILAC, containing strains of *Enterococcus faecalis*, *Clostridium butyricum*, *Bacillus mesentericus* and *Bacillus coagulans* [70], Lakcid L, containing *Lacticaseibacillus rhamnosus* 573L/1, 573L/2, 573L/3 [71], *Lactobacil-*

lus casei and Lactobacillus acidophilus strains CERELA [72] and *Lacticaseibacillus rhamnosus* 19070-2 and *L. reuteri* DSM 12,246 [73]. However, several studies did not report the strains used, which decreased the quality and reproducibility of the studies.

The probiotic strain *Lacticaseibacillus rhamnosus* GG, previously known as *Lactobacillus rhamnosus* GG (LGG), is a gram-positive lactobacillus, known to promote immunological responses and influence the intestinal microbiota by producing both a biofilm that can mechanically protect the mucosa, and different soluble factors beneficial to the gut by enhancing intestinal crypt survival, diminishing apoptosis of the intestinal epithelium, and preserving cytoskeletal integrity [91]. The ESPGHAN recommends LGG as an adjuvant therapy for gastrointestinal infections in children [77,80]. It was used in a large multi-centre European trial [74] with patients from Poland, Egypt, Croatia, Italy, Slovenia, the Netherlands, Greece, Israel, the United Kingdom, and Portugal. Administering the oral rehydration solution containing LGG to children with acute diarrhoea was found safe and resulted in a shorter duration of diarrhoea, less chance of a protracted course, and faster discharge from the hospital. There is also a large cohort of other studies using the same strain LGG that also confirms this effect [61,63,67,75]. A study by Szajewska et al. [92] that investigated the prevention of nosocomial diarrhoea found that supplementation with *Lacticaseibacillus rhamnosus* GG resulted in a reduced risk of nosocomial diarrhoea in children. A systematic review also confirms the reduction in the duration of rotavirus-induced diarrhoea, where a higher dose was efficient [93]. Another important factor to consider is the possible effect of rotavirus immunisation on the effectiveness of LGG, as noted in the meta-analysis [94], where the authors concluded that rotavirus immunisation affected the efficacy of LGG for the treatment of children with acute diarrhoea, which could be one of the underlying reasons for the mixed results. However, other reviews conclude that probiotics as adjuvants in vaccination should be considered in future studies, especially in the elderly and in children, where vaccine effectiveness and duration of immunisation really matter [38,95].

The probiotic yeast *Saccharomyces cerevisiae* var. *boulardii* is the only yeast used in clinical practice and is recommended for the prevention of antibiotic-associated diarrhoea and acute gastroenteritis in children as an adjunct [79,96] The mechanisms of action include inhibition of growth and invasion of pathogens by interfering with pathogen attachment, production of small peptides that inhibit endotoxins, as well as stimulation of short-chain fatty acids, especially butyrate, that restore intestinal functions and immunoregulation. However, the effect of *Saccharomyces cerevisiae* var. *boulardii* against common viruses responsible for diarrhoea, such as the rotavirus, adenovirus or norovirus, is still very limited, and further research is advocated [96]. *Saccharomyces cerevisiae* var. *boulardii* was efficient in the treatment of rotavirus gastroenteritis in children in six clinical studies noted in Table 1 [58,60,64–66,72].

In a small clinical study conducted in the Republic of Korea by Shin and co-authors [58], 50 hospitalized children with rotavirus enteritis were divided into three groups. The first group received a novel strain *Lactiplantibacillus plantarum* LRCC5310; however, neither the concentration of the probiotic nor the dosage was specified. Group II was the control group that did not receive any probiotics, and group III received a probiotic containing the *Saccharomyces cerevisiae* species according to the treatment policy of the hospital. Group III was retrospectively analysed through medical records. The novel strain LRCC5310 improved clinical symptoms and was comparable to, or more effective than the probiotic containing a *Saccharomyces cerevisiae* species. Several rotavirus genotypes were detected in stools, including: G9P8, G1P8, G1P18, G3P8, G2P4, G4P6, and G9P4. The rotavirus titre was significantly reduced in patients that received the novel strain LRCC5310 compared to those who did not take any probiotic formulations (Group II). Intake of LRCC5310 was found to be effective in the suppression of viral symptoms, as well as in prognosis and treatment, via virus titre reduction. The authors did not discuss the mechanisms involved, but the most likely mechanisms of the antiviral effect of the probiotic was due to modulation of the intestinal microbiota and the improvement of immune function, as several *Lactiplantibacillus plantarum* strains have exhibited enhancement of immune activity during infectious and inflammatory conditions, as well as improving lower gastrointestinal symptoms and modulation of intestinal microbiota after dysbiosis due to infections [97–104]. Although some probiotic traits are strain-specific, other core traits are in fact species-specific [31].

The study by Lee and co-authors [55] investigated the antiviral influence of a multi-strain probiotic against viral gastroenteritis in paediatric patients. Nine of the twenty-nine patients had a rotavirus infection. A six-species supplement containing *Bifidobacterium longum, Bifidobacterium lactis, Lactobacillus acidophilus, Lacticaseibacillus rhamnosus, Lactiplantibacillus plantarum,* and *Pediococcus pentosaceus* (strains not specified) proved effective in statistically significantly reducing the duration of diarrhoea in the probiotic group. Similarly, another multi-species probiotic, containing *Lactobacillus acidophilus, Lacticaseibacillus rhamnosus, Bifidobacterium longum,* and *Saccharomyces boulardii* (strains not specified) was also efficient [66,68]. Supplementation with bifidobacteria, including the probiotic *Bifidobacterium bifidum* Bb12, has been shown to protect against rotavirus infection, as children receiving this probiotic had a statistically significant lower concentration of the rotavirus-specific IgA antibody compared to the control group [105]. The well-known probiotic VSL#3 was also used in a study by Dubey et al. [69], conducted in India, and found a statistically significant lower duration and frequency of rotavirus diarrhoea in the probiotic group compared to the control group. Interestingly, the authors report that the statistically significant differences were still observed on day 4, but by day 8 the control group also spontaneously improved, and the results became comparable with the probiotic group. The antiviral effect of the multi-strain probiotic Bifilac was also found [70]. However, the author does not specify the composition of the supplement in the clinical trial.

Huang et al. [62] found that supplementation with a three-strain probiotic containing *Enterococcus faecalis* T-110, *Clostridium butyricum* TO-A, and *Bacillus mesentericus* TO-A resulted in a significant decrease in the duration of severe diarrhoea in the probiotic group compared to the placebo in children with infectious gastroenteritis. In the patients with rotavirus, a statistically significant decrease in gastroenteritis (Vesikari score) and diarrhoea frequency was also observed in the probiotic group. According to the authors of this study, the three strains acted symbiotically to facilitate the proliferation of the others. The dosage in this study was different compared to other clinical trials as the probiotic was given three times daily, whereas other clinical studies supplemented their patients once or twice a day.

Some of the clinical studies were not double-blind, but either single-blind [65,68] or open-labelled [61,62], which enhances the possibility of bias due to knowledge of the patient's treatment group [106].

Besides probiotics, prebiotics [107], synbiotics [108], postbiotics [109], or even fermented foods [110] could have positive effects for rotavirus diarrhoea due to enhancement of the natural intestinal microbiota, as a combination of probiotics and prebiotics (synbiotics) could have a synergistic effect; in some cases, heat-killed probiotics or postbiotics could even be safer than viable microorganisms. Some human and animal studies have addressed these effects [84,111–113], opening the possibility for more well-designed clinical studies.

5. Studies with No Antiviral Effect of Probiotics against Rotavirus Infections

On the other hand, several other studies using the same strains or other strains did not find statistically significant differences after probiotic administration. *Lacticaseibacillus rhamnosus* GG did not appear to enhance short-term recovery following acute diarrhoeal illness in children in five clinical studies [114–117]. In a large study by Freedman et al. [118], no significant differences were found in paediatric patients aged between 3 and 48 months with diarrhoea after a five-day supplementation with 4.0×10^9 cfu twice daily of either *Lacticaseibacillus rhamnosus* R0011 and *Lactobacillus helveticus* R0052 or *Lacticaseibacillus rhamnosus* GG, regardless of whether there were gastroenteritis-causing pathogens (e.g., adenovirus, norovirus, rotavirus, or bacteria). More results from the same clinical study (NCT01853124) were also published and showed no indication that probiotic administration lessened the burden of disease, regardless of the etiologic pathogen group (i.e., virus, bacteria, or parasite) or specific viral aetiologies (i.e., adenovirus, norovirus, or rotavirus) [119–122]. Perhaps the duration of supplementation with this probiotic was too short to exhibit a positive immunological effect as other clinical studies using the same strain achieved significant differences, such as Sindhu et al. [63] where the probiotic was consumed for 4 weeks, and an immunological effect was found in the probiotic group. Fewer children with rotavirus diarrhoea on LGG had repeated diarrhoeal episodes. Although no differences were found in the duration of diarrhoea, the immunological effect was evident. The dosage could also have been a factor, as Aggarwahl et al. and Basu et al. [61,67] both reported a statistically significant shorter duration of diarrhoea in children with watery diarrhoea after supplementation with LGG for five days at a dosage of 10^{10} CFU daily, whilst the dosage in the Freedman et al. study was 8.0×10^9 CFU daily. No significant differences were also found in the immunogenicity of the rotavirus vaccine given to infants in a poor urban community in India after supplementation with the probiotic *Lacticaseibacillus rhamnosus* GG in a study by Lazarus et al. [123]; however, among probiotic recipients, the abundance of lactobacilli in stools showed a modest association with rotavirus shedding after the first dose of the vaccine, consistent with the concept that probiotic bacteria may promote vaccine virus replication and the immune response.

A study conducted in Vietnam [124] using *Lactobacillus acidophilus* (4.0×10^8 CFU twice daily) also did not yield any significant differences in the duration of rotavirus diarrhoea compared to the placebo. The strain used was not specified and perhaps it was not a probiotic strain; or the dosage used was one log-step lower than other studies, which could have caused the lack of significant differences, as an appropriate strain and adequate administered amount of a probiotic are necessary to achieve a health benefit [31].

No significant differences were found in the studies conducted in Poland by Urbanska et al. [125] and Wanke et al. [126], using *Limosilactobacillus reuteri* DSM 17,938 for preventing nosocomial diarrhoea in children, including rotavirus infection. *Limosilactobacillus reuteri* DSM 17,938 has otherwise shown to be effective in the prevention and treatment of infantile colic and regurgitation and gastrointestinal disorders [127–129].

Several other clinical studies also did not show a statistically significant anti-rotavirus effect after intervention with *Bacillus coagulans* [130], *Lacticaseibacillus casei* ST11 [131]. A study using *Bifidobacterium animalis* subsp. *lactis* Bb-12 and *Streptococcus thermophilus* TH4 also did not find any differences in the treatment of gastroenteritis compared to the placebo; however, a decrease in rotavirus shedding was observed [86].

One of the possible reasons for the lack of effect of probiotics in these clinical trials, even though the same strains were used that previously exhibited a health benefit, could have been the quality of the production procedures of the probiotic strains. Lyophilisation, and the form of probiotics including lyophilized or heat-dried powders in capsule or powder form, can also influence the shelf-life and general quality of the probiotic [132]. Depending on production, some probiotics need to be stored in the refrigerator and others do not. Finally, the stability of the product must remain during storage, as an adequate concentration of viable probiotics must be persevered for the whole shelf-life. The probiotics used in clinical studies come from various commercial markets, and since many are foodstuffs or dietary supplements—not medicinal products—the quality may not always be assured or controlled [132,133]. All these factors can indirectly influence the reality of the results of clinical studies.

6. Conclusions, Limitations, and Future Directions

The effect of probiotics on enteric virus infections has been studied for years, and there is still much research to be done in the line of the microbiota–host–pathogen interactions. Although probiotics have shown promising results in the prevention of viral AGE, we still need an effective weapon to prevent the high mortality rate in early childhood in low-income countries. At least for rotavirus infection, effective vaccines are available to make progress in lowering the disease burden.

In probiotics studies, we need to be cautious as results are inconsistent—sometimes the same probiotic strain was not effective, whilst in other clinical studies it was, showing the importance of correct dosages, the duration of treatment, and quality of the probiotic.

With careful consideration of strains, dosages, and durations of supplementation, probiotics appear to be a safe and effective adjuvant in the treatment of rotavirus diarrhoea via modulation of the immune system and the intestinal microbiota. However, more clinical studies with different probiotics, perhaps combined with prebiotics to achieve a synergistic effect to optimally influence the restoration of the intestinal microbiota after dysbiosis due to diarrhoea, are warranted.

Author Contributions: All authors contributed to the content, drafting and writing of this document. All authors have read and agreed to the published version of the manuscript.

Funding: This research received no external funding.

Data Availability Statement: Not applicable.

Conflicts of Interest: The authors declare no conflict of interest.

References

1. Bányai, K.; Estes, M.K.; Martella, V.; Parashar, U.D. Viral gastroenteritis. *Lancet* **2018**, *392*, 175–186. [CrossRef] [PubMed]
2. Kotwal, G.; Cannon, J.L. Environmental persistence and transfer of enteric viruses. *Curr. Opin. Virol.* **2014**, *4*, 37–43. [CrossRef] [PubMed]
3. Hamza, I.A.; Jurzik, L.; Überla, K.; Wilhelm, M. Methods to detect infectious human enteric viruses in environmental water samples. *Int. J. Hyg. Environ. Health* **2011**, *214*, 424–436. [CrossRef] [PubMed]
4. Lundgren, O.; Svensson, L. Pathogenesis of Rotavirus diarrhea. *Microbes Infect.* **2001**, *3*, 1145–1156. [CrossRef]
5. Matthijnssens, J.; Ciarlet, M.; McDonald, S.M.; Attoui, H.; Banyai, K.; Brister, J.R.; Buesa, J.; Esona, M.D.; Estes, M.K.; Gentsch, J.R.; et al. Uniformity of rotavirus strain nomenclature proposed by the Rotavirus Classification Working Group (RCWG). *Arch. Virol.* **2011**, *156*, 1397–1413. [CrossRef]
6. Burnett, E.; Parashar, U.D.; Tate, J.E. Rotavirus Infection, Illness, and Vaccine Performance in Malnourished Children: A Review of the Literature. *Pediatr. Infect. Dis. J.* **2021**, *40*, 930–936. [CrossRef]
7. Burke, R.M.; Tate, J.E.; Kirkwood, C.D.; Steele, A.D.; Parashar, U.D. Current and new rotavirus vaccines. *Curr. Opin. Infect. Dis.* **2019**, *32*, 435–444. [CrossRef]
8. Armah, G.E.; Sow, O.S.; Breiman, R.F.; Dallas, M.J.; Tapia, M.D.; Feikin, D.R.; Binka, F.N.; Steele, A.D.; Laserson, K.F.; Ansah, A.N.; et al. Efficacy of pentavalent rotavirus vaccine against severe rotavirus gastroenteritis in infants in developing countries in sub-Saharan Africa: A randomised, double-blind, placebo-controlled trial. *Lancet* **2010**, *376*, 606–614. [CrossRef]
9. Zaman, K.; Anh, D.D.; Victor, J.C.; Shin, S.; MBBS, Y.; Dallas, M.J.; Podder, G.; Thiem, V.D.; Mai, L.T.P.; Luby, S.P.; et al. Efficacy of pentavalent rotavirus vaccine against severe rotavirus gastroenteritis in infants in developing countries in Asia: A randomised, double-blind, placebo-controlled trial. *Lancet* **2010**, *376*, 615–623. [CrossRef]
10. Desselberger, U. Differences of Rotavirus Vaccine Effectiveness by Country: Likely Causes and Contributing Factors. *Pathogens* **2017**, *6*, 65. [CrossRef]
11. Steele, A.; Victor, J.; Carey, M.; Tate, J.; Atherly, D.; Pecenka, C.; Diaz, Z.; Parashar, U.; Kirkwood, C. Experiences with rotavirus vaccines: Can we improve rotavirus vaccine impact in developing countries? *Hum. Vaccines Immunother.* **2019**, *15*, 1215–1227. [CrossRef] [PubMed]
12. Kandasamy, S.; Vlasova, A.N.; Fischer, D.; Kumar, A.; Chattha, K.S.; Rauf, A.; Shao, L.; Langel, S.N.; Rajashekara, G.; Saif, L.J. Differential Effects of *Escherichia coli* Nissle and *Lactobacillus rhamnosus* Strain GG on Human Rotavirus Binding, Infection, and B Cell Immunity. *J. Immunol.* **2016**, *196*, 1780–1789. [CrossRef] [PubMed]
13. Kim, A.H.; Hogarty, M.P.; Harris, V.C.; Baldridge, M.T. The Complex Interactions Between Rotavirus and the Gut Microbiota. *Front. Cell. Infect. Microbiol.* **2021**, *10*, 586751. [CrossRef] [PubMed]
14. Stevens, E.J.; Bates, K.A.; King, K.C. Host microbiota can facilitate pathogen infection. *PLoS Pathog.* **2021**, *17*, e1009514. [CrossRef]
15. Gonzalez-Ochoa, G.; Flores-Mendoza, L.K.; Icedo-Garcia, R.; Gomez-Flores, R.; Tamez-Guerra, P. Modulation of rotavirus severe gastroenteritis by the combination of probiotics and prebiotics. *Arch. Microbiol.* **2017**, *199*, 953–961. [CrossRef]
16. Villena, J.; Vizoso-Pinto, M.G.; Kitazawa, H. Intestinal Innate Antiviral Immunity and Immunobiotics: Beneficial Effects against Rotavirus Infection. *Front. Immunol.* **2016**, *7*, 563. [CrossRef]
17. Wang, Y.; Moon, A.; Huang, J.; Sun, Y.; Qiu, H.-J. Antiviral Effects and Underlying Mechanisms of Probiotics as Promising Antivirals. *Front. Cell. Infect. Microbiol.* **2022**, *12*, 724. [CrossRef]
18. Cani, P.D. Human gut microbiome: Hopes, threats and promises. *Gut* **2018**, *67*, 1716–1725. [CrossRef]
19. Maldonado Galdeano, C.; Cazorla, S.I.; Lemme Dumit, J.M.; Vélez, E.; Perdigón, G. Beneficial Effects of Probiotic Consumption on the Immune System. *Ann. Nutr. Metab.* **2019**, *74*, 115–124. [CrossRef]

20. Fijan, S.; Frauwallner, A.; Langerholc, T.; Krebs, B.; Younes, J.A.T.H.; Heschl, A.; Turk, D.M.; Rogelj, I. Efficacy of Using Probiotics with Antagonistic Activity against Pathogens of Wound Infections: An Integrative Review of Literature. *BioMed Res. Int.* **2019**, *2019*, 7585486. [CrossRef]
21. Hooper, L.V.; Littman, D.R.; MacPherson, A.J. Interactions Between the Microbiota and the Immune System. *Science* **2012**, *336*, 1268–1273. [CrossRef] [PubMed]
22. Clarke, G.; Stilling, R.M.; Kennedy, P.J.; Stanton, C.; Cryan, J.F.; Dinan, T.G. Minireview: Gut Microbiota: The Neglected Endocrine Organ. *Mol. Endocrinol.* **2014**, *28*, 1221–1238. [CrossRef] [PubMed]
23. Ettinger, S. Diet, Gut Microbiome, and Cognitive Decline. *Curr. Nutr. Rep.* **2022**. Epub ahead of print. [CrossRef] [PubMed]
24. Foster, J.; Zhou, L. Psychobiotics and the gut–brain axis: In the pursuit of happiness. *Neuropsychiatr. Dis. Treat.* **2015**, *11*, 715–723. [CrossRef] [PubMed]
25. Lee, J.E.; Walton, D.; O'Connor, C.P.; Wammes, M.; Burton, J.P.; Osuch, E.A. Drugs, Guts, Brains, but Not Rock and Roll: The Need to Consider the Role of Gut Microbiota in Contemporary Mental Health and Wellness of Emerging Adults. *Int. J. Mol. Sci.* **2022**, *23*, 6643. [CrossRef]
26. Salem, I.; Ramser, A.; Isham, N.; Ghannoum, M.A. The Gut Microbiome as a Major Regulator of the Gut-Skin Axis. *Front. Microbiol.* **2018**, *9*, 1459. [CrossRef]
27. Shah, T.; Shah, Z.; Baloch, Z.; Cui, X. The role of microbiota in respiratory health and diseases, particularly in tuberculosis. *Biomed. Pharmacother.* **2021**, *143*, 112108. [CrossRef]
28. Taghinezhad-S, S.; Keyvani, H.; Bermúdez-Humarán, L.G.; Donders, G.G.G.; Fu, X.; Mohseni, A.H. Twenty years of research on HPV vaccines based on genetically modified lactic acid bacteria: An overview on the gut-vagina axis. *Cell. Mol. Life Sci.* **2020**, *78*, 1191–1206. [CrossRef]
29. Desselberger, U. Significance of the Gut Microbiome for Viral Diarrheal and Extra-Intestinal Diseases. *Viruses* **2021**, *13*, 1601. [CrossRef]
30. Harper, A.; Vijayakumar, V.; Ouwehand, A.C.; ter Haar, J.; Obis, D.; Espadaler, J.; Binda, S.; Desiraju, S.; Day, R. Viral Infections, the Microbiome, and Probiotics. *Front. Cell. Infect. Microbiol.* **2021**, *10*, 596166. [CrossRef]
31. Hill, C.; Guarner, F.; Reid, G.; Gibson, G.R.; Merenstein, D.J.; Pot, B.; Morelli, L.; Canani, R.B.; Flint, H.J.; Salminen, S.; et al. Expert consensus document: The International Scientific Association for Probiotics and Prebiotics consensus statement on the scope and appropriate use of the term probiotic. *Nat. Rev. Gastroenterol. Hepatol.* **2014**, *11*, 506–514. [CrossRef] [PubMed]
32. Bagga, D.; Reichert, J.L.; Koschutnig, K.; Aigner, C.S.; Holzer, P.; Koskinen, K.; Moissl-Eichinger, C.; Schöpf, V. Probiotics drive gut microbiome triggering emotional brain signatures. *Gut Microbes* **2018**, *9*, 486–496. [CrossRef] [PubMed]
33. Chamberlain, R.; Lau, C. Probiotics are effective at preventing Clostridium difficile-associated diarrhea: A systematic review and meta-analysis. *Int. J. Gen. Med.* **2016**, *9*, 27–37. [CrossRef] [PubMed]
34. Liu, Y.; Alookaran, J.J.; Rhoads, J.M. Probiotics in Autoimmune and Inflammatory Disorders. *Nutrients* **2018**, *10*, 1537. [CrossRef]
35. Wahab, S.; Almaghaslah, D.; Mahmood, S.E.; Ahmad, F.; Alsayegh, A.A.; Abu Haddash, Y.M.; Rahman, M.A.; Ahamd, I.; Ahmad, W.; Khalid, M.; et al. Pharmacological Efficacy of Probiotics in Respiratory Viral Infections: A Comprehensive Review. *J. Pers. Med.* **2022**, *12*, 1292. [CrossRef]
36. Rawashdeh, M.; Al-Zoubi, B.; Aliwat, M.B.; Burayzat, S.; Alhindawi, E.; Al-Matti, A.A.; Altamimi, E. National Consensus for the Management of Acute Gastroenteritis in Jordanian Children: Consensus Recommendations Endorsed by the Jordanian Paediatric Society. *Int. J. Pediatr.* **2022**, *2022*, 4456232. [CrossRef]
37. George, S.; Aguilera, X.; Gallardo, P.; Farfán, M.; Lucero, Y.; Torres, J.P.; Vidal, R.; O'Ryan, M. Bacterial Gut Microbiota and Infections During Early Childhood. *Front. Microbiol.* **2022**, *12*, 793050. [CrossRef]
38. Ghosh, S.; Malik, Y.S.; Kobayashi, N. Therapeutics and Immunoprophylaxis Against Noroviruses and Rotaviruses: The Past, Present, and Future. *Curr. Drug Metab.* **2018**, *19*, 170–191. [CrossRef]
39. Omatola, C.A.; Olaniran, A.O. Rotaviruses: From Pathogenesis to Disease Control—A Critical Review. *Viruses* **2022**, *14*, 875. [CrossRef]
40. Afchangi, A.; Latifi, T.; Jalilvand, S.; Marashi, S.M.; Shoja, Z. Combined use of lactic-acid-producing bacteria as probiotics and rotavirus vaccine candidates expressing virus-specific proteins. *Arch. Virol.* **2021**, *166*, 995–1006. [CrossRef]
41. Ahmadi, E.; Alizadeh-Navaei, R.; Rezai, M.S. Efficacy of probiotic use in acute rotavirus diarrhea in children: A systematic review and meta-analysis. *Casp. J. Intern. Med.* **2015**, *6*, 187–195.
42. Di, J.-B.; Gai, Z.-T. Protective efficacy of probiotics on the treatment of acute rotavirus diarrhea in children: An updated meta-analysis. *Eur. Rev. Med. Pharmacol. Sci.* **2020**, *24*, 9675–9683. [PubMed]
43. Mizutani, T.; Aboagye, S.Y.; Ishizaka, A.; Afum, T.; Mensah, G.I.; Asante-Poku, A.; Asandem, D.A.; Parbie, P.K.; Abana, C.Z.-Y.; Kushitor, D.; et al. Gut microbiota signature of pathogen-dependent dysbiosis in viral gastroenteritis. *Sci. Rep.* **2021**, *11*, 13945. [CrossRef] [PubMed]
44. Gizatulina, S.S.; Birger, M.O.; Nikovskaia, M.I.; Mastiukova, I.N.; Potashova, L.A. Mikroflora kishechnika u deteĭ rannego vozrasta s rotavirusnoĭ infektsieĭ [Intestinal microflora in young children with rotavirus infection]. *Zh Mikrobiol. Epidemiol. Immunobiol.* **1992**, *3*, 29–30.
45. Wilkins, L.J.; Monga, M.; Miller, A.W. Defining Dysbiosis for a Cluster of Chronic Diseases. *Sci. Rep.* **2019**, *9*, 12918. [CrossRef]
46. Britton, R.A.; Versalovic, J. Probiotics and Gastrointestinal Infections. *Interdiscip. Perspect. Infect. Dis.* **2008**, *2008*, 290769. [CrossRef]

47. Alvarez-Vieites, E.; López-Santamarina, A.; Miranda, J.M.; Mondragón, A.D.C.; Lamas, A.; Cardelle-Cobas, A.; Nebot, C.; Franco, C.M.; Cepeda, A. Influence of the Intestinal Microbiota on Diabetes Management. *Curr. Pharm. Biotechnol.* **2020**, *21*, 1603–1615. [CrossRef]
48. Mahooti, M.; Miri, S.M.; Abdolalipour, E.; Ghaemi, A. The immunomodulatory effects of probiotics on respiratory viral infections: A hint for COVID-19 treatment? *Microb. Pathog.* **2020**, *148*, 104452. [CrossRef]
49. Belguesmia, Y.; Bendjeddou, K.; Kempf, I.; Boukherroub, R.; Drider, D. Heterologous Biosynthesis of Five New Class II Bacteriocins From Lactobacillus paracasei CNCM I-5369 With Antagonistic Activity Against Pathogenic Escherichia coli Strains. *Front. Microbiol.* **2020**, *11*, 1198. [CrossRef]
50. Tabata, T.; Petitt, M.; Puerta-Guardo, H.; Michlmayr, D.; Wang, C.; Fang-Hoover, J.; Harris, E.; Pereira, L. Zika Virus Targets Different Primary Human Placental Cells, Suggesting Two Routes for Vertical Transmission. *Cell Host Microbe* **2016**, *20*, 155–166. [CrossRef]
51. Umair, M.; Jabbar, S.; Zhaoxin, L.; Jianhao, Z.; Abid, M.; Khan, K.-U.R.; Korma, S.A.; Alghamdi, M.A.; El-Saadony, M.T.; El-Hack, M.E.A.; et al. Probiotic-Based Bacteriocin: Immunity Supplementation Against Viruses. An Updated Review. *Front. Microbiol.* **2022**, *13*, 876058. [CrossRef] [PubMed]
52. Wachsman, M.B.; Castilla, V.; De Ruiz Holgado, A.P.; De Torres, R.A.; Sesma, F.; Coto, C.E. Enterocin CRL35 inhibits late stages of HSV-1 and HSV-2 replication in vitro. *Antivir. Res.* **2003**, *58*, 17–24. [CrossRef] [PubMed]
53. Butel, M.-J. Probiotics, gut microbiota and health. *Med. Mal. Infect.* **2014**, *44*, 1–8. [CrossRef] [PubMed]
54. de Vrese, M.; Offick, B. Chapter 14—Probiotics and Prebiotics: Effects on Diarrhea. In *Bioactive Foods in Promoting Health*; Watson, R.R., Preedy, V.R., Eds.; Academic Press: Boston, MA, USA, 2016; pp. 205–227.
55. Lee, D.K.; Park, J.E.; Kim, M.J.; Seo, J.G.; Lee, J.H.; Ha, N.J. Probiotic bacteria, B. longum and L. acidophilus inhibit infection by rotavirus in vitro and decrease the duration of diarrhea in pediatric patients. *Clin. Res. Hepatol. Gastroenterol.* **2015**, *39*, 237–244. [CrossRef]
56. Ramamurthy, T.; Kumari, S.; Ghosh, A. Diarrheal disease and gut microbiome. *Prog. Mol. Biol. Transl. Sci.* **2022**, *192*, 149–177.
57. Xiong, L.; Li, Y.; Li, J.; Yang, J.; Shang, L.; He, X.; Liu, L.; Luo, Y.; Xie, X. Intestinal microbiota profiles in infants with acute gastroenteritis caused by rotavirus and norovirus infection: A prospective cohort study. *Int. J. Infect. Dis.* **2021**, *111*, 76–84. [CrossRef]
58. Shin, D.Y.; Yi, D.Y.; Jo, S.; Lee, Y.M.; Kim, J.-H.; Kim, W.; Park, M.R.; Yoon, S.M.; Kim, Y.; Yang, S.; et al. Effect of a new Lactobacillus plantarum product, LRCC5310, on clinical symptoms and virus reduction in children with rotaviral enteritis. *Medicine* **2020**, *99*, e22192. [CrossRef]
59. Park, M.S.; Kwon, B.; Ku, S.; Ji, G.E. The Efficacy of Bifidobacterium longum BORI and Lactobacillus acidophilus AD031 Probiotic Treatment in Infants with Rotavirus Infection. *Nutrients* **2017**, *9*, 887. [CrossRef]
60. Das, S.; Gupta, P.K.; Das, R.R. Efficacy and Safety of *Saccharomyces boulardii* in Acute Rotavirus Diarrhea: Double Blind Randomized Controlled Trial from a Developing Country. *J. Trop. Pediatr.* **2016**, *62*, 464–470. [CrossRef]
61. Aggarwal, S.; Upadhyay, A.; Shah, D.; Teotia, N.; Agarwal, A.; Jaiswal, V. Lactobacillus GG for treatment of acute childhood diarrhoea: An open labelled, randomized controlled trial. *Indian J. Med Res.* **2014**, *139*, 379–385.
62. Huang, Y.-F.; Liu, P.-Y.; Chen, Y.-Y.; Nong, B.-R.; Huang, I.-F.; Hsieh, K.-S.; Chen, K.-T. Three-Combination Probiotics Therapy in Children With Salmonella and Rotavirus Gastroenteritis. *J. Clin. Gastroenterol.* **2014**, *48*, 37–42. [CrossRef] [PubMed]
63. Sindhu, K.N.C.; Sowmyanarayanan, T.V.; Paul, A.; Babji, S.; Ajjampur, S.S.R.; Priyadarshini, S.; Sarkar, R.; Balasubramanian, K.A.; Wanke, C.A.; Ward, H.D.; et al. Immune Response and Intestinal Permeability in Children With Acute Gastroenteritis Treated With Lactobacillus rhamnosus GG: A Randomized, Double-Blind, Placebo-Controlled Trial. *Clin. Infect. Dis.* **2014**, *58*, 1107–1115. [CrossRef] [PubMed]
64. Corrêa, N.B.; Penna, F.J.; Lima, F.M.; Nicoli, J.R.; Filho, L.A. Treatment of Acute Diarrhea With Saccharomyces boulardii in Infants. *J. Pediatr. Gastroenterol. Nutr.* **2011**, *53*, 497–501. [CrossRef]
65. Dalgic, N.; Sancar, M.; Bayraktar, B.; Pullu, M.; Hasim, O. Probiotic, zinc and lactose-free formula in children with rotavirus diarrhea: Are they effective? *Pediatr. Int.* **2011**, *53*, 677–682. [CrossRef]
66. Grandy, G.; Medina, M.; Soria, R.; Terán, C.G.; Araya, M. Probiotics in the treatment of acute rotavirus diarrhoea. A randomized, double-blind, controlled trial using two different probiotic preparations in Bolivian children. *BMC Infect. Dis.* **2010**, *10*, 253. [CrossRef]
67. Basu, S.; Paul, D.K.; Ganguly, S.; Chatterjee, M.; Chandra, P.K. Efficacy of High-dose Lactobacillus rhamnosus GG in Controlling Acute Watery Diarrhea in Indian Children. *J. Clin. Gastroenterol.* **2009**, *43*, 208–213. [CrossRef]
68. Teran, C.G.; Teran-Escalera, C.N.; Villarroel, P. Nitazoxanide vs. probiotics for the treatment of acute rotavirus diarrhea in children: A randomized, single-blind, controlled trial in Bolivian children. *Int. J. Infect. Dis.* **2009**, *13*, 518–523. [CrossRef] [PubMed]
69. Dubey, A.P.; Rajeshwari, K.; Chakravarty, A.; Famularo, G. Use of VSL♯3 in the Treatment of Rotavirus Diarrhea in Children. *J. Clin. Gastroenterol.* **2008**, *42*, S126–S129. [CrossRef] [PubMed]
70. Narayanappa, D. Randomized double blinded controlled trial to evaluate the efficacy and safety of Bifilac in patients with acute viral diarrhea. *Indian J. Pediatr.* **2008**, *75*, 709–713. [CrossRef]
71. Szymański, H.; Pejcz, J.; Jawień, M.; Chmielarczyk, A.; Strus, M.; Heczko, P.B. Treatment of acute infectious diarrhoea in infants and children with a mixture of three Lactobacillus rhamnosus strains—A randomized, double-blind, placebo-controlled trial. *Aliment. Pharmacol. Ther.* **2006**, *23*, 247–253. [CrossRef]

72. Gaón, D.; García, H.; Winter, L.; Rodríguez, N.; Quintás, R.; Gonzalez, S.N.; Oliver, G. Effect of Lactobacillus strains and Saccharomyces boulardii on persistent diarrhea in children. *Medicina* **2003**, *63*, 293–298. [PubMed]
73. Rosenfeldt, V.; Michaelsen, K.F.; Jakobsen, M.; Larsen, C.N.; Møller, P.L.; Pedersen, P.; Tvede, M.; Weyrehter, H.; Valerius, N.H.; Paerregaard, A. Effect of probiotic Lactobacillus strains in young children hospitalized with acute diarrhea. *Pediatr. Infect. Dis. J.* **2002**, *21*, 411–416. [CrossRef] [PubMed]
74. Guandalini, S.; Pensabene, L.; Abu Zikri, M.; Dias, J.A.; Casali, L.G.; Hoekstra, H.; Kolacek, S.; Massar, K.; Micetic–Turk, D.; Papadopoulou, A.; et al. Lactobacillus GG Administered in Oral Rehydration Solution to Children with Acute Diarrhea: A Multicenter European Trial. *J. Craniofacial Surg.* **2000**, *30*, 54–60. [CrossRef] [PubMed]
75. Guarino, A.; Canani, R.B.; Spagnuolo, M.I.; Albano, F.; Di Benedetto, L. Oral Bacterial Therapy Reduces the Duration of Symptoms and of Viral Excretion in Children with Mild Diarrhea. *J. Pediatr. Gastroenterol. Nutr.* **1997**, *25*, 516–519. [CrossRef]
76. Zheng, J.; Wittouck, S.; Salvetti, E.; Franz, C.M.A.P.; Harris, H.M.B.; Mattarelli, P.; O'Toole, P.W.; Pot, B.; Vandamme, P.; Walter, J.; et al. A taxonomic note on the genus Lactobacillus: Description of 23 novel genera, emended description of the genus Lactobacillus Beijerinck 1901, and union of Lactobacillaceae and Leuconostocaceae. *Int. J. Syst. Evol. Microbiol.* **2020**, *70*, 2782–2858. [CrossRef]
77. Faure, C. Role of Antidiarrhoeal Drugs as Adjunctive Therapies for Acute Diarrhoea in Children. *Int. J. Pediatr.* **2013**, *2013*, 612403. [CrossRef]
78. Guarino, A.; Ashkenazi, S.; Gendrel, D.; Vecchio, A.L.; Shamir, R.; Szajewska, H. European Society for Pediatric Gastroenterology, Hepatology, and Nutrition/European Society for Pediatric Infectious Diseases Evidence-Based Guidelines for the Management of Acute Gastroenteritis in Children in Europe. *J. Craniofacial Surg.* **2014**, *59*, 132–152. [CrossRef]
79. Hojsak, I.; Fabiano, V.; Pop, T.L.; Goulet, O.; Zuccotti, G.V.; Çokuğraş, F.; Pettoello-Mantovani, M.; Kolacek, S. Guidance on the use of probiotics in clinical practice in children with selected clinical conditions and in specific vulnerable groups. *Acta Paediatr.* **2018**, *107*, 927–937. [CrossRef]
80. Hojsak, I.; Szajewska, H.; Canani, R.B.; Guarino, A.; Indrio, F.; Kolacek, S.; Orel, R.; Shamir, R.; Vandenplas, Y.; van Goudoever, J.B.; et al. Probiotics for the Prevention of Nosocomial Diarrhea in Children. *J. Pediatr. Gastroenterol. Nutr.* **2018**, *66*, 3–9. [CrossRef]
81. Holscher, H.D.; Czerkies, L.A.; Cekola, P.; Litov, R.; Benbow, M.; Santema, S.; Alexander, D.D.; Perez, V.; Sun, S.; Saavedra, J.M.; et al. *Bifidobacterium lactis* Bb12 Enhances Intestinal Antibody Response in Formula-Fed Infants. *J. Parenter. Enter. Nutr.* **2012**, *36*, 106S–117S. [CrossRef]
82. Szymański, H.; Chmielarczyk, A.; Strus, M.; Pejcz, J.; Jawień, M.; Kochan, P.; Heczko, P.B. Colonisation of the gastrointestinal tract by probiotic L. rhamnosus strains in acute diarrhoea in children. *Dig. Liver Dis.* **2006**, *38*, S274–S276. [CrossRef] [PubMed]
83. Xie, Y.-M.; Gao, S.; Wang, L.-Y.; Wang, Z.-L. Therapeutic effect of probiotics and oral IgY as supplementary drugs in the treatment of pediatric rotavirus enteritis: A comparative study. *Zhongguo Dang Dai Er Ke Za Zhi* **2013**, *15*, 1000–1005. [PubMed]
84. Binns, C.; Lee, A.; Harding, H.; Gracey, M.; Barclay, D. The CUPDAY Study: Prebiotic-probiotic milk product in 1–3-year-old children attending childcare centres. *Acta Paediatr.* **2007**, *96*, 1646–1650. [CrossRef] [PubMed]
85. Fang, S.-B.; Lee, H.-C.; Hu, J.-J.; Hou, S.-Y.; Liu, H.-L.; Fang, H.-W. Dose-dependent effect of Lactobacillus rhamnosus on quantitative reduction of faecal rotavirus shedding in children. *J. Trop. Pediatr.* **2009**, *55*, 297–301. [CrossRef] [PubMed]
86. Mao, M.; Yu, T.; Xiong, Y.; Wang, Z.; Liu, H.; Gotteland, M.; Brunser, O. Effect of a lactose-free milk formula supplemented with bifidobacteria and streptococci on the recovery from acute diarrhoea. *Asia Pac. J. Clin. Nutr.* **2008**, *17*, 30–34.
87. Nopchinda, S.; Varavithya, W.; Phuapradit, P.; Sangchai, R.; Suthutvoravut, U.; Chantraruksa, V.; Haschke, F. Effect of bifidobacterium Bb12 with or without Streptococcus thermophilus supplemented formula on nutritional status. *J. Med Assoc. Thail.* **2002**, *85* (Suppl. 4), S1225–S1231.
88. Zarin, I.; Ali, M.A.; Paul, S.K.; Mazid, R.; Amin, S.E. Efficacy of Probiotics in Treatment of Acute Rotavirus and Non Rotavirus Watery Diarrhoea in Children Admitted in Mymensingh Medical College Hospital. *Mymensingh. Med. J.* **2022**, *31*, 49–54.
89. Liu, X.L.; Li, M.L.; Ma, W.X.; Xia, S.L.; Xu, B.L. Clinical trial on the prevention of diarrhea by oral BIFICO for infants aged 1–6 years. *Zhonghua Shi Yan He Lin Chuang Bing Du Xue Za Zhi* **2013**, *27*, 277–279.
90. Le Luyer, B.; Makhoul, G.; Duhamel, J.-F. Étude multicentrique, contrôlée en double insu d'une formule adaptée enrichie en Saccharomyces boulardii dans le traitement des diarrhées aiguës du nourrisson. *Arch. Pédiatrie* **2010**, *17*, 459–465. [CrossRef]
91. Capurso, L. Thirty Years of Lactobacillus rhamnosus GG. *J. Clin. Gastroenterol.* **2019**, *53*, S1–S41. [CrossRef]
92. Szajewska, H.; Kotowska, M.; Mrukowicz, J.Z.; Armánska, M.; Mikolajczyk, W. Efficacy of Lactobacillus GG in prevention of nosocomial diarrhea in infants. *J. Pediatr.* **2001**, *138*, 361–365. [CrossRef] [PubMed]
93. Li, Y.-T.; Xu, H.; Ye, J.-Z.; Wu, W.-R.; Shi, D.; Fang, D.-Q.; Liu, Y.; Li, L.-J. Efficacy of Lactobacillus rhamnosus GG in treatment of acute pediatric diarrhea: A systematic review with meta-analysis. *World J. Gastroenterol.* **2019**, *25*, 4999–5016. [CrossRef] [PubMed]
94. Vecchio, A.L.; Nunziata, F.; Bruzzese, D.; Conelli, M.; Guarino, A. Rotavirus immunisation status affects the efficacy of *Lacticaseibacillus rhamnosus* GG for the treatment of children with acute diarrhoea: A meta-analysis. *Benef. Microbes* **2022**, *13*, 283–294. [CrossRef] [PubMed]
95. Peroni, D.; Morelli, L. Probiotics as Adjuvants in Vaccine Strategy: Is There More Room for Improvement? *Vaccines* **2021**, *9*, 811. [CrossRef] [PubMed]
96. Canani, R.B.; Cucchiara, S.; Cuomo, R.; Pace, F.; Papale, F. Saccharomyces boulardii: A summary of the evidence for gastroenterology clinical practice in adults and children. *Eur. Rev. Med Pharmacol. Sci.* **2011**, *15*, 809–822.

97. Chong, H.-X.; Yusoff, N.A.A.; Hor, Y.Y.; Lew, L.C.; Jaafar, M.H.; Choi, S.-B.; Yusoff, M.S.; Wahid, N.; Bin Abdullah, M.F.I.L.; Zakaria, N.; et al. Lactobacillus plantarum DR7 improved upper respiratory tract infections via enhancing immune and inflammatory parameters: A randomized, double-blind, placebo-controlled study. *J. Dairy Sci.* **2019**, *102*, 4783–4797. [CrossRef] [PubMed]
98. Cunningham-Rundles, S.; Ahrné, S.; Bengmark, S.; Johann-Liang, R.; Marshall, F.; Metakis, L.; Califano, C.; Dunn, A.M.; Grassey, C.; Hinds, G.; et al. Probiotics and immune response. *Am. J. Gastroenterol.* **2000**, *95* (Suppl. 1), S22–S25. [CrossRef]
99. Han, Y.; Kim, B.; Ban, J.; Lee, J.; Kim, B.J.; Choi, B.S.; Hwang, S.; Ahn, K.; Kim, J. A randomized trial of *Lactobacillus plantarum* CJLP133 for the treatment of atopic dermatitis. *Pediatr. Allergy Immunol.* **2012**, *23*, 667–673. [CrossRef]
100. Jung, K.; Kim, A.; Lee, J.-H.; Cho, D.; Seo, J.; Jung, E.S.; Kang, H.-J.; Roh, J.; Kim, W. Effect of Oral Intake of *Lactiplantibacillus plantarum* APsulloc 331261 (GTB1^{TM}) on Diarrhea-Predominant Irritable Bowel Syndrome: A Randomized, Double-Blind, Placebo-Controlled Study. *Nutrients* **2022**, *14*, 2015. [CrossRef]
101. Lee, A.; Lee, Y.J.; Yoo, H.J.; Kim, M.; Chang, Y.; Lee, D.S.; Lee, J.H. Consumption of Dairy Yogurt Containing Lactobacillus paracasei ssp. paracasei, Bifidobacterium animalis ssp. lactis and Heat-Treated Lactobacillus plantarum Improves Immune Function Including Natural Killer Cell Activity. *Nutrients* **2017**, *9*, 558. [CrossRef]
102. Mañé, J.; Pedrosa, E.; Lorén, V.; Gassull, A.M.; Espadaler, J.; Cuñé, J.; Audivert, S.; Bonachera, A.M.; Cabré, E. A mixture of Lactobacillus plantarum CECT 7315 and CECT 7316 enhances systemic immunity in elderly subjects. A dose-response, double-blind, placebo-controlled, randomized pilot trial. *Nutr. Hosp.* **2011**, *26*, 228–235. [PubMed]
103. Yang, B.; Yue, Y.; Chen, Y.; Ding, M.; Li, B.; Wang, L.; Wang, Q.; Stanton, C.; Ross, R.P.; Zhao, J.; et al. Lactobacillus plantarum CCFM1143 Alleviates Chronic Diarrhea via Inflammation Regulation and Gut Microbiota Modulation: A Double-Blind, Randomized, Placebo-Controlled Study. *Front. Immunol.* **2021**, *12*, 746585. [CrossRef] [PubMed]
104. Shin, C.M.; Choi, Y.J.; Lee, D.H.; Moon, J.S.; Kim, T.-Y.; Kim, Y.-K.; Lee, W.-H.; Yoon, H.; Park, Y.S.; Kim, N. Validity and safety of ID-JPL934 in lower gastrointestinal symptom improvement. *Sci. Rep.* **2021**, *11*, 13046. [CrossRef] [PubMed]
105. Phuapradit, P.; Varavithya, W.; Vathanophas, K.; Sangchai, R.; Podhipak, A.; Suthutvoravut, U.; Nopchinda, S.; Chantraruksa, V.; Haschke, F. Reduction of rotavirus infection in children receiving bifidobacteria-supplemented formula. *J. Med Assoc. Thail.* **1999**, *82* (Suppl. 1), S43–S48.
106. Kahan, B.C.; Cro, S.; Doré, C.J.; Bratton, D.J.; Rehal, S.; Maskell, A.N.; Rahman, N.; Jairath, V. Reducing bias in open-label trials where blinded outcome assessment is not feasible: Strategies from two randomised trials. *Trials* **2014**, *15*, 456. [CrossRef] [PubMed]
107. Gibson, G.R.; Hutkins, R.; Sanders, M.E.; Prescott, S.L.; Reimer, R.A.; Salminen, S.J.; Scott, K.; Stanton, C.; Swanson, K.S.; Cani, P.D.; et al. Expert consensus document: The International Scientific Association for Probiotics and Prebiotics (ISAPP) consensus statement on the definition and scope of prebiotics. *Nat. Rev. Gastroenterol. Hepatol.* **2017**, *14*, 491–502. [CrossRef]
108. Swanson, K.S.; Gibson, G.R.; Hutkins, R.; Reimer, R.A.; Reid, G.; Verbeke, K.; Scott, K.P.; Holscher, H.D.; Azad, M.B.; Delzenne, N.M.; et al. The International Scientific Association for Probiotics and Prebiotics (ISAPP) consensus statement on the definition and scope of synbiotics. *Nat. Rev. Gastroenterol. Hepatol.* **2020**, *17*, 687–701. [CrossRef]
109. Salminen, S.; Collado, M.C.; Endo, A.; Hill, C.; Lebeer, S.; Quigley, E.M.M.; Sanders, M.E.; Shamir, R.; Swann, J.R.; Szajewska, H.; et al. The International Scientific Association of Probiotics and Prebiotics (ISAPP) consensus statement on the definition and scope of postbiotics. *Nat. Rev. Gastroenterol. Hepatol.* **2021**, *18*, 649–667. [CrossRef]
110. Marco, M.L.; Sanders, M.E.; Gänzle, M.; Arrieta, M.C.; Cotter, P.D.; De Vuyst, L.; Hill, C.; Holzapfel, W.; Lebeer, S.; Merenstein, D.; et al. The International Scientific Association for Probiotics and Prebiotics (ISAPP) consensus statement on fermented foods. *Nat. Rev. Gastroenterol. Hepatol.* **2021**, *18*, 196–208. [CrossRef]
111. Bruno, C.; Paparo, L.; Pisapia, L.; Romano, A.; Cortese, M.; Punzo, E.; Canani, R.B. Protective effects of the postbiotic deriving from cow's milk fermentation with L. paracasei CBA L74 against Rotavirus infection in human enterocytes. *Sci. Rep.* **2022**, *12*, 6268. [CrossRef]
112. Morales-Ferré, C.; Azagra-Boronat, I.; Massot-Cladera, M.; Tims, S.; Knipping, K.; Garssen, J.; Knol, J.; Franch, A.; Castell, M.; Pérez-Cano, F.J.; et al. Preventive Effect of a Postbiotic and Prebiotic Mixture in a Rat Model of Early Life Rotavirus Induced-Diarrhea. *Nutrients* **2022**, *14*, 1163. [CrossRef] [PubMed]
113. A Pedone, C.; Arnaud, C.C.; Postaire, E.R.; Bouley, C.F.; Reinert, P. Multicentric study of the effect of milk fermented by Lactobacillus casei on the incidence of diarrhoea. *Int. J. Clin. Pr.* **2000**, *54*, 568–571.
114. Mastretta, E.; Longo, P.; Laccisaglia, A.; Balbo, L.; Russo, R.; Mazzaccara, A.; Gianino, P. Effect of Lactobacillus GG and Breast-feeding in the Prevention of Rotavirus Nosocomial Infection. *J. Craniofacial Surg.* **2002**, *35*, 527–531. [CrossRef] [PubMed]
115. Misra, S.; Sabui, T.K.; Pal, N.K. A Randomized Controlled Trial to Evaluate the Efficacy of Lactobacillus GG in Infantile Diarrhea. *J. Pediatr.* **2009**, *155*, 129–132. [CrossRef]
116. Ritchie, B.K.; Brewster, D.R.; Tran, C.D.; Davidson, G.P.; McNeil, Y.; Butler, R.N. Efficacy of Lactobacillus GG in Aboriginal Children With Acute Diarrhoeal Disease: A Randomised Clinical Trial. *J. Pediatr. Gastroenterol. Nutr.* **2010**, *50*, 619–624. [CrossRef]
117. Salazar-Lindo, E.; Miranda-Langschwager, P.; Campos-Sanchez, M.; Chea-Woo, E.; Sack, R.B. Lactobacillus caseistrain GG in the treatment of infants with acute watery diarrhea: A randomized, double-blind, placebo controlled clinical trial [ISRCTN67363048]. *BMC Pediatr.* **2004**, *4*, 18. [CrossRef]
118. Freedman, S.B.; Williamson-Urquhart, S.; Farion, K.J.; Gouin, S.; Willan, A.R.; Poonai, N.; Hurley, K.; Sherman, P.M.; Finkelstein, Y.; Lee, B.E.; et al. Multicenter Trial of a Combination Probiotic for Children with Gastroenteritis. *N. Engl. J. Med.* **2018**, *379*, 2015–2026. [CrossRef]

119. Freedman, S.B.; Williamson-Urquhart, S.; Schuh, S.; Sherman, P.M.; Farion, K.J.; Gouin, S.; Willan, A.R.; Goeree, R.; Johnson, D.W.; Black, K.; et al. Impact of emergency department probiotic treatment of pediatric gastroenteritis: Study protocol for the PROGUT (Probiotic Regimen for Outpatient Gastroenteritis Utility of Treatment) randomized controlled trial. *Trials* **2014**, *15*, 170. [CrossRef]
120. Freedman, S.B.; Xie, J.; Nettel-Aguirre, A.; Pang, X.L.; Chui, L.; Williamson-Urquhart, S.; Schnadower, D.; Schuh, S.; Sherman, P.M.; Lee, B.E.; et al. A randomized trial evaluating virus-specific effects of a combination probiotic in children with acute gastroenteritis. *Nat. Commun.* **2020**, *11*, 2533. [CrossRef]
121. Freedman, S.B.; Finkelstein, Y.; Pang, X.-L.; Chui, L.; Tarr, I.P.; VanBuren, J.M.; Olsen, C.; Lee, E.B.; Hall-Moore, A.C.; Sapien, R.; et al. Pathogen-Specific Effects of Probiotics in Children With Acute Gastroenteritis Seeking Emergency Care: A Randomized Trial. *Clin. Infect. Dis.* **2021**, *75*, 55–64. [CrossRef]
122. Freedman, S.B.; Horne, R.; Johnson-Henry, K.; Xie, J.; Williamson-Urquhart, S.; Chui, L.; Pang, X.-L.; Lee, B.; Schuh, S.; Finkelstein, Y.; et al. Probiotic stool secretory immunoglobulin A modulation in children with gastroenteritis: A randomized clinical trial. *Am. J. Clin. Nutr.* **2021**, *113*, 905–914. [CrossRef] [PubMed]
123. Lazarus, R.P.; John, J.; Shanmugasundaram, E.; Rajan, A.K.; Thiagarajan, S.; Giri, S.; Babji, S.; Sarkar, R.; Kaliappan, P.S.; Venugopal, S.; et al. The effect of probiotics and zinc supplementation on the immune response to oral rotavirus vaccine: A randomized, factorial design, placebo-controlled study among Indian infants. *Vaccine* **2018**, *36*, 273–279. [CrossRef] [PubMed]
124. Chau, T.T.H.; Chau, N.N.M.; Le, N.T.H.; The, H.C.; Vinh, P.V.; To, N.T.N.; Ngoc, N.M.; Tuan, H.M.; Ngoc, T.L.C.; Kolader, M.-E.; et al. A Double-blind, Randomized, Placebo-controlled Trial of Lactobacillus acidophilus for the Treatment of Acute Watery Diarrhea in Vietnamese Children. *Pediatr. Infect. Dis. J.* **2018**, *37*, 35–42. [CrossRef] [PubMed]
125. Urbańska, M.; Gieruszczak–Białek, D.; Szymański, H.; Szajewska, H. Effectiveness of Lactobacillus reuteri DSM 17938 for the Prevention of Nosocomial Diarrhea in Children. *Pediatr. Infect. Dis. J.* **2016**, *35*, 142–145. [CrossRef]
126. Wanke, M.; Szajewska, H. Lack of an effect of Lactobacillus reuteri DSM 17938 in preventing nosocomial diarrhea in children: A randomized, double-blind, placebo-controlled trial. *J. Pediatr.* **2012**, *161*, 40–43.e41. [CrossRef]
127. Dargenio, V.; Cristofori, F.; Dargenio, C.; Giordano, P.; Indrio, F.; Celano, G.; Francavilla, R. Use of *Limosilactobacillus reuteri* DSM 17938 in paediatric gastrointestinal disorders: An updated review. *Benef. Microbes* **2022**, *13*, 221–242. [CrossRef]
128. Zermiani, A.P.d.R.B.; Soares, A.L.P.P.d.P.; Moura, B.L.d.S.G.d.; Miguel, E.R.A.; Lopes, L.D.G.; Santana, N.D.C.S.; Santos, T.D.S.; Demarchi, I.G.; Teixeira, J.J. Evidence of Lactobacillus reuteri to reduce colic in breastfed babies: Systematic review and meta-analysis. *Complement. Ther. Med.* **2021**, *63*, 102781. [CrossRef]
129. Foster, J.P.; Dahlen, H.G.; Fijan, S.; Badawi, N.; Schmied, V.; Thornton, C.; Smith, C.; Psaila, K. Probiotics for preventing and treating infant regurgitation: A systematic review and meta-analysis. *Matern. Child Nutr.* **2021**, *18*, e13290. [CrossRef]
130. Dutta, P.; Mitra, U.; Dutta, S.; Rajendran, K.; Saha, T.K.; Chatterjee, I.K. Randomised controlled clinical trial of Lactobacillus sporogenes (Bacillus coagulans), used as probiotic in clinical practice, on acute watery diarrhoea in children. *Trop. Med. Int. Health* **2011**, *16*, 555–561. [CrossRef]
131. Sarker, S.A.; Sultana, S.; Fuchs, G.J.; Alam, N.H.; Azim, T.; Brussow, H.; Hammarstrom, L. *Lactobacillus paracasei* Strain ST11 Has No Effect on Rotavirus but Ameliorates the Outcome of Nonrotavirus Diarrhea in Children From Bangladesh. *Pediatrics* **2005**, *116*, e221–e228. [CrossRef]
132. Zawistowska-Rojek, A.; Zaręba, T.; Tyski, S. Microbiological Testing of Probiotic Preparations. *Int. J. Environ. Res. Public Health* **2022**, *19*, 5701. [CrossRef] [PubMed]
133. Lorbeg, P.M.; Golob, M.; Kramer, M.; Treven, P.; Matijašić, B.B. Evaluation of Dietary Supplements Containing Viable Bacteria by Cultivation/MALDI-TOF Mass Spectrometry and PCR Identification. *Front. Microbiol.* **2021**, *12*, 700138. [CrossRef] [PubMed]

microorganisms

Article

Probiotic Characteristics and Antimicrobial Potential of a Native *Bacillus subtilis* Strain Fa17.2 Rescued from Wild *Bromelia* sp. Flowers

Gabriela N. Tenea *, Gabriela Lizeth Gonzalez and Jose Luis Moreno

Biofood and Nutraceutics Research and Development Group, Faculty of Engineering in Agricultural and Environmental Sciences, Technical University of the North, Av. 17 de Julio s-21, Barrio El Olivo, Ibarra 100150, Ecuador; glgonzalezc@utn.edu.ec (G.L.G.); jlmoreno@utn.edu.ec (J.L.M.)

* Correspondence: gntenea@utn.edu.ec

Abstract: In the present study, we identified the *Bacillus subtilis* strain annotated Fa17.2 isolated from *Bromelia* flower inflorescences collected from the subtropical humid mesothermal region, Santo Domingo de Los Tsachilas Province, Ecuador. The probiotic capacity and antimicrobial potential against four foodborne pathogens were assessed. The cell culture of Fa17.2 is highly resistant to synthetic gastric acid (pH 2.5, 3.0, and 3.5), bile salts (0.3%), tolerating different sodium chloride concentrations (1, 3, and 5%), and growth conditions (15 °C and 45 °C), suggesting its potential probiotic features. The isolate showed no antibiotic resistance and was considered safe as no hemolysis was detected on sheep blood agar. The optimum medium for bacterial growth and the release of antimicrobial compounds was MRS with 10% glucose. The active components released in the neutralized crude extract (NCE) were insensitive to organic solvents, surfactants, and nonproteolytic enzymes and sensitive to proteolytic enzymes suggesting their proteinaceous nature. The antimicrobial activity was enhanced by heat and maintained active over a wide range of pH (2.0–8.0). Moreover, the crude extract (CE) showed inhibitory activity against several Gram-negative and Gram-positive bacteria. The molecular weight of partially purified precipitated bacteriocin-like substances (BLISs) was about 14 kDa in 20% Tricine-SDS-PAGE. The CE obtained from Fa17.2 inhibits the growth of four foodborne pathogens, *Staphylococcus aureus*, *Escherichia coli*, *Kosaconia cowanii*, and *Shigella dysenteriae*, which implies its potential as an antimicrobial producer strain.

Keywords: probiotic; *Bacillus subtilis*; antimicrobials; foodborne pathogens

Citation: Tenea, G.N.; Gonzalez, G.L.; Moreno, J.L. Probiotic Characteristics and Antimicrobial Potential of a Native *Bacillus subtilis* Strain Fa17.2 Rescued from Wild *Bromelia* sp. Flowers. *Microorganisms* **2022**, *10*, 860. https://doi.org/10.3390/microorganisms10050860

Academic Editor: Sabina Fijan

Received: 24 March 2022
Accepted: 18 April 2022
Published: 21 April 2022

1. Introduction

Worldwide, millions of individuals suffer from gastrointestinal problems, most of them due to the consumption of contaminated food and water [1]. Ecuador makes no exception. Throughout 2019, foodborne pathologies reached 19,487 cases, while in 2021 a decrease in about 60% was reported [2]; however, there are still cases due to lack of knowledge of health standards, and handling and conservation of food by sellers, to guarantee the quality of the products that are marketed. This decrease may be related to the pandemic disease of the coronavirus, as street sales were blocked. Nonetheless, in most cities, the lack of an appropriate structure in the retail markets might be the main cause of the contamination; however, the products failed to reach the required quality [3]. In addition, the use of beneficial microorganisms such as probiotics in foods that contain antimicrobial substances is very limited in the Ecuadorian market. Probiotics are defined as live microorganisms that are administered to hosts in adequate amounts to improve human health [4]. However, only a few products contain such microorganisms as *Lactobacillus rhamnosus* GG (ATCC53100), a commercial strain [5].

Customer demand for high-quality, free of chemicals or antibiotics containing foods is increasing; therefore, the identification of natural alternatives using beneficial microorganisms, or their derivatives, can be a solution. In the last decade, several researchers have investigated the use of antimicrobials fabricated by food-grade microorganisms, such as peptides or proteins with antimicrobial activity that is secreted into the extracellular matrix during the metabolic process of various bacteria, which can prevent the increase in single or combined pathogens; they are easily degraded by enzymes, thus do not affect the human body pathogenically [6]. Although many bacterial species generate antimicrobials, only a few are applied to foods as biological preservatives [7].

One of the best-studied Gram-positive bacteria, *Bacillus subtilis*, an aerobic or facultatively anaerobic bacterium, is considered a model of cell differentiation and industrial exploitation [8]. Currently, various commercial formulations contain bacilli as active ingredients, thanks to their ability to colonize, reproduce easily, and their high stability concerning endosporogenesis; this last characteristic is especially essential, as it allows them to survive in stressful situations, such as abiotic conditions, that facilitate long term production and storage [9]. These species remain widely distributed in a wide diversity of habitats, integrated freshwater, rhizosphere, marine, and terrestrial ecosystems, and their species are commonly associated with plants [10]. *Bacillus subtilis* are recognized as safe and reliable probiotic strains that are non-pathogenic to humans and animals [11]. Antimicrobial metabolites are generated during their growth and reproduction [12].

Several registered strains have been marketed as probiotic supplements for human consumption in Asia, Europe, and the US [13]. Feed supplementation with spores can provide numerous benefits including animal improvement in digestibility and immune modulation [14]. The spores are metabolically quiescent and should be in a metabolically active state to perform certain probiotic functions such as secretion of antimicrobial compounds and enzymes, and synthesis of short-chain fatty acids [15,16]. *B. subtilis* gained more interest to be used as a probiotic, and their consumption in foods is believed to be associated with numerous health benefits, such as increased immunity, reduced bone loss in postmenopausal women, and antiallergic effects [17]. In addition, *Bacillus* isolates are well-known for producing a wide range of antimicrobial compounds, including lipopeptides and BLIS [18]. The main types of antimicrobial compounds from *B. subtilis* comprise peptides such as lantibiotics and lantibiotic-like peptides, and non-peptide compounds such as polyketides, an amino sugar, and phospholipid [19].

The increase in the value of biological diversity and the exceptional richness of tropical forests improve the chances of the systematic use of genetic resources or their derivatives in various biotechnological processes. Ecuador has not yet used its genetic resources effectively. There are no studies that quantify values in the bioproducts market. In the case of biological products, research and development activities are concentrated in multinational companies. From free of board (FOB), the annual import values of these products are close to USD 250 million; therefore, there is a huge opportunity to enter this market with innovative technologies, such as probiotics, nutraceuticals, and derivatives developed by the researchers and transferred to different productive sectors of the Andean region [20]. Given the extensive changes made by Ecuadorian government policies, some natural areas such as subtropical forests have been considered relevant genetic resources for biotechnological research. However, to detect and characterize new bacterial species, the bacterial microbiota associated with flowers and fruits was studied [21]. Particularly, lactic acid bacteria (LAB) associated with these micro-niches were investigated [21]. We taught that the strains associated with these extreme niches may allow selecting more robust strains with broad antimicrobial capacity against foodborne pathogens as well as native strains with probiotic potential. Throughout the selection process for isolates showing antimicrobial capacity against at least two Gram-negative bacteria (*Salmonella enterica* and *E. coli*), one bacillus showing sporulation "escapes" along with other isolated lactobacilli from the selection on MRS media. Moreover, these isolates during cultivation were characterized by a "particular flower-fragrance" which might be linked with the secretion of some volatile compounds that need further attention. At this point, we speculate that this feature might be connected to the *Bromelia* flower origin. Due to its comparable inhibitory activity with lactobacilli, we selected this isolate for further taxonomic identification and evaluated its probiotic capacity and antimicrobial potential against some foodborne pathogenic bacteria. Therefore, we performed various in vitro studies to test their tolerance to intestinal gastric acid, bile salts, and sodium chloride, as well as different growth temperatures, and hemolysis and antibiotic susceptibility for safety issues. In addition, the effect of medium composition on the production of antimicrobial substances as well as the antimicrobial spectrum was evaluated against several Gram-positive and Gram-negative bacteria. Moreover, the nature of these antimicrobials was evaluated in vitro along with their sensitivity to various pH, heat, inorganic, and organic treatments. Moreover, the partially precipitated BLIS were analyzed by Tricine-SDS-PAGE to estimate their molecular weight.

2. Materials and Methods

2.1. Sampling, Bacterial Isolation, and Identification

Samples consisting of flower inflorescences of *Bromelia* sp. were collected aseptically from a subtropical humid mesothermal region of Santo Domingo de Los Tsachilas Province, 43 km away from Quito, the capital city. Samples were packaged in clean bags, then stored at 4 °C for further analysis. The isolation and selection procedures were performed as described earlier [21]. One isolate assigned Fa17.2, showing spore formation, was selected based on its capacity to inhibit *Salmonella enterica* subsp. *enterica* ATCC 51741 and *E. coli* ATCC25922. The BBL Crystal Gram-positive identification system (cat # 245010, BD Company, Franklin Lakes, NJ, USA), a miniaturized identification method using 29 enzymatic and biochemical substrates, was used for genera classification according to the manufacturer's instructions. Moreover, 16S rRNA gene sequencing was used for taxonomical classification following a standard procedure (Macrogen Inc., Seoul, Korea). The microorganism culture was preserved by deep freezing in glycerol solution before use in further analyses.

2.2. In Vitro Probiotic Feature Assessment

2.2.1. Survival under Gastric Juice Conditions

Survival was determined using 8 log CFU/mL of the overnight culture of Fa17.2 by the plate-agar method using the MRS agar medium (MRS, Difco, Detroit, MI, USA). Briefly, after incubation at 37 °C (with shaking 200 rpm), the bacterial cells were harvested at 5000× *g* for 5 min at 4 °C, the biomass was rinsed twice with sterile Ringer's solution (pH 7.2) and resuspended in synthetic gastric juice solution with the established pH of 2.5, 3.0, and 3.5 followed by incubation for 4 h at 37 °C. The cell viability was determined at intervals of 1 h by counting the cells on the MRS agar. The gastric juice was formulated as follows: glucose (3.5 g/L), NaCl (2.05 g/L), KH_2PO_4 (0.60 g/L), $CaCl_2$ (0.11 g/L), and KCl (0.37 g/L), adjusted to corresponding pH using 1 M HCl. After autoclavation at 121 °C for 15 min, porcine bile (0.05 g/L), lysozyme (0.1 g/L), and pepsin (13.3 mg/L) were added as stock solutions before analysis [22]. Components were obtained from Sigma Aldrich (Sigma Aldrich, St. Louis, MO, USA). The % of cell survival was calculated as follows: ((cell counts at initial incubation time − cell counts at the final incubation time)/cell counts at the initial time) × 100). The results were compared with a probiotic reference strain, *Lactobacillus acidophilus* ATCC4846 (LA).

2.2.2. Survival under Bile Conditions

In the case of bile, the overnight Fa17.2 cell culture (8 log CFU/mL) was incubated in MRS broth containing 0.3% bile salt (oxgall, *w*/*v*) at 37 °C for 4 h (with shaking, 200 rpm). The cell viability was determined by plating 100 µL bacterial cells on MRS agar (MRS, Difco, Detroit, MI, USA). The % of cell survival was calculated as indicated in Section 2.2.1. No modified MRS broth was used as control and the experiment was run in triplicates starting from different batches of culture. The results were compared with the probiotic LA reference strain.

2.2.3. Optimum Temperature and Growth Tolerance in the Presence of Sodium Chloride

Overnight culture (8 log CFU/mL) of Fa17.2 was inoculated in tubes containing MRS broth and incubated at 15 °C and 45 °C for 24 h (with shaking, 200 rpm), and the absorbance at 605 nm was measured at the initial and final incubation time. Similarly, the tolerance in the presence of sodium chloride was evaluated upon the inoculation of the overnight culture of Fa17.2 in broth medium containing 1%, 3%, and 5% sodium chloride (*w*/*v*) for 24 h at 37 °C. Cell growth was monitored for each treatment and the effect of sodium chloride on cell survival was determined using the plate-agar method. The % of cell survival was determined and is described in Section 2.2.1. No modified MRS broth was used as control and the experiment was run in triplicate starting from individual batches of bacterial culture. The results were compared with the probiotic LA reference strain.

2.2.4. Hemolysis Test

The hemolytic activity of the isolate was determined as previously described [23]. The Columbia agar containing 5% (*w*/*v*) sheep blood was used. After incubation at 37 °C for 48 h, the hemolytic activity was evaluated and classified based on the lysis of red blood cells in the medium around the colonies: the green zones around colonies (α-hemolysis), clear zones around colonies (β-hemolysis), and no zone around colonies (γ-hemolysis). The strain is considered safe if γ-hemolysis was detected.

2.2.5. Antibiotic Susceptibility

Susceptibility to several antibiotics was determined using commercial disks of ampicillin, gentamicin, kanamycin, amoxicillin/clavulanic acid, tetracycline, and cefuroxime at the concentrations recommended by the Scientific Committee on Animal Nutrition (disks provided by Merck, Darmstadt, Germany) by the disk diffusion assay as described in [21]. The experiment was run in triplicate starting with different batches of bacteria culture and the disks were verified by *E. coli* ATCC25922, a reference strain for quality control. Using a similar approach, the minimum inhibitory concentration (MIC) distribution within the bacillus group was measured using the E-test (Biomerieux, Durham, NC, USA, E-test) assay following the manufacturer's instructions. The microbiological breakpoints reported by the FEEDAP document were used to categorize bacilli as susceptible or resistant. The strains showing a MIC higher than the EFSA breakpoint were considered resistant [24].

2.3. *Characterization of Antimicrobial Substances Produced by Fa17.2 In Vitro*

2.3.1. Preparation of CE and Antimicrobial Assay

CE from the target strain was obtained as described in [25]. In brief, the CE was recovered by centrifugation (13,000× *g* for 20 min at 4 °C) and filtration using a 0.22 µm porosity syringe filter (# STF020025H, ChemLab Group, Fort Smith, AR, USA) of an overnight culture of Fa17.2. The indicator strain (100 µL) grown in broth medium (7 log CFU/mL) was mixed with 3.5 mL of soft MRS agar (0.75%), overlaid on nutrient agar plates, and incubated at 37 °C for 2 h. The CE (100 µL) were transferred onto wells (6 mm) on overlaid agar, incubated at 37 °C, and subsequently examined for inhibition zones at 48 h. The experiments were run in triplicate and the mean value of the inhibition zone was determined. As indicator microorganisms: *S. aureus* ATCC1026, *S. dysenteriae* UTNFa37-1, *K. cowanii* B2Sh1 (laboratory isolate), and *E. coli* ATCC25922 were used. Each experiment was performed in triplicates starting from individual bacterial cultures.

2.3.2. Effect of Medium Composition on Antimicrobials Production

The effect of medium composition on bacterial growth and the release of antimicrobial compounds was evaluated. The following media were tested: (1) MRSS: MRS broth supplemented with sucrose (5, 10, 20, 40, and 50%); (2) MRSG: MRS broth supplemented with glucose ((5, 10, 20, 40, and 50%); (3) and MRSGly: MRS broth supplemented with glycerol (5, 10, 20, 40, and 50%). As a control, MRS broth without additional nutrients was used. The Fa17.2 strain was inoculated individually in each medium combination for 24 h and the CE obtained as mentioned above was used in the agar-well diffusion assay. As an indicator, the microorganisms described in Section 2.3.1 were used. Each experiment was performed in triplicates starting from individual bacterial cultures.

2.3.3. Evaluation of Antimicrobial Spectrum of Inhibition

The antimicrobial activity of the CE was obtained after growing the bacteria in the optimum medium detected for each target bacteria. The results were compared with the antimicrobial activity obtained from bacteria grown in MRS media (−) with no additional nutrients. The percentage of antimicrobial activity change was calculated as follows: ((average diameter of inhibition zone of CE obtained from the optimum media—average diameter of inhibition zone of CE obtained from MRS (−)/average diameter of inhibition zone of CE obtained from MRS media (−)). The bacteriocinogenic strain *L. plantarum* ATCC8014 (LP) was used as a reference. The indicator strains of Gram-positive: *S. aureus* ATCC1026, *Lactococcus lactis* ATCC11474, *L. acidophilus* ATCC4358, *Bifidobacterium brevis* ATCC15700, and *Streptococcus thermophilus* ATCC19298; and Gram-negative: *Shigella* ssp. UTNShg1 (laboratory strain), *S. enterica* subsp. *enterica* ATCC51741, *Salmonella* ssp. UTNSm2 (laboratory strain), *S. sonnei* ATCC25931, *S. dysenteriae* UTNFa37-1 (laboratory strain), *E. coli* ATCC25922, *E. coli* ssp. UTNEc2 (laboratory strain), and *K. cowanii* B2Sh1 (laboratory strain) were used.

2.3.4. Estimation of the Chemical Nature of CE

The CE were submitted to different treatments as described previously [25]. Briefly, aliquots of CE were treated 10 min at 80 °C and pH 6.0 to rule out the effect of acids on the antimicrobial activity (NCE). Moreover, NCE was treated with catalase enzyme (1 mg/mL) to prevent the possible inhibitory of hydrogen peroxidase. Furthermore, NCE was independently treated with proteinase K, pepsin, lysozyme, and α-chymotrypsin (Sigma-Aldrich Corporation, St. Louis, MO, USA) at the final concentration of 1 mg/mL, incubated for 2 h at 37 °C and 5 min at 100 °C for enzyme inactivation. All experiments were run in triplicate using *S. aureus* ATCC1026, *S. dysenteriae* UTNFa37-1, *K. cowanii* B2Sh1, and *E. coli* ATCC25922 as indicator strains. The control for all experiments was the sterile MRS medium.

2.3.5. Effect of Heat, pH, and Detergents on Antimicrobial Activity

Aliquots of CE were incubated for 10, 30, and 60 min at 60, 80, 90, and 100 °C as well as 15 min at 121 °C (autoclaving). In another batch, aliquots of CE were adjusted at pH 2.0, 4.0, 6.0, and 8.0, incubated for 3 h at room temperature. In addition, the effect of Triton X-100 (BDH Chemicals Ltd., Poole, UK), sodium dodecyl sulphate ((SDS) Sigma-Aldrich Corporation, St. Louis, MO, USA), and ethylenediaminetetraacetic acid ((EDTA) Sigma-Aldrich Corporation, St. Louis, MO, USA) at the final concentration of 1 mg/mL was evaluated. All experiments were run in triplicate using *S. aureus* ATCC1026, *S. dysenteriae* UTNFa37-1, *K. cowanii* B2Sh1, and *E. coli* ATCC25922 as indicator strains. The control for all experiments was the sterile MRS medium.

2.4. Bacteriocin Molecular Size Approximation

To obtain the precipitated bacteriocin, the CE was treated with 80% ammonium sulfate, incubated overnight at 4 °C, and centrifuged at 10,000× *g* for 30 min. The bacteriocin was recovered in ammonium acetate 25 mM (pH 6.5) and desalted using a midi dialysis kit (cat # PURD10005-1KT, Sigma-Aldrich, St. Louis, MO, USA) pre-equilibrated with phosphate buffer (pH 7.0) and stored at −80 °C before use. The bacteriocin molecular weight was determined by the Tricine-SDS-PAGE method using pre-casted acrylamide gels (4–20%) and a mini-vertical electrophoresis system (Expedeon, Abcam, Cambridge, MA, USA). The broad range protein molecular marker (cat # V8491, Promega, Madison, WI, USA) was used for molecular weight determination. The gel was stained with InstantBlue ready-to-use stain (Expedeon, Abcam, Cambridge, MA, USA) for 2 h and distained with a solution of 30% methanol (v/v) and glacial acetic acid, 10% (v/v) until the bands became clear.

2.5. Statistical Analysis

The means were calculated from repeated measurements performed three times. For the antimicrobial activity, the effect of the medium and enzymes, one-way analysis of variance (ANOVA), and Tukey's post hoc test were used to determine significant differences between the means. For the effect of heat, and detergents, the ANOVA with a split-split-plot experimental design was performed. Then, Duncan's multiples tests and Least Significant Difference with Bonferroni correction (LSD) were applied to determine significant differences between the means. The statistical significance used was $p < 0.05$ (SPSS version 10.0.6, IBM, Armonk, NY, USA).

3. Results and Discussion

Based on the 16S rRNA gene sequences, a comparative sequence analysis, and a biochemical characteristics analysis, the new isolate assigned Fa17.2 belongs to the genus *Bacillus* with 99% identity to *Bacillus subtilis*. The strain was registered at GenBank with the accession number KY046251.1 (1 November 2016). Using multiple sequences alignment with Jalview (version 2.11.2.0) [26], the average distance was calculated from the percentage of identity between the sequences of some *Bacillus* strains retrieved from the database and the contig sequence of the target strain (Fa17.2). (Figure 1). The closest genome to Fa17.2 was *B. subtilis* strain *subtilis* (MN611449.1).

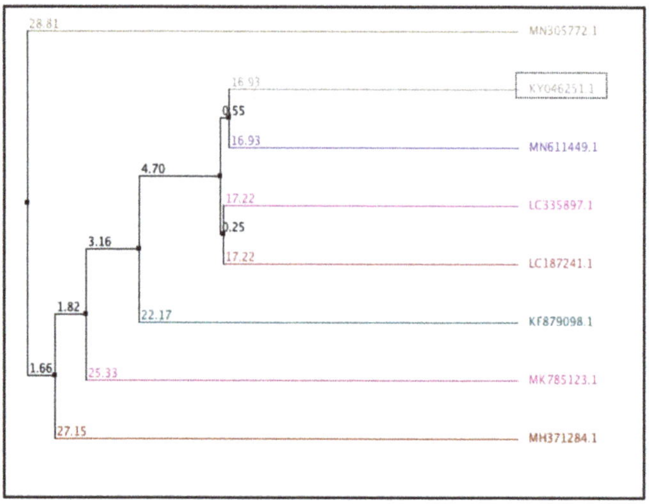

Figure 1. Average distance calculated based on percentage of similarity between sequences. Bacillus strains from the database and the contig 16S rRNA sequence of Fa17.2. Trees were calculated based on a measure of similarity between each pair of sequences in the alignment: PID. The percentage identity between the two sequences at each aligned position. The number on the branch is the bootstrap value that indicates the extent of relatedness between two subjects. Legend: KY046251.1: *B. subtilis* strain Fa17.2; MH371284.1: *B. subtilis* strain C1; MK785123.1: *B. vallismortis* strain VS-5; LC335897.1: *B. subtilis* PH; KF879098.1: *Bacillus* spp. BAB-2797; MN305772.1: *B. subtilis* strain OTG009; MN611449.1: *B. subtilis* strain *subtilis*; LC187241.1: *Lactobacillus murinus* strain LAP1.

3.1. Assessment of Probiotic Characteristics of B. subtilis Strain Fa17.2

3.1.1. Tolerance to Gastric Juice and Bile Salts

To exert their probiotic potential, the new bacterial isolates should present resistance to gastric acid and bile [27]. Figure 2 shows the cell viability (%) of both strains after 4 h of incubation with gastric juice at pH 2.5, 3.0, and 3.5. The initially inoculated population at the time 0 (hours) and during 4 h of incubation is shown in Table S1. At pH2.5, the registered percentage of decrease was 6.31% for *B. subtilis* Fa17.2 and 14.66% for the reference probiotic strain *L. acidophilus* ATCC4846. Similarly, cell viability decreases of 5.71% and 12.43% were registered at pH 3.0, and 6.79% and 11.75% at pH 3.5 for Fa17.2 and LA, respectively, upon 4 h of incubation. Although a percentage decrease in the growth of *Bacillus* spp. was evidenced at pH 3.0, by increasing the pH (4.0), an increase in cell viability was observed. A similar study indicates that the vegetative cells of *B. subtilis* DET6 and *B. megaterium* JHT3 had poor resistance to artificial gastric acid, whereas the spore cells were resistant to gastric conditions [28]. In addition, the resistance to synthetic gastric acid of two *B. subtilis* CBD2 and KMKW4 strains isolated from Korean fermented foods were demonstrated [10]. At 4 h, a significant increase ($p > 0.05$) in the cell population of Fa17.2 was observed when incubated with 0.3% bile salts (oxgall, w/v), while the cell population of the reference LA strain decreases (Table S1). In a previous analysis, several selected LAB strains isolated from wild fruits and flowers showed high bile resistance with a significant increase in the cell population at 4 h of incubation, thus suggesting that bile might stimulate cell growth [21]. In another study, an increase in *B. subtilis* P223 vegetative cells in bile salts (0.3%) was registered [29]. Based on these results, we suggest that *B. subtilis* Fa17.2 can resist acidic gastric conditions; therefore, it might pass through the intestinal gut, an important criterion when selecting potential probiotic strains. Moreover, the bile stress did not affect the growth of Fa17.2 cells, while the reference commercial probiotic cells were less tolerant. We suggested that this property is species-specific and might be connected to the origin (tropical flower), but this statement needs further investigation.

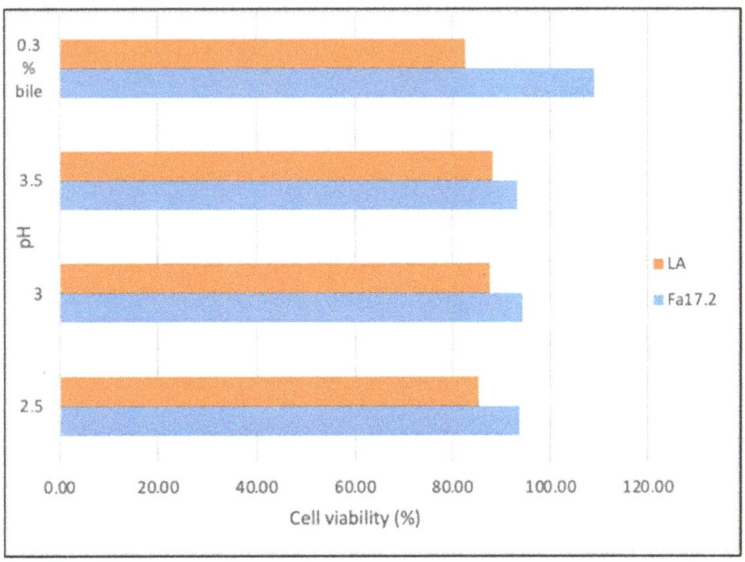

Figure 2. Cell viability (%) upon incubation with gastric juice at different pHs and bile salt (0.3%). Legend: Fa17.2: *B. subtilis* Fa17.2; LA: *L. acidophilus* ATCC4846.

3.1.2. Tolerance to Sodium Chloride and High-Temperature Growth Conditions

Temperature is one of the most important factors that affect the growth and survival of microorganisms, it varies between different genera and reflects the optimal temperature range of its natural habitat [28]. In this study, the results indicated that both strains grow at 15 °C and 45 °C and tolerate different concentrations of sodium chloride (Table S2). At 15 °C a lower decrease in cell population was observed for the native Fa17.2 strain compared with reference LA, while at 45 °C, the strains showed a comparable tolerance profile (Figure 3). In a similar study, less tolerance at high temperature (43 °C) and salt concentrations (2, 4, 7, and 10%) was noticed for different *Bacillus* strains [29]. Our results agreed with previous studies showing that sodium chloride tolerance might be strain-dependent [21,29].

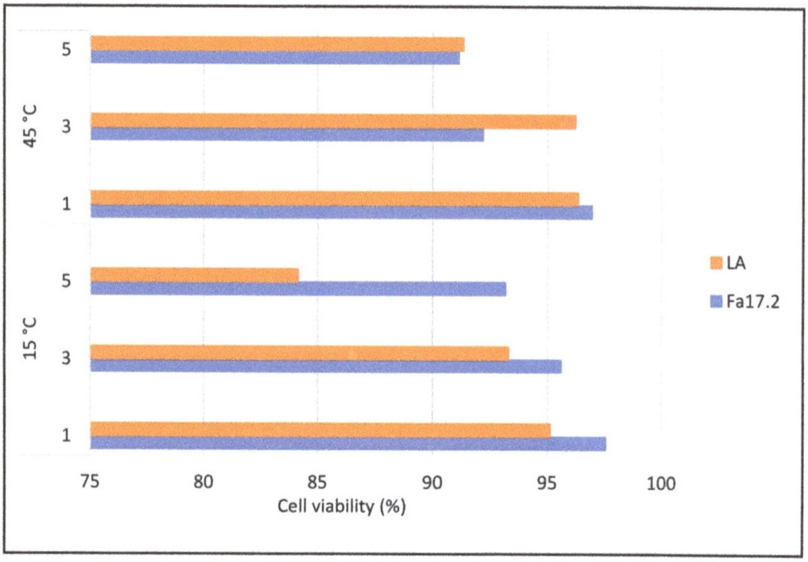

Figure 3. Cell viability (%) in different concentrations of NaCl upon 24 h of incubation at 15 °C and 45 °C. Legend: Fa17.2: *B. subtilis* Fa17.2; LA: *L. acidophilus* ATCC4846.

3.1.3. Antibiotic Susceptibility and Pathogenicity

Antibiotic susceptibility of bacterial strains intended for use as probiotics must be confirmed for safety proof [30]. Antibiotic tolerance can help balance the gut microflora after antibiotic administration [31]. The main concern about using *Bacillus* spp. as probiotics is due to their ability to transfer antibiotic resistant genes [32]. In addition, *Bacillus* spp. do not belong to the commensal microbiota of the digestive tract; however, several strains of the genus are integrated into food supplies [33]. Moreover, some strains of *Bacillus* are used as feed additives, biomass for animal feed consumption, or enzyme/vitamin production [34], and many species have been added to the EFSA QPS list [35]. Based on disk diffusion agar assay results, the strain Fa17.2 was sensible to all antibiotics except gentamycin (data not shown). According to EFSA [24], *Bacillus* strains are listed as resistant to all antibiotics except ampicillin. The selected Fa17.2 does not show ampicillin resistance. In a similar study, *B. clausii* ATCC700160 and *B. subtilis* P223 strains were found resistant to streptomycin [29]. The microbiological breakpoints reported by the FEEDAP were used to categorize bacilli as susceptible or resistant [24]. In this study, the E-test assay confirmed that the strain was sensible to the antibiotics tested. In addition, the resistance showed by Fa17.2 to gentamycin was not confirmed by the E-test analysis (Table S3). Moreover, the strain did not show any hemolysis on sheep blood agar, indicating that the strain is not pathogenic.

3.2. Assessment of Inhibitory Capacity and Characterization of Antimicrobial Substances Produced by Fa17.2

3.2.1. Culture Medium Optimization to Enhance the Bacteria Growth and Antimicrobials Production

The agar diffusion analysis was carried out to identify the effect of cell growth in several culture media on the inhibitory activity against four indicator strains. The average mean values registered against all indicator strains are presented in Tables S4–S7. The results indicated that all media tested displayed favorable inhibitory effects against all indicator strains. Nonetheless, the greater inhibitory effect of CE Fa17.2 against *S. aureus* ATCC1026 was detected after growing in MRS media supplemented with 10% and 20% glucose (Table S4). Comparable results were obtained against *E. coli* ATCC25922 when growing Fa17.2 in MRS supplemented with 10% glucose (Table S5). Moreover, the results indicated that against the B2Sh1 strain the optimum media for Fa17.2 to exert maximum inhibitory activity were MRS supplemented with 20% glycerol and 10% glucose (Table S6). The MRS media supplemented with 10% glucose had a favorable effect on the production of antimicrobial compounds for the Fa17.2 strain against *S. dysenteriae* Fa37-1 (Table S7). Likewise, the reference strain LP showed greater activity in medium supplemented with 20% glycerol against all tested indicator strains. Supplementation of MRS with different concentrations of carbohydrates provided an increase in yield which varies depending on the type of sugar used. Although several media compositions showed differences in the growth of the cells and coupled with antimicrobial activity against the four pathogens, the optimum media chosen for the target strain were MRS with 10%. Sucrose also served as a carbon source for optimal growth; however, did not show the same antimicrobial effect against all indicator bacteria, while glycerol appears to be deficient in growth. The optimum media for the bacteriocin production of LP were MRS with 20% glycerol. Early research showed the positive effect of a medium supplemented with different sugars (glucose, sucrose, and xylose) on the *B. subtilis* growth, and an enhanced antimicrobial activity was correlated with the accumulation of cell biomass [36]. In another study, Monteiro et al. [37] found that for *B. subtilis*, glucose exerted an inhibitory effect on spore production if its concentration exceeded 20 g/L. However, the results agreed with other studies indicating that the inhibitory efficiency depends on the medium composition and pathogen.

3.2.2. Inhibitory Spectrum

In this study, the inhibitory activity of CE obtained from Fa17.2 and the reference LP strains was evaluated against several indicator strains. The percentage of inhibitory activity changes (%) is shown in Figure 4. Although both strains showed inhibitory activity against all tested indicator strains, the highest activity was registered by the LP strain against *Shigella* UTNShg1. Within the Fa17.2 group, the most sensitive strains were *S. enterica* ATCC51741, *S. aureus* ATCC1026, *E. coli* ATCC25922, and *Salmonella* UTNSm2 ($p > 0.05$). Lower activity was detected against *Lactobacillus* and *Bifidobacterium* strains. An early study indicated a higher inhibitory capacity of a cell-free supernatant extracted from the *B. subtilis* KKU213 strain against several Gram-positive pathogens such as *B. cereus*, *S. aureus*, and *L. monocytogenes*, with the low levels of activity against lactic acid bacteria, *E. faecalis* BT2, and MG30 [38]. In another study, antibacterial substances produced by *B. subtilis* LFB112 isolated from Chinese herbs were effective against Gram-positive and Gram-negative bacteria involved in domestic animal diseases, including *E. coli*, *S. pullarum*, *Pseudomonas* ssp., *Clostridium perfringens*, *Micrococcus luteus*, *S. bovis*, and *S. aureus* [38]. Due to its nature as an endospore former strain, the ability of *B. subtilis* to resist harsh environments and produce cocktails of antimicrobial substances such as bacteriocins and lipopeptides is an advantage to stimulate the inhibition of pathogenic bacteria. In the industry, the control of these microorganisms is of vital importance, so these strains represent a great option to be used as natural biopreservatives [39]. Moreover, endospore formation allows it to withstand extreme stresses and offers biological solutions to formulation conservation problems when produced on an industrial scale [40]. Genome sequencing highlighted the genus *Bacillus* as an unexpected source of antimicrobial compounds including surfactin, fengycin, iturin, mycosubtilins, and bacillomycins, which are amphiphilic, membrane-active biosurfactants [41]. These molecules might contribute to the overall inhibitory action. Further investigations are required to identify the antimicrobial molecules produced by the target Fa17.2 as well as their mode of action against pathogenic strains.

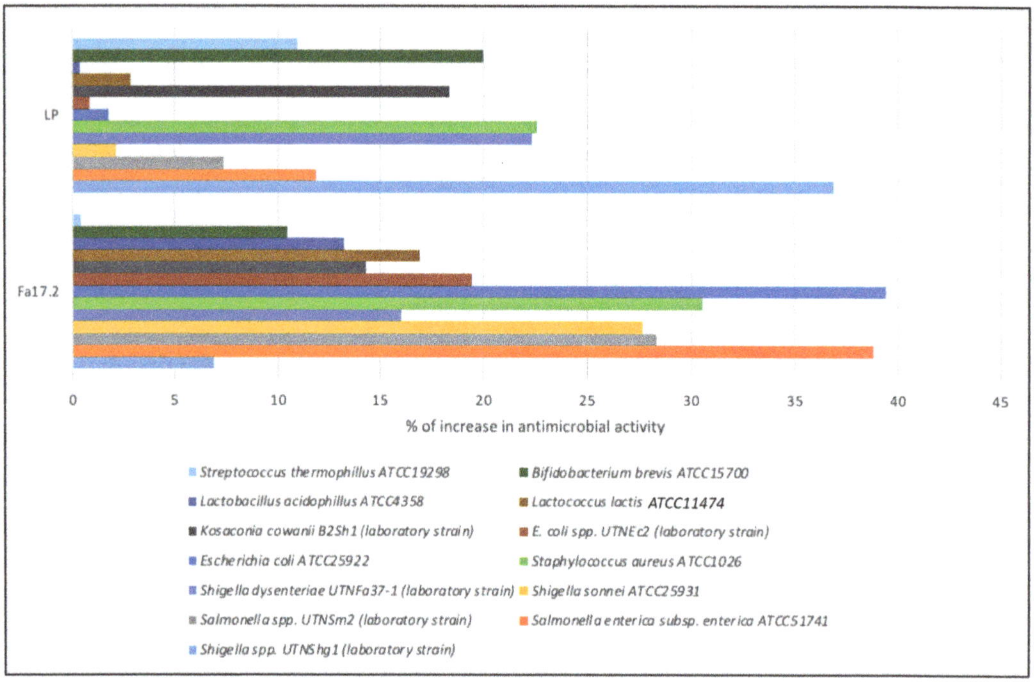

Figure 4. Percentage of increase in the inhibitory activity of CE obtained from Fa17.2 and LP after growth in optimum media versus control (MRS control).

3.2.3. Detection of the Nature of Antimicrobial Substances

Table 1 shows a comparison of the averages of the diameter of the inhibition zone obtained at 48 h of incubation for different treatments of NCE with proteolytic and non-proteolytic enzymes against the indicator bacteria under study. However, the treatment with proteinase K (1 mg/mL) resulted in completely abolishing the antimicrobial activity, suggesting the protein-like nature of compounds released in the CE. The treatment with alpha-chymotrypsin, pepsin, and trypsin resulted in a little decrease in inhibitory activity ($p < 0.05$), while the treatment with lysozyme did not show any change in the antimicrobial activity. Thus, the substances released in the CE might not be affected by the presence of these enzymes in the medium. The treatment with catalase resulted in the reduction in the inhibitory effect indicating that the activity might be hydrogen peroxide dependent. Similar results were observed in the case of *B. subtilis* strain RLID 12.1 isolated from the soil when a decrease in activity was shown after treatment with proteinase K (10 mg/mL), while the activity was partially lost upon exposure to pronase E, trypsin, amylase, and lipase [18]. The activity was maintained after lysozyme treatment, indicating that the protein might be glycosylated, while treatment with lipase and α-amylase can explain the lack of carbohydrate or lipid moieties. Overall, our data indicated the presence of antimicrobial substances (protein-like) in the bacterial CE of Fa17.2.

Table 1. Effect of enzymes on the antimicrobial activity.

Strains	Indicator Strains	NCE + Enzymes (1 mg/mL)					NCE (Control)
		α Chymotrypsin	Lysozyme	Proteinase K	Catalase	Pepsin	
Fa17.2	*S. aureus* ATCC1026	9.33 ± 0.1 b	10.33 ± 0.1 a	6.01 ± 0.1 c	9.33 ± 0.2 b	9.67 ± 0.2 b	10.33 ± 0.1 a
	S. dysenteriae UTNFa37-1	9.67 ± 0.6 b	10.33 ± 0.1 a	6.01 ± 0.1 b	9.33 ± 0.2 b	9.67 ± 0.2 b	10.33 ± 0.1 a
	K. cowanii B2Sh1	8.33 ± 0.2 c	11.20 ± 0.2 a	6.01 ± 0.1 d	9.33 ± 0.2 b	8.67 ± 0.2 bc	11.33 ± 0.2 a
	E. coli ATCC25922	9.33 ± 0.4 b	10.67 ± 0.1 a	6.01 ± 0.1 c	9.33 ± 0.2 b	9.33 ± 0.2 b	10.67 ± 0.1 a
LP	*S. aureus* ATCC1026	9.67 ± 0.2 b	10.33 ± 0.2 a	6.01 ± 0.1 d	8.33 ± 0.2 c	9.67 ± 0.1 b	10.33 ± 0.2 a
	S. dysenteriae UTNFa37-1	8.67 ± 0.6 bc	10.33 ± 0.2 a	6.01 ± 0.1 d	9.33 ± 0.2 b	8.33 ± 0.2 c	10.33 ± 0.2 a
	K. cowanii B2Sh1	8.33 ± 0.2 b	9.67 ± 0.2 a	6.01 ± 0.1 c	8.67 ± 0.1 ab	8.33 ± 0.2 b	9.67 ± 0.2 a
	E. coli ATCC25922	8.33 ± 0.2 b	10.33 ± 0.2 a	6.01 ± 0.1 c	8.67 ± 0.1 b	8.67 ± 0.1 b	10.33 ± 0.2 a

Data are mean ± standard error. Values in the same row with small letters are significantly different versus NCE ($p < 0.05$); NCE (control): neutralized CE (10 min heat at 80 °C and pH 6.0); CE: crude-extract.

3.2.4. Effectiveness of Inhibitory Activity upon Heat, pH, and Detergent Exposure

The antimicrobial substance was found to be heat stable at all temperatures and times tested, this feature is important when selecting bacteriocinogenic producer strains intended to be used as preservation agents in processed foods. In this study, the statistical analysis revealed that the effectiveness of inhibitory activity was influenced by both temperature-time and pathogen-temperature interactions (Figure 5). Analysis from the split-split-plot design (where main plot: pathogen; sub-plot 1: temperature; and sub-plot 2: incubation time) indicated that the activity was maintained with the incubation time at all four temperatures tested, with a significant increase versus control recorded after 30 and 60 min of incubation (LSD with Bonferroni correction) (Figure 5A). The greatest activity was registered against *K. cowanii* B2Sh1 at all temperature tested indicating that the antimicrobial effectiveness is pathogen-dependent (Figure 5B). Such an increase was not observed in the case of the reference LP strain (data not shown). We hypothesized that the increased inhibitory activity after heat exposure might follow the same path as the thermal process-induced chemical reaction between active elements such as the amino and carbonyl groups, known as the Maillard reaction [42]. Previous studies reported the efficacy of Maillard reaction products with inhibitory action against pathogens, these properties might be linked with the high molecular weight of the proteins released which can bind chemical elements such as iron, copper, or zinc, increasing the antimicrobial effect [43]. Considering the tropical microenvironment origin of the raw material, this might be a significant finding as other studies did not mention such property of heat-time inducing inhibitory activity of *B. subtilis*. At the autoclavation temperature (121 °C for 15 min) the activity was maintained against the four indicator strains under study (data not shown), suggesting the benefit of these molecules if tested as preservatives in association with thermal processing foods. A statistically significant increase ($p < 0.05$) in activity was observed in highly acidic conditions (pH 2.0) towards all indicator strains. Table 2 shows the diameter of the inhibition zone registered against indicator bacteria at different pH treatments of CE. The results indicated that the acidity stimulates the antimicrobial activity, due to the increase in bacteriocin solubility or due to the ability of acids to pass beyond the target cell membranes acidifying the cytoplasm and increasing its permeability [44]. At pH 4.0 and 6.0, the activity was maintained, while a significant decrease was registered at pH 8.0. Nonetheless, the data obtained from CE neutralization only provide a preliminary indication of the active ingredients. Other experiments, integrated CE pH control and acidification of growing cultures, and further investigation is need it to better verify the potential role of organic acids. In agreement with other studies, we suggest that organic acids, if present, may have potentiated the activity of other antimicrobial metabolites, which can trigger acidification and/or acid-mediated cell membrane variation to exert an apparent antagonistic effect [45]. In addition to their pH minimizing characteristics, the antimicrobial effect of organic acids might reflect a specific mode of action that can subjectively be independent of pH. For example, acids permeabilizes the outer membrane of Gram-negative species, causing structural alterations in the phospholipid components [46]. However, the resistance of BLIS to different treatments such as acidity and temperature is an important characteristic of a probiotic strain, as this resistance might enhance the strain capacity to pass through the digestive tract, adhere to, and colonize the host gut [47]. In addition, a significant increase in antimicrobial activity ($p < 0.05$) relative to the untreated counterpart was observed when adding EDTA and SDS for both strains (Figure 6). The positive effect of EDTA and SDS on the inhibitory activity against Gram-negative species was previously described [48]. This activity was linked to the increase in outer membrane permeability beyond extracting cations (Ca^{2+}, Mg^{2+}); thus, allowing bacteriocins to reach the cytoplasmic membrane. Similarly, in this study, the effectiveness of inhibitory activity was positively influenced by the treatment with SDS and chelating EDTA agent. Likewise, a slight decrease in activity was observed when CE was treated with Triton-X100. Comparable results were obtained with the reference to the LP strain (data not shown). In a similar study, no such increase in activity was observed when treating the cell-free supernatant from *B. subtilis* RLID 12.1 with EDTA, Triton-X100, or SDS, indicating that such effect might be bacteriocin-dependent [18]. In conclusion, our results indicated that the efficiency of bacteriocin-like substances of the selected Fa17.2 strain was positively regulated by heat, acidic condition, and chelating agents, these features might help for further identification of the mode of action against multidrug-resistant pathogens.

(A)

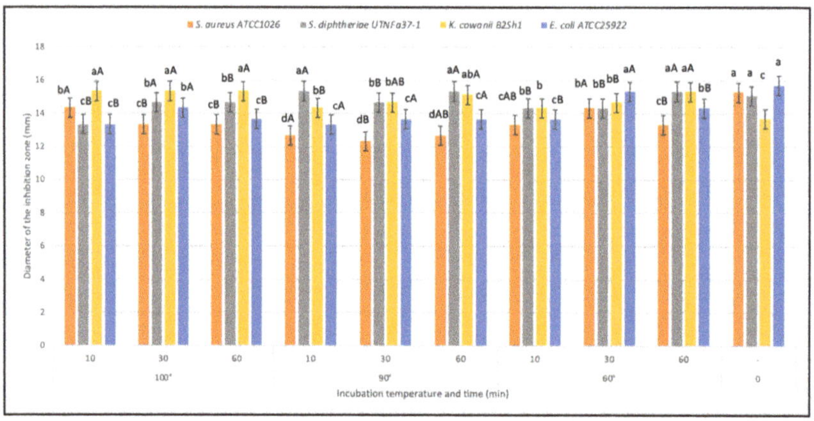

(B)

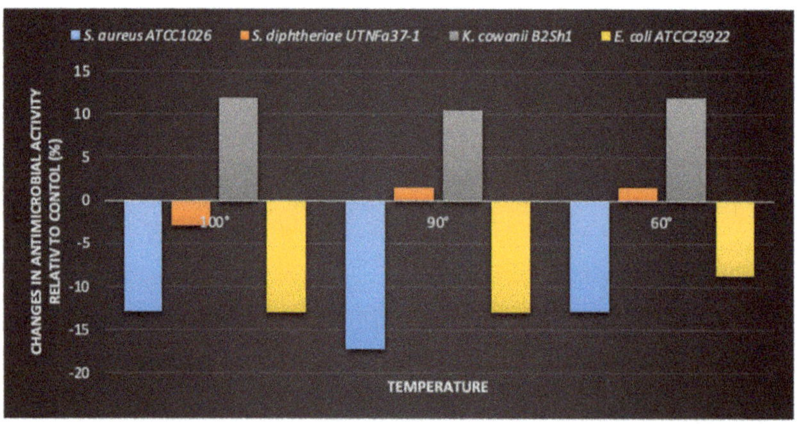

Figure 5. Effect of heat on bacteriocin activity. (**A**) Diameter of the inhibition zone (mm) at different temperatures and incubation time. Bars are the means ± standard error. Values with different letters

are significantly different $p < 0.05$. Small letters show the difference between temperature-incubation time and control (LSD with Bonferroni correction); capital letters show the differences within the incubation time (Duncan's test). (**B**) The influence of pathogen in the inhibitory activity. The changes (%) in antimicrobial activity relative to the control are shown.

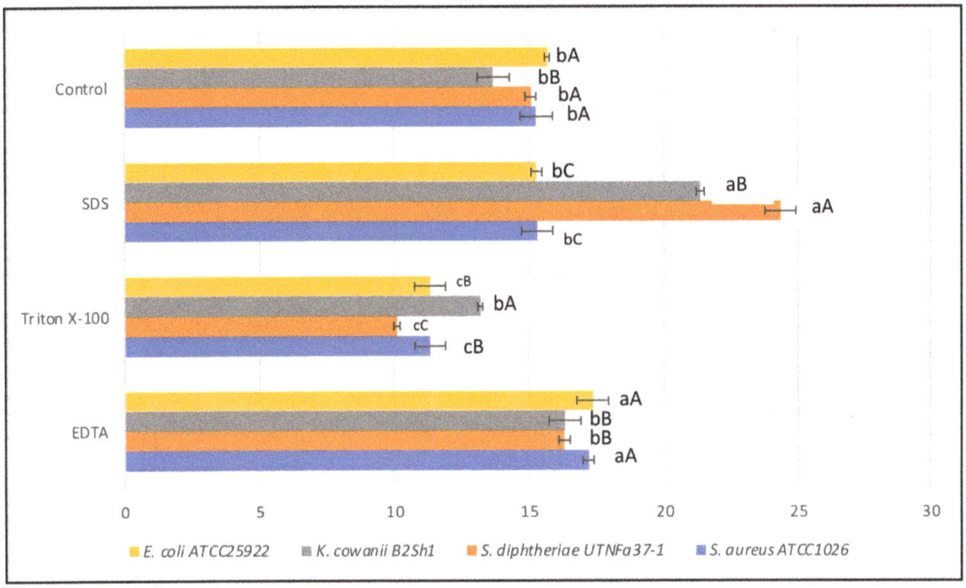

Figure 6. Effect of EDTA, Triton-X100, SDS on antimicrobial activity of CE obtained from Fa17.2. Diameter of the inhibition zone (mm) is shown. Bars are the means ± standard error. Values with different letters are significantly different $p < 0.05$. Small letters show the difference between treatment-pathogen and control (LSD with Bonferroni correction); capital letters indicate the differences within pathogen (Duncan's test).

Table 2. Antimicrobial activity (inhibition zone expressed in mm) of the CE at different pH against indicator bacteria.

Samples	Indicator Strains	Diameter of the Inhibition Zone (mm)				
		pH				Control CE (No Treatment)
		2.0	4.0	6.0	8.0	
Fa17.2	*S. aureus* ATCC1026	16.1 ± 0.1^{eA}	14.6 ± 1.0^{dC}	9.3 ± 1.0^{abD}	8.1 ± 0.1^{bE}	15.3 ± 0.6^{B}
	S. dysenteriae UTNFa37-1	18.3 ± 0.6^{cA}	15.1 ± 0.1^{cdB}	9.1 ± 0.1^{bC}	8.1 ± 0.1^{bD}	15.1 ± 0.2^{B}
	K. cowanii B2Sh1	16.1 ± 0.1^{eA}	14.6 ± 1.0^{dB}	9.1 ± 0.1^{bD}	8.1 ± 0.1^{bE}	13.7 ± 0.6^{C}
	E. coli ATCC25922	15.3 ± 0.6^{fA}	14.1 ± 0.1^{dB}	9.7 ± 1.0^{aD}	9.1 ± 0.1^{abD}	13.3 ± 0.1^{C}
LP	*S. aureus* ATCC1026	17.1 ± 0.1^{dA}	16.1 ± 0.1^{bB}	9.7 ± 1.0^{aC}	9.3 ± 0.6^{aC}	15.7 ± 0.6^{B}
	S. dysenteriae UTNFa37-1	26.1 ± 0.6^{aA}	17.3 ± 1.0^{aB}	9.7 ± 1.0^{aD}	8.1 ± 0.1^{bE}	16.1 ± 0.1^{C}
	K. cowanii B2Sh1	20.7 ± 0.6^{bA}	15.3 ± 1.0^{cB}	9.1 ± 0.1^{bC}	8.1 ± 0.1^{bD}	15.1 ± 0.1^{B}
	E. coli ATCC25922	21.7 ± 1.0^{bA}	15.3 ± 0.1^{cB}	9.1 ± 0.1^{bD}	7.1 ± 0.1^{cE}	12.3 ± 0.6^{C}

Data are mean ± standard error. Values in the same column with small different letters are statistically different ($p < 0.05$). Values in the same row with capital letters are significantly different versus control (no treatment).

3.3. Molecular Weight Estimation of BLIS Substances

Members of the genus *Bacillus* are known to produce a wide arsenal of antimicrobial substances, including peptide and lipopeptide antibiotics, and bacteriocins [49]. Some bacteriocins were earlier characterized [50]. The molecular weight of the band in Tricine-SDS-PAGE was estimated to be about 14 kDa (Figure 7). The size was larger than previously characterized *B. subtilis* L-Q11 of 3.5 kDa [51] and 5 kDa of *B. subtilis* RLID 12.1 [18]. Bacteriocins with antimicrobial activity, greater than 10 kDa of class III, were already detected in different *Bacillus* species [49]. For example, baciamin, bacisubin, CAMT2, and Bac14B showed high antifungal and antimicrobial activity [52–55]. Further biochemical and molecular characterization after complete purification will be undertaken.

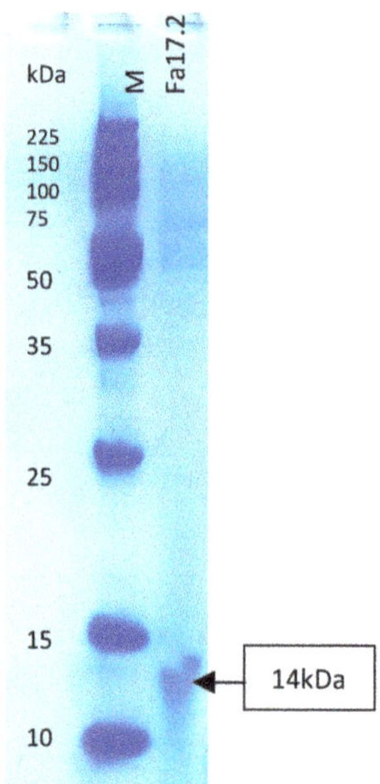

Figure 7. Tricine-SDS-PAGE of the partial purified bacteriocin of Fa17.2. Legend: M: molecular marker (low molecular range marker Promega); Fa17.2 purified peptide extract from *B. subtilis* Fa17.2 strain.

4. Conclusions

To the best of our knowledge, this is the first study showing the presence of *B. subtilis* in tropical *Bromelia* sp. inflorescence showing strain-specific probiotic and antimicrobial strength. The Fa17.2 strain exerted high tolerance to artificial gastric acid, bile, and sodium chloride and was sensible to various antibiotics. The selected strain tolerates high temperatures thus, unlike many other probiotic strains can resist food processing. To a lesser extent, the results were superior to the probiotic LA strain. The Fa17.2 strain was considered safe as no hemolysis was detected in sheep blood agar. In addition, it generates highly thermostable antimicrobials with characteristics very similar to the bacteriocinogenic LP strain. The CE produced by the selected strain was effective against several indicator microorganisms including Gram-negative bacteria, thus using CE as antimicrobial components can be an interesting plan to combat spoilage microorganisms in foods. The antimicrobial activity was stimulated by heat and remains active over a wide pH range. The molecular weight of the partially purified BLIS was about 14 kDa, but more research is required to determine its chemical composition. The present characterization revealed interesting properties of *B. subtilis* strain Fa17.2 with potential applications for biological control of pathogenic strains. Additional research should be aimed at identifying the molecular mechanism of pathogen inhibition. Similarly, the selected microbiome associated with such a microenvironment (tropical flowers and fruits) must be further tailored as a unique probiotic consortium inoculum harboring several species that can confer multifunctional characteristics on the raw matrix with which they interact. Taken together, the data obtained from this research represent the starting point of a demanding study aimed to select beneficial microorganisms from native raw materials that should be further exploited in the food market as an innovative strategy to maintain or improve the quality of the products, guaranteeing food security in the region, while at the same time implementing sustainable solutions in developing countries.

Supplementary Materials: The following are available online at: https://www.mdpi.com/article/10.3390/microorganisms10050860/s1, Table S1. Cell viability (log CFU/mL) registered during incubation with the gastric juice at different pH and 0.3% bile (w/v, oxgall). Table S2. The cell viability (log CFU/mL) was registered during incubation with NaCl at different concentrations and temperatures for 24 h. Table S3. Antibiotic susceptibility of the B. subtilis Fa17.2 Table S4. Diameter of the zone of inhibition (mm) produced by CE of Fa17.2 and LP in different media against S. aureus ATCC1026. Table S5. Diameter of the zone of inhibition (mm) produced by CE of Fa17.2 and LP in different media against E. coli ATCC25922. Table S6. Diameter of the zone of inhibition (mm) produced by CE of Fa17.2 and LP in different media enriched with different carbon sources against K. cowanii B2Sh1. Table S7. Diameter of the zone of inhibition (mm) produced by CE of Fa17.2 and LP in different media against S. dysenteriae UTNFa37-1.

Author Contributions: G.N.T., G.L.G. and J.L.M. contributed to the formal investigation and analysis. G.L.G. and J.L.M. performed the statistics. G.N.T. contributes to conceptualization, methodology, data curation, supervision, project administration, funding acquisition, writing, review, and editing. All authors have read and agreed to the published version of the manuscript.

Funding: This research was funded by Technical University of the North, Grant No. 2929/2019.

Institutional Review Board Statement: Not applicable.

Informed Consent Statement: Not applicable.

Data Availability Statement: Not applicable.

Acknowledgments: The authors are thankful to Technical University of the North for funding this research.

Conflicts of Interest: The authors declare no conflict of interest.

References

1. Prestinaci, F.; Pezzotti, P.; Pantosti, A. Antimicrobial resistance: A global multifaceted phenomenon. *Pathog. Glob. Health* **2015**, *109*, 309–318. [CrossRef]
2. Ministry of Public Health, Ecuador. SIVE Surveillance Subsystem—Water and Foodborne Diseases Alert Ecuador, 2021, SE 18. Available online: https://www.salud.gob.ec/wp-content/uploads/2021/05/Etas-SE-18.pdf (accessed on 10 February 2021).
3. Tenea, G.N.; Olmedo, D. Antimicrobial cocktail combining specific peptide extracts from native probiotic bacteria hamper adulteration of ready-to-eat mango wedges. *Appl. Sci.* **2021**, *11*, 2246. [CrossRef]
4. FAO; WHO. *Probiotics in Food: Health and Nutritional Properties and Guidelines for Evaluation*; FAO Food and Nutrition Paper; World Health Organization and Food Agriculture Organization of the United Nations: Rome, Italy, 2006.
5. Capurso, L. Thirty years of *Lactobacillus rhamnosus* GG: A Review. *J. Clin. Gastroenterol.* **2019**, *53*, S1–S41. [CrossRef] [PubMed]
6. Park, K.M.; Yoon, S.G.; Choi, T.H.; Kim, H.J.; Park, K.J.; Koo, M. The bactericidal effect of a combination of food-grade compounds and their application as alternative antibacterial agents for food contact surfaces. *Foods* **2020**, *9*, 59. [CrossRef] [PubMed]
7. Kimelman, H.; Shemesh, M. Probiotic bifunctionality of *Bacillus subtilis*—Rescuing lactic acid bacteria from desiccation and antagonizing pathogenic *Staphylococcus aureus*. *Microorganisms* **2019**, *7*, 407. [CrossRef] [PubMed]
8. Urdaci, M.C.; Pinchuk, I. Antimicrobial activity of *Bacillus* probiotics. In *Bacterial Spore Formers–Probiotics and Emerging Applications*; Horizon Bioscience: Norfolk, UK, 2004; pp. 171–182.
9. Su, Y.; Liu, C.; Fang, H.; Zhang, D. Bacillus subtilis: A universal cell factory for industry, agriculture, biomaterials and medicine. *Microb. Cell Fact.* **2020**, *19*, 173. [CrossRef] [PubMed]
10. Kim, B.J.; Hong, J.H.; Jeong, Y.S.; Jung, H.K. Evaluation of two *Bacillus subtilis* strains isolated from Korean fermented food as probiotics against loperamide-induced constipation in mice. *J. Korean Soc. Appl. Biol. Chem.* **2014**, *57*, 797–806. [CrossRef]
11. Lu, Z.; Guo, W.; Liu, C. Isolation, identification and characterization of novel *Bacillus subtilis*. *J. Vet. Med. Sci.* **2018**, *80*, 427–433. [CrossRef]
12. Luo, X.Z.; Shi, B.H.; Zheng, H.; Wei-Bin, W.U.; Shi, Q.Q. Physical and chemical properties and antimicrobial spectrum of subtilin from *Bacillus subtilis* FB123. *J. Microbiol.* **2008**, *28*, 64–67.

13. Foster, L.M.; Tompkins, T.A.; Dahl, W.J. A comprehensive post-market review of studies on a probiotic product containing *Lactobacillus helveticus* R0052 and *Lactobacillus rhamnosus* R0011. *Benef. Microb.* **2011**, *2*, 319–334. [CrossRef]
14. Bernardeau, M.; Lehtinen, M.J.; Forssten, S.D.; Nurminen, P. Importance of the gastrointestinal life cycle of *Bacillus* for probiotic functionality. *J. Food Sci. Technol.* **2017**, *54*, 2570–2584. [CrossRef]
15. Cutting, S.M. *Bacillus* probiotics. *Food Microbiol.* **2011**, *28*, 214–220. [CrossRef]
16. Ilinskaya, O.N.; Ulyanova, V.V.; Yarullina, D.R.; Gataullin, I.G. Secretome of intestinal *Bacilli*: A natural guard against pathologies. *Front Microbiol.* **2017**, *8*, 1666. [CrossRef]
17. Lefevre, M.; Racedo, S.M.; Denayrolles, M.; Ripert, G.; Desfougères, T.; Lobach, A.R.; Simon, R.; Pélerin, F.; Jüsten, P.; Urdaci, M.C. Safety assessment of *Bacillus subtilis* CU1 for use as a probiotic in humans. *Regul. Toxicol. Pharmacol.* **2017**, *83*, 54–65. [CrossRef]
18. Ramachandran, R.; Chalasani, A.G.; Lal, R.; Roy, U. A broad-spectrum antimicrobial activity of *Bacillus subtilis* RLID12.1. *Sci. World J.* **2014**, *2014*, 968487. [CrossRef]
19. Stein, T. *Bacillus subtilis* antibiotics: Structures, syntheses and specific functions. *Mol. Microbiol.* **2005**, *56*, 845–857. [CrossRef]
20. Commercial Bulletin. Ministry of Production, Foreign Trade, Investment and Fisheries 2021, Twelfth Edition. 2021. Available online: https://www.produccion.gob.ec/wp-content/uploads/2021/12/VFBoletinComercioExteriorDiciembre2021-final.pdf?fbclid=IwAR3dpQeWuwUgQF8IOVsb7SSmLCbMPBHZkKhpxXqMA7HPEY61wWhN (accessed on 10 January 2022).
21. Benavidez, A.B.; Ulcuango, M.; Yepez, L.; Tenea, G.N. Assessment of the in vitro bioactive properties of lactic acid bacteria isolated from native ecological niches of Ecuador. *Rev. Arg. Microbiol.* **2016**, *48*, 236–244. [CrossRef]
22. Corcoran, B.M.; Stanton, C.; Fitzgerald, G.F.; Ross, R.P. Survival of probiotic lactobacilli in acidic environments is enhanced in the presence of metabolizable sugars. *Appl. Environ. Microbiol.* **2005**, *71*, 3060–3067. [CrossRef]
23. Yasmin, I.; Saeed, M.; Khan, W.A.; Khaliq, A.; Chughtai, M.; Iqbal, R.; Tehseen, S.; Naz, S.; Liaqat, A.; Mehmood, T.; et al. In vitro probiotic potential and safety evaluation (hemolytic, cytotoxic activity) of *Bifidobacterium* strains isolated from raw camel milk. *Microorganisms* **2020**, *8*, 354. [CrossRef]
24. EFSA. Panel on additives and products or substances used in animal feeds. Guidance on the assessment of bacterial susceptibility to antimicrobials of human and veterinary importance. *EFSA J.* **2012**, *10*, 2740. [CrossRef]
25. Garzon, K.; Ortega, C.; Tenea, G.N. Characterization of bacteriocin-producing lactic acid bacteria isolated from native fruits of Ecuadorian Amazon. *Pol. J. Microbiol.* **2017**, *66*, 473–481. [CrossRef]
26. Waterhouse, A.M.; Procter, J.B.; Martin, D.M.; Clamp, M.; Barton, G.J. Jalview version 2-a multiple sequence alignment editor and analysis workbench. *Bioinformatics* **2009**, *25*, 1189–1191. [CrossRef]
27. Lee, N.K.; Han, K.J.; Son, S.H.; Eom, S.J.; Lee, S.K.; Paik, H.D. Multifunctional effect of probiotic *Lactococcus lactis* KC24 isolated from kimchi. *LWT-Food Sci. Technol.* **2015**, *64*, 1036–1041. [CrossRef]
28. Patel, A.K.; Ahire, J.J.; Pawar, S.P.; Chaudhari, B.L.; Chincholkar, S.B. Comparative accounts of probiotic characteristics of *Bacillus* spp. isolated from food wastes. *Food Res. Int.* **2009**, *42*, 505–510. [CrossRef]
29. Jeon, H.L.; Lee, N.K.; Yang, S.J.; Kim, W.S.; Paik, H.D. Probiotic characterization of *Bacillus subtilis* P223 isolated from kimchi. *Food Sci. Biotech.* **2017**, *26*, 1641–1648. [CrossRef]
30. Bevilacqua, A.; Altieri, C.; Carbo, M.R.; Sinigaglia, M.; Ouoba, L.I.I. Characterization of lactic acid bacteria isolated from Italian Belladi Cerignola table olives: Selection of potential multifunctional starter cultures. *J. Food Sci.* **2010**, *25*, 536–544. [CrossRef]
31. Gueimonde, M.; Sánchez, B.G.; de Los Reyes-Gavilán, C.; Margolles, A. Antibiotic resistance in probiotic bacteria. *Front. Microbiol.* **2013**, *4*, 202. [CrossRef]
32. Lee, N.K.; Kim, W.S.; Paik, H.D. *Bacillus* strains as human probiotics: Characterization, safety, microbiome, and probiotic carrier. *Food Sci. Biotechnol.* **2019**, *28*, 1297–1305. [CrossRef]
33. Sarkar, P.K.; Hasenack, B.; Nout, M.J. Diversity and functionality of *Bacillus* and related genera isolated from spontaneously fermented soybeans (Indian Kinema) and locust beans (African Soumbala). *Int. J. Food Microbiol.* **2002**, *77*, 175–186. [CrossRef]
34. Hong, H.A.; Duc, L.H.; Cutting, S.M. The use of bacterial spore formers as probiotics. *FEMS Microbiol. Rev.* **2005**, *29*, 813–835. [CrossRef]
35. EFSA. Scientific opinion on the maintenance of the list of QPS biological agents intentionally added to food and feed. *EFSA J.* **2012**, *10*, 3020. [CrossRef]
36. Khardziani, T.; Kachlishvili, E.; Sokhadze, K.; Elisashvili, V.; Weeks, R.; Chikindas, M.L.; Chistyakov, V. Elucidation of *Bacillus subtilis* KATMIRA 1933 potential for spore production in submerged fermentation of plant raw materials. *Probiotics Antimicrob. Proteins* **2017**, *9*, 435–443. [CrossRef] [PubMed]
37. Monteiro, S.M.; Clemente, J.J.; Carrondo, M.J.; Cunha, A. Enhanced spore production of *Bacillus subtilis* grown in a chemically defined medium. *Ai Mag.* **2014**, *4*, 444–454. [CrossRef]
38. Khochamit, N.; Siripornadulsil, S.; Sukon, P.; Siripornadulsil, W. Antibacterial activity and genotypic-phenotypic characteristics of bacteriocin-producing *Bacillus subtilis* KKU213: Potential as a probiotic strain. *Microbiol. Res.* **2015**, *170*, 36–50. [CrossRef] [PubMed]
39. Xie, J.; Zhang, R.; Shang, C.; Guo, Y. Isolation and characterization of a bacteriocin produced by an isolated *Bacillus subtilis* LFB112 that exhibits antimicrobial activity against domestic animal pathogens. *Afr. J. Biotechnol.* **2009**, *8*, 5611–5619.
40. Yi, Y.; Zhang, Z.; Zhao, F.; Liu, H.; Yu, L.; Zha, J.; Wang, G. Probiotic potential of *Bacillus velezensis* JW: Antimicrobial activity against fish pathogenic bacteria and immune enhancement effects on *Carassius auratus*. *Fish Shellfish Immunol.* **2018**, *78*, 322–330. [CrossRef]

41. Sharma, A.; Satyanarayana, T. Comparative genomics of *Bacillus* species and its relevance in industrial microbiology. *Genom. Insights* **2013**, *6*, 25–36. [CrossRef]
42. Hiramoto, S.; Itoh, K.; Shizuuchi, S.; Kawachi, Y.; Morishita, Y.; Nagase, M.; Suzuki, Y.; Nobuta, Y.; Sudou, Y.; Nakamura, O.; et al. Melanoidin, a food protein-derived advanced Maillard reaction product, suppresses *Helicobacter pylori* in vitro and in vivo. *Helicobacter* **2004**, *9*, 429–435. [CrossRef]
43. Kukuminato, S.; Koyama, K.; Koseki, S. Antibacterial properties of melanoidins produced from various combinations of Maillard reaction against pathogenic bacteria. *Microbiol. Spectr.* **2021**, *9*, e0114221. [CrossRef]
44. Banerjee, S.P.; Dora, K.C.; Chowdhury, S. Detection, partial purification and characterization of bacteriocin produced by *Lactobacillus brevis* FPTLB3 isolated from freshwater fish: Bacteriocin from *Lb. brevis* FPTLB3. *J. Food Sci. Technol.* **2013**, *50*, 17–25. [CrossRef]
45. Arena, M.P.; Silvain, A.; Normanno, G.; Grieco, F.; Drider, D.; Spano, G.; Fiocco, D. Use of *Lactobacillus plantarum* strains as a bio-control strategy against food-borne pathogenic microorganisms. *Front. Microbiol.* **2016**, *7*, 464. [CrossRef]
46. Alakomi, H.L.; Skyttä, E.; Saarela, M.; Mattila-Sandholm, T.; Latva-Kala, K.; Helander, I.M. Lactic acid permeabilizes gram-negative bacteria by disrupting the outer membrane. *Appl. Environ. Microbiol.* **2000**, *66*, 2001–2005. [CrossRef]
47. Goyal, C.; Malik, R.K.; Pradhan, D. Purification and characterization of a broad spectrum bacteriocin produced by a selected *Lactococcus lactis* strain 63 isolated from Indian dairy products. *J. Food Sci. Technol.* **2018**, *55*, 3683–3692. [CrossRef]
48. Gálvez, A.; Abriouel, H.; López, R.L.; Ben Omar, N. Bacteriocin-based strategies for food biopreservation. *Int. J. Food Microbiol.* **2007**, *120*, 51–70. [CrossRef]
49. Abriouel, H.; Franz, C.M.; Ben Omar, N.; Gálvez, A. Diversity and applications of *Bacillus* bacteriocins. *FEMS Microbiol. Rev.* **2011**, *35*, 201–232. [CrossRef]
50. Fuchs, S.W.; Jaskolla, T.W.; Bochmann, S.; Kötter, P.; Wichelhaus, T.; Karas, M.; Stein, T.; Entian, K.D. Entianin, a novel subtilin-like lantibiotic from *Bacillus subtilis* subsp. *spizizenii* DSM 15029T with high antimicrobial activity. *Appl. Environ. Microbiol.* **2011**, *77*, 1698–1707. [CrossRef]
51. Qin, Y.; Wang, Y.; He, Y.; Zhang, Y.; She, Q.; Chai, Y.; Li, P.; Shang, Q. Characterization of subtilin L-Q11, a novel class I bacteriocin synthesized by *Bacillus subtilis* L-Q11 isolated from orchard soil. *Front. Microbiol.* **2019**, *10*, 484. [CrossRef]
52. Hammami, I.; Jaouadi, B.; Bacha, A.B.; Rebai, A.; Bejar, S.; Nesme, X.; Rhouma, A. *Bacillus subtilis* bacteriocin Bac 14B with a broad inhibitory spectrum: Purification, amino acid sequence analysis, and physicochemical characterization. *Biotechnol. Bioprocess Eng.* **2012**, *17*, 41–49. [CrossRef]
53. Wong, J.H.; Hao, J.; Cao, Z.; Qiao, M.; Xu, H.; Bai, Y.; Ng, T.B. An antifungal protein from *Bacillus amyloliquefaciens*. *J. Appl. Microbiol.* **2008**, *105*, 1888–1898. [CrossRef]
54. An, J.; Zhu, W.; Liu, Y.; Zhang, X.; Sun, L.; Hong, P.; Wang, I.; Xu, C.; Xu, D.; Liu, H. Purification and characterization of a novel bacteriocin CAMT2 produced by *Bacillus amyloliquefaciens* isolated from marine fish *Epinephelus areolatus*. *Food Control.* **2015**, *51*, 278–282. [CrossRef]
55. Liu, Y.; Chen, Z.; Ng, T.B.; Zhang, J.; Zhou, M.; Song, F.; Lu, F.; Liu, Y. Bacisubin, an antifungal protein with ribonuclease and hemagglutinating activities from *Bacillus subtilis* strain B-916. *Peptides* **2007**, *28*, 553–559. [CrossRef] [PubMed]

microorganisms

MDPI

Article

Leuconostoc mesenteroides Strains Isolated from Carrots Show Probiotic Features

Emily Schifano [1], Alberta Tomassini [2,3], Adele Preziosi [1], Jorge Montes [1], Walter Aureli [4], Patrizia Mancini [5], Alfredo Miccheli [3,6] and Daniela Uccelletti [1,*]

1 Department of Biology and Biotechnology "C. Darwin", Sapienza University of Rome, Piazzale Aldo Moro 5, 00185 Rome, Italy; emily.schifano@uniroma1.it (E.S.); adele.preziosi@uniroma1.it (A.P.); jmontes1677@gmail.com (J.M.)
2 Department of Physiology and Pharmacology "Vittorio Erspamer", Sapienza University of Rome, 00185 Rome, Italy; alberta.tomassini@uniroma1.it
3 NMR-Based Metabolomics Laboratory (NMLab), Sapienza University of Rome, Piazzale Aldo Moro 5, 00185 Rome, Italy; alfredo.miccheli@uniroma1.it
4 R&D, Aureli Mario S.S. Agricola, Via Mario Aureli 7, 67050 Ortucchio, Italy; produzione@aurelimario.com
5 Department of Experimental Medicine, University of Rome, Viale Regina Elena 324, 00161 Rome, Italy; patrizia.mancini@uniroma1.it
6 Department of Environmental Biology, Sapienza University of Rome, Piazzale Aldo Moro 5, 00185 Rome, Italy
* Correspondence: daniela.uccelletti@uniroma1.it

Abstract: Lactic acid bacteria (LAB) share several beneficial effects on human organisms, such as bioactive metabolites' release, pathogens' competition and immune stimulation. This study aimed at determining the probiotic potential of autochthonous lactic acid bacteria isolated from carrots. In particular, the work reported the characterization at the species level of four LAB strains deriving from carrots harvested in Fucino highland, Abruzzo (Italy). Ribosomal 16S DNA analysis allowed identification of three strains belonging to *Leuconostoc mesenteroides* and a *Weissella soli* strain. In vitro and in vivo assays were performed to investigate the probiotic potential of the different isolates. Among them, *L. mesenteroides* C2 and *L. mesenteroides* C7 showed high survival percentages under in vitro simulated gastro-intestinal conditions, antibiotic susceptibly and the ability to inhibit in vitro growth against *Salmonella enterica* serovar *Typhimurium*, *Listeria monocytogenes*, *Pseudomonas aeruginosa* and *Staphylococcus aureus* pathogens. In parallel, the simple model *Caenorhabditis elegans* was used for in vivo screenings. *L. mesenteroides* C2 and *L. mesenteroides* C7 strains significantly induced pro-longevity effects, protection from pathogens' infection and innate immunity stimulation. Overall, these results showed that some autochthonous LAB from vegetables such as carrots have functional features to be considered as novel probiotic candidates.

Keywords: probiotic; carrots; *Caenorhabditis elegans*; pathogen resistance

Citation: Schifano, E.; Tomassini, A.; Preziosi, A.; Montes, J.; Aureli, W.; Mancini, P.; Miccheli, A.; Uccelletti, D. *Leuconostoc mesenteroides* Strains Isolated from Carrots Show Probiotic Features. *Microorganisms* **2021**, *9*, 2290. https://doi.org/10.3390/microorganisms9112290

Academic Editor: Sabina Fijan

Received: 21 October 2021
Accepted: 2 November 2021
Published: 4 November 2021

1. Introduction

The Food and Agriculture Organization of the United Nations (FAO) and the World Health Organization (WHO) defined probiotics as live microorganisms that, when administered in adequate amounts, confer a health benefit on the host [1]. According to this statement, probiotics must be safe, and not exert pathogenic effects or show antibiotic resistance genes that could be transferred. Moreover, probiotic strains should be resistant to gastrointestinal conditions, such as stomach acid pH and bile acids, produce antimicrobial compounds and compete with pathogens by stimulating immunity [2]. Furthermore, probiotics' efficacy should be confirmed in human studies.

Among the various microbial species associated with food, some of them may share probiotic features. The main source of probiotics used in humans is represented by dairy foods, but increasing evidence has highlighted the importance to select probiotics from other sources, such as fresh fruits and vegetables [3]. Indeed, the availability of commercial milk-based products limits their consumption by people who are intolerant or allergic to lactose. Therefore, fruits and vegetables offer healthy alternatives thanks to their large distribution and nutritive value. Among them, carrot (*Daucus carota* L.), as well as being rich in minerals and antioxidants, is reported to be a reservoir of carotenoids, vitamins and fiber [4–6]. Many studies on carrots have focused on cultivation, breeding, tissue culture, nutrient content and carotenoid synthesis regulation, while few works deal with microbial composition in terms of potential probiotic bacteria [7,8]. Indeed, the most common probiotics, isolated from fruits and vegetables, include different strains belonging to the lactic acid bacteria (LAB) group. This heterogeneous group of Gram-positive and non-spore-forming bacteria are normally present in food products, involved in numerous fermentation processes and some of them are widely used in industrial processes [9]. The major representatives of this microorganism group are *Lactobacillus, Streptococcus, Leuconostoc, Pediococcus, Propionibacterium, Enterococcus, Bifidobacterium* and *Weissella* genera [10]. Although the *Weissella genus* is found in multiple habitats, many species were isolated from different foods, such as fermented crop products, meat and fish, along with *Leuconostoc* species. Moreover, many of them produce exopolysaccharides, influencing the adhesion to substrates and affecting the structure of fermented foods.

Since the direct evaluation of probiotic potentials in vivo is often expensive and timewasting, the use of simple and inexpensive model systems is needed. *Caenorhabditis elegans* is a powerful in vivo model to screen for probiotic bacteria. Nematodes feed only on microorganisms, which reach the intestine, influencing nematodes' physiology [11]. Among its many advantages, the possibility to easily monitor anti-aging markers or innate immunity pathways could be used for the screening of microorganisms to identify new probiotic strains and to explore the possible molecular pathways involved. Indeed, several foodborne LAB were reported to exert positive effects in worms, and the mechanisms correlated with innate immunity and lifespan extension have been elucidated [12]. Recently, different *Lactobacillus* strains, isolated from vegetables or dairy products, were reported to increase nematode viability, delay the aging process and protect against foodborne *S. enterica* serovar typhimurium LT2 or *L. monocytogenes* OH pathogens [13,14]. Moreover, it has been demonstrated that *Bifidobacterium* isolates can also exert beneficial effects on *C. elegans* health and lifespan [15].

This study aimed at determining the probiotic potential of four lactic acid bacteria strains isolated from carrots. Tolerance to gastrointestinal conditions, antibiotic susceptibility and antagonism toward human pathogenic microorganisms were evaluated in vitro. The different isolates were tested in vivo using the *C. elegans* animal model to analyze possible beneficial effects on worm lifespan, gut colonization, the aging process and pathogen resistance.

2. Materials and Methods

2.1. Species Isolation and Identification

Carrots were provided by Aureli Mario S.S. Agricola (Ortucchio, AQ, Italy). The carrot cultivar (*Daucus carota* L., Nantese Dordogne, Syngenta seeds) was grown in Fucino highland (Abruzzo, Italy) and harvested at commercial maturity, as indicated by the supplier's geneticists. Epidermis and shallow flesh of five carrots (about 20 g) were homogenized with mortar under aseptic conditions and diluted in sterile H_2O_{dd}. Dilution aliquots were plated on De Man Rogosa Sharpe (MRS) medium for 24–48 h at 30 °C, anaerobically. After that, morphologically different colonies were streaked on new MRS plates and grown at 30 °C to isolate purified strains. Each strain was then inoculated in MRS broth anaerobically and, after growth, a stock at −80 °C was carried out.

For bacterial identification, DNA was extracted and amplified according to Schifano et al. [16]. The primer pairs F8 (5′-AGAGTTTGATCCTGGCTCAG-3′) and R1492 (5′-GGTTACCTTGTTACGACTT-3′) were used to amplify the 16S rDNA region of LAB isolates. FASTA sequences of the amplified region from each LAB isolate were submitted to GenBank, and the associated accession numbers are reported in the Results Section.

2.2. Growth Conditions of Bacterial Isolates

Bacterial strains isolated from carrots and used in this study were *Leuconostoc mesenteroides* C1, *L. mesenteroides* C2, *L. mesenteroides* C7 and *Weissella soli* T4. The LAB strains described in this work were grown in MRS medium at 30 °C under anaerobic conditions. Commercial probiotic strain *L. rhamnosus* GG ATCC® 53103™ (LGG), used as the LAB reference strain, was grown at 37 °C anaerobically. For *C. elegans* experiments, *Escherichia coli* OP50 strain was used as standard food. For in vitro and in vivo resistance to pathogens, *Pseudomonas aeruginosa* ATCC 15692, *Staphylococcus aureus* ATCC 25923, *S. enterica* serovar typhimurium LT2 and *L. monocytogenes* OH were used. *E. coli* OP50 and pathogen strains were grown in Luria-Bertani (LB) broth at 37 °C overnight, under shaking.

2.3. Resistance to Lysozyme, Acid pH and Bile Salts

Isolates were grown in MRS broth overnight at 30 °C. For the lysozyme tolerance assay, 10 mL of overnight culture was centrifuged at 5000 rpm at 4 °C and suspended in the same volume of SES buffer (0.22 g/L $CaCl_2$, 6.2 g/L NaCl, 2.2 g/L KCl, 1.2 g/L $NaHCO_3$) containing 0.1 mg/mL of lysozyme (Sigma-Aldrich, St. Louis, MO, USA) [17]. After 30 min and 2 h of incubation at 37 °C, 100 μL of each suspension was plated on MRS agar plates and further incubated at 37 °C for 24 h, under anaerobic conditions. SES without lysozyme was used as a control.

The acid tolerance assay was performed according to [18], with some modifications. A 1 mL aliquot of overnight culture (10^9 cfu/mL) was inoculated into 10 mL of sterile phosphate-buffered saline (9 g/L NaCl, 9 g/L Na_2HPO_4 $2H_2O$, 1.5 g/L KH_2PO_4) adjusted to pH 2.5 and pH 3.0 with 8M HCl. pH 5 buffer was used as a control. The tubes were incubated at 37 °C and the viable organisms were recovered after 3 h of incubation on MRS agar incubated for 48 h at 30 °C. For resistance to bile salts, the same protocol was performed, using phosphate-buffered saline with 0.3% bile salts (Sigma-Aldrich).

The viability was measured as percent viability = $[(CFU_{treated}/mL)/(CFU_{untreated}/mL)] \times 100$. The untreated value corresponds to plate counts of inoculated bacteria in control phosphate-buffered saline, and the treated value corresponds to the bacterial counts obtained after incubation in simulated GI conditions.

2.4. Antibiotic Resistance

For the susceptibility test, antibiotic discs (Biolab Zrt., Budapest, Hungary) were used. The experiment was performed as described in [13]. Briefly, 100 μL of overnight cultures of different isolates or LGG were plated onto MRS agar plates; then, the antibiotic discs were gently placed on the plates and incubated under anaerobic conditions for 24 h at 37 °C. The zones of inhibition were measured from the center of the disc, recorded and compared with those of the reference strain.

2.5. Antimicrobial Activity

The agar diffusion test was performed using, as indicator strains, *P. aeruginosa* ATCC 15692, *S. aureus* ATCC 25923, *S. enterica* serovar typhimurium LT2 and *L. monocytogenes* OH. To evaluate the antagonistic activity of LAB isolates against the different pathogens, 100 μL each of LAB overnight cultures was spotted onto MRS agar and coated with 5 mL of TSA soft agar (0.7%), previously inoculated with 500 μL of each pathogen indicator strain. Plates were incubated at 37 °C for 24 h. The antagonist activity was recorded as the diameter (mm) of growth inhibition halo around each spot.

2.6. *C. elegans* Strain and Lifespan Assay

The wild-type *C. elegans* strain, Bristol N2, was grown at 16 °C on Nematode Growth Medium (NGM) plates plated with *E. coli* OP50. Fertile N2 adults were placed to lay embryos for 8 h on peptone-free NGM plates, plated with LAB strains, LGG or *E. coli* OP50, and then sacrificed. For the preparation of the bacterial lawns, overnight cultures were centrifuged for 15 min at 6000 rpm. The pellet was weighed and suspended in M9 buffer in order to obtain a final concentration of 400 mg/mL. Then, 25 µL of each type of bacterial lawn was plated on mNGM, as described in [19]. When the progeny became fertile (t0), 60 worms per condition were transferred to new plates plated with fresh bacterial cultures and monitored daily. A worm was considered dead when it did not respond to touch.

2.7. *Fertility Assay*

As described in [20], synchronized worms obtained as above were incubated at 16 °C on mNGM plates plated with different strains, allowing embryo laying. Animals were transferred onto new plates every day and the number of progeny was documented until the mother worms became infertile.

2.8. *Colonization Assay of* C. elegans *Gut*

For each condition, 10 L4 larvae or 8-day-old adults were washed in M9 buffer and lysed, as described in [21]. Whole worm lysates and serial dilutions were plated onto MRS-agar plates. The number of colony-forming units (CFU) was counted after 24 h of incubation at 37 °C, anaerobically. Instead, OP50-fed worm lysates were plated onto LB-agar and incubated at 37 °C.

2.9. *Aging Markers' Analysis*

For the pharyngeal pumping rate, the number of grinder contractions was counted under a Zeiss Axiovert 25 microscope in 10-day-old adult animals fed different bacteria from embryo hatching. Ten worms were analyzed for each treatment, during a period of 30 s. The locomotion ability of nematodes was analyzed by body bending counting after 30 s. In particular, as described in [14], 10 worms for each treatment were washed in M9 buffer to remove bacteria, and then placed in 10 µL of M9 buffer to facilitate the locomotion measure. For lipofuscin accumulation analysis, 10-day-old adult worms, after washes in M9 buffer, were placed onto a 3% agar pad containing 20 mM of sodium azide. Afterwards, nematodes were observed with the Axio Observer Z1 inverted microscope, equipped with an ApoTome.2 System (Carl Zeiss Inc., Oberkochen, Germany). Digital images were acquired with the AxioCam MRm high-resolution digital camera (Zeiss) and processed with the AxioVision 4.8.2 software (Zeiss). ApoTome optical sectioning images of animals were recorded under a 40 Å~/0.75 objective (Zeiss). Median fluorescence intensity was analyzed using the ImageJ software, measuring the ratio of pixels per area of the worm.

2.10. *Resistance to Pathogens in* C. elegans

For the killing assay, 35 mm NGM plates were seeded with 60 µL of *L. mesenteroides* C2 or *L. mesenteroides* C7 mixed with different pathogens, in a 1:1 ratio. *Pseudomonas aeruginosa* ATCC 15692 and *Staphylococcus aureus* ATCC 25923 were chosen as representative pathogens of Gram-negative and Gram-positive bacteria, respectively. The assay was performed as described in [22]. Synchronous L4 larvae were placed onto the different co-cultures and incubated at 25 °C. Lifespan was monitored daily and worms fed with pathogen alone were taken as the control. A worm was considered dead when it failed to respond to touch.

2.11. *Real-Time qPCR*

At the stage of 1-day-old adults, 200 worms for each condition were lysed and total RNA was extracted as described in [23]. *pmk-1*, *skn-1*, *daf-16*, *sod-3* and *hsf-1* mRNA levels were analyzed. The differences between the mean CT value of each sample and the CT value of the housekeeping gene (*act-1*) were calculated. Primers used in this study are reported in Table 1. The experiment was carried out in triplicate.

Table 1. Primers for real-time qPCR analysis.

hsf-1	FOR REV	5′-ATGACTCCACTGTCCCAAGG 5′-TCTTGCCGATTGCTTTCTCT
pmk-1	FOR REV	5′-AAATGACTCGCCGTGATTTC 5′-CATCGTGATAAGCAGCCAGA
sod-3	FOR REV	5′-AGAACCTTCAAAGGAGCTGATG 5′-CCGCAATAGTGATGTCAGAAAG
act-1	FOR REV	5′-GAGCGTGGTTACTCTTTCA 5′-CAGAGCTTCTCCTTGATGTC
skn-1	FOR REV	5′-GTTCCCAACATCCAACTACG 5′-TGGAGTCTGACCAGTGGATT
daf-16	FOR REV	5′-TCAAGACCTCAAAGCCAATCAACTC 5′-ACGAGAAAGAAGGAGTAAGAGGAGG

2.12. Statistical Analysis

All experiments were performed at least in triplicate. Data are presented as mean ± SD. The statistical significance was determined by Student's *t* test or one-way ANOVA coupled with a Bonferroni post-test (GraphPad Prism 5.0 software, GraphPad Software Inc., La Jolla, CA, USA). Differences with *p*-values < 0.05 were considered significant and were indicated as follows: * $p < 0.05$, ** $p < 0.01$ and *** $p < 0.001$.

3. Results

3.1. Bacteria Isolation

In this study, potential probiotic properties of four bacterial isolates were evaluated in vitro and in vivo. Different bacterial colonies were isolated from carrots and identified at the molecular level by the amplification of 16S rDNA. The sequences obtained from sequencing were compared with those in the BLAST database, so that three strains belonging to *Leuconostoc mesenteroides* and a *Weissella soli* strain were identified. The Gram-positive strains isolated were identified as *L. mesenteroides* C1 (accession number OK513088), *L. mesenteroides* C2 (accession number OK513089), *L. mesenteroides* C7 (accession number OK513090) and *Weissella soli* T4 (accession number OK513091).

3.2. In Vitro Tests

3.2.1. Resistance to Lysozyme, Low pH and Bile Salts

To perform a selection of possible probiotic candidates, their resistance to the extreme conditions of the gastrointestinal tract, such as low pH in stomach and bile in the upper intestine, was evaluated. The high concentration of lysozyme present in the mouth represents the first barrier. Figure 1A reports bacteria survival data after 30 and 120 min of treatment with 1 mg/mL of lysozyme. All LAB strains showed high resistance to lysozyme after 30 min of incubation, with percentages of survival of 100%. Notably, *L. mesenteroides* C2 was able to resist even after 120 min of incubation with lysozyme, which can be considered a severe treatment. On the other hand, *W. soli* T4 and *L. mesenteroides* C7 showed a percent survival of 60% and 80% respectively, after 2 h of incubation. *L. mesenteroides* C1, instead, did not resist lysozyme treatment. Tolerance to low pH conditions was performed to simulate microbial flux along the mammalian gastrointestinal tract. As shown in Figure 1B, 3 h of incubation in pH 2.5 exerted a strong reduction (about 80%) of *L. mesenteroides* C1, *L. mesenteroides* C7 and *W. soli* T4 counts, as compared to the control. This result suggested that these strains were not able to endure acidic environments. Notably, *L. mesenteroides* C2 cell recovery showed a higher ability to survive in low pH conditions, similarly to the probiotic LGG strain. Similar results were obtained after testing the ability of different strains to resist at pH 3.0.

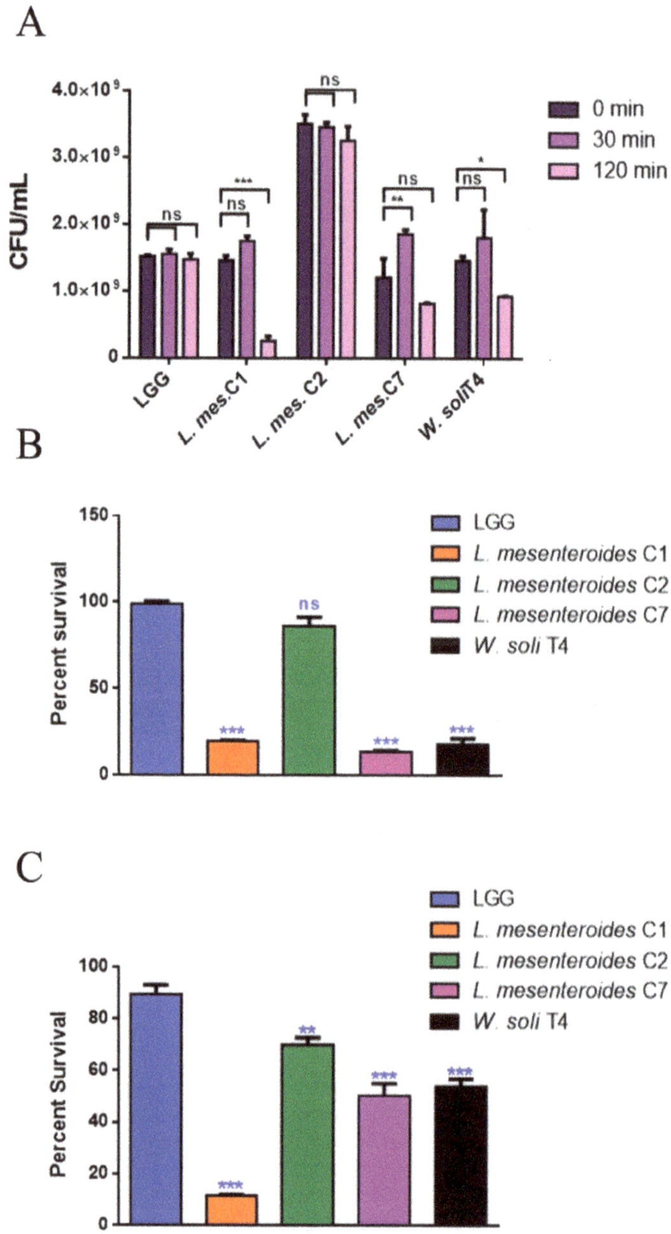

Figure 1. In vitro tolerance to lysozyme, pH 2.5 and 0.3% bile salts. (**A**) Cell counts of viable bacteria recovered at the initial time point (t0), following 30 or 120 min of incubation in 1 mg/mL of lysozyme SES buffer. (**B**) Recovery of viable bacteria after 3 h of incubation in phosphate buffer adjusted to pH 2.5 or (**C**) 0.3% bovine bile salts. LGG was taken as the LAB reference strain. Columns represent the mean ± SD of three independent experiments. Statistical analysis was performed by one-way ANOVA, followed by the Bonferroni post-test. Asterisks indicate significant differences (* $p < 0.05$; ** $p < 0.01$; *** $p < 0.001$), ns: not significant.

The human bile concentration ranges from 0.3% to 0.5%, and a high bile tolerance improves probiotics' colonization in the host GI tract [24]. To investigate the ability of the different isolates to survive in the presence of bile, the percent survival in the presence of 0.3% bile was tested. As shown in Figure 1C, *L. mesenteroides* C2, *L. mesenteroides* C7 and *W. soli* T4 strains showed a growth percentage above 50% in the presence of bile. As expected, the reference probiotic strain LGG exhibited a resistance of about 90%. On the other hand, *L. mesenteroides* C1 was not able to resist the treatment, showing a decrease in viability of about 90%.

3.2.2. Antibiotic Susceptibility and Antagonistic Activity to Pathogens

Antibiotic susceptibility was determined by the disc diffusion assay, analyzing a panel of 20 antibiotics. Among them, there are inhibitors of synthesis of cell wall, DNA and RNA, proteins and inhibitors of membrane function. As reported in Table 1, LAB isolates showed an antibiotic susceptibility pattern very similar to that of the LGG control strain. *L. mesenteroides* C2 displayed resistance to only five antibiotics, and in some cases, the inhibition halo was larger as compared to the control (Table 2).

Table 2. Resistance to antibiotics of different isolates. The zones of inhibition were measured from the center of the disc and recorded in mm ± SD. Absence of inhibition halo was indicated as (+), ns: not significant.

Antibiotic	**Amount on Disc (μg)**	**Zone of Inhibition (mm)**					
		LGG	***L. mesenteroides* C1**	***L. mesenteroides* C2**	***L. mesenteroides* C7**	***W. soli* T4**	***p*-Value**
Amikacin	30	4 ± 0.03	5 ± 0.06	10 ± 0.08	4 ± 0.05	4 ± 0.03	$p < 0.001$
Ampicillin	10	7 ± 0.06	7 ± 0.05	6 ± 0.10	5 ± 0.20	5 ± 0.05	$p < 0.01$
Aztreonam	30	+	+	+	+	+	ns
Carbenicillin	100	8 ± 0.08	7 ± 0.03	5 ± 0.09	7 ± 0.03	5 ± 0.08	$p < 0.01$
Cefalotin	30	+	3 ± 0.10	+	4 ± 0.08	+	$p < 0.01$
Cefotaxime	30	7 ± 0.08	+	7 ± 0.20	+	+	$p < 0.001$
Cefuroxime	30	5 ± 0.02	+	12 ± 0.08	+	+	$p < 0.001$
Clindamycin	2	8 ± 0.12	9 ± 0.35	4 ± 0.08	10 ± 0.10	9 ± 0.30	$p < 0.01$
Chloramphenicol	30	8 ± 0.08	8 ± 0.35	14 ± 0.10	8 ± 0.20	10 ± 0.15	$p < 0.01$
Erythromycin	15	8 ± 0.03	7 ± 0.08	8 ± 0.15	7 ± 0.09	8 ± 0.08	ns
Fosfomycin	50	+	+	+	+	+	ns
Gentamicin	10	4 ± 0.05	5 ± 0.12	+	5 ± 0.15	5 ± 0.10	$p < 0.05$
Mezlocillin	75	14 ± 0.05	10 ± 0.15	13 ± 0.20	8 ± 0.20	11 ± 0.10	$p < 0.001$
Oxacillin	1	+	+	4 ± 0.03	+	+	$p < 0.05$
Penicillin	10	13 ± 0.10	9 ± 0.20	14 ± 0.20	9 ± 0.05	11 ± 0.09	$p < 0.001$
Rinfampicin	30	13 ± 0.02	11 ± 0.09	8 ± 0.10	11 ± 0.08	10 ± 0.08	$p < 0.01$
Streptomycin	25	4 ± 0.12	5 ± 0.08	+	4 ± 0.06	4 ± 0.05	$p < 0.01$
Tetracycline	30	14 ± 0.15	10 ± 0.30	7 ± 0.05	8 ± 0.06	7 ± 0.10	$p < 0.001$
Tobramycin	10	4 ± 0.08	4 ± 0.05	4 ± 0.09	3 ± 0.05	4 ± 0.05	ns
Vancomycin	30	+	+	+	+	+	ns

The antagonistic activity was evaluated through the agar double-layer diffusion test, against four pathogen test strains: the Gram-positive *Staphylococcus aureus* and *Listeria monocytogenes*, and the Gram-negative *Pseudomonas aeruginosa* and *S. enterica* serovar typhimurium LT2. The antagonistic activity was less variable among the different isolates, as shown by inhibition halo diameters on all four pathogen test strains (Table 3). Indeed, the inhibition halo diameters produced by the different isolates were comparable to that of the probiotic strain LGG.

Table 3. Antagonistic activity in vitro. The diameter of inhibition halos was recorded in mm and the data were expressed as average ± SD.

Pathogen	LGG	*L. mesenteroides* C1	*L. mesenteroides* C2	*L. mesenteroides* C7	*W. soli* T4	*p*-Value
S. aureus	40 ± 0.08	38 ± 0.4	35 ± 0.5	35 ± 0.1	35 ± 0.5	$p < 0.01$
L. monocytogenes	30 ± 0.2	29 ± 0.2	30 ± 0.8	31 ± 0.6	30 ± 0.5	ns
P. aeruginosa	40 ± 0.07	38 ± 0.5	33 ± 0.5	40 ± 0.1	35 ± 0.6	$p < 0.01$
S. enterica	30 ± 0.3	31 ± 0.3	30 ± 0.5	28 ± 0.5	30 ± 0.08	ns

3.3. *In Vivo Tests*

3.3.1. Effects on *C. elegans* Lifespan and Colonization Capability

In vivo screening of the four strains was performed in the *C. elegans* model system, to test possible beneficial effects exerted by LAB. For this purpose, worms were separately fed each of the isolated strains starting from embryo hatching, using animals fed LGG or standard *E. coli* OP50 as control populations. Among the tested strains, *L. mesenteroides* C2 and *L. mesenteroides* C7 induced a relevant increase in *C. elegans* viability (Figure 2A), showing similar survival as compared to those fed the probiotic strain LGG. Indeed, 50% of viability was recorded at days 17, 18 and 21 in *L. mesenteroides* C1, *L. mesenteroides* C2 and *L. mesenteroides* C7 fed nematodes respectively, in comparison with day 22 in LGG-fed nematodes. On the other hand, only *W. soli* T4 showed a similar worm lifespan when compared to the control OP50 diet, with 50% of viability recorded at days 14 and 12, respectively. To test possible effects on fertility, progeny production was evaluated. Like probiotic LGG, the brood size of worms fed different isolates showed a reduction of about 60% of the progeny number compared to the OP50 control (Figure 2B). Afterwards, the gut colonization capability was explored by plating worm lysates at different time points and by CFU counting. Results highlighted the increase of all bacterial strains along the lifespan (Figure 2C). At the L4 stage, the CFU number relative to *L. mesenteroides* C1 resulted to be about 2-fold higher than that relative to controls. At the stage of 8 days of adulthood, instead, the gut colonization capability of *L. mesenteroides* C1 was about 80% lower than that of probiotic LGG. On the other hand, in adult worms, *L. mesenteroides* C2 resulted to be able to colonize *C. elegans* gut, similarly to the LGG control. Along worms' lifespan, *L. mesenteroides* C7 and *W. soli* T4 showed a colonization capability similar to the OP50 strain.

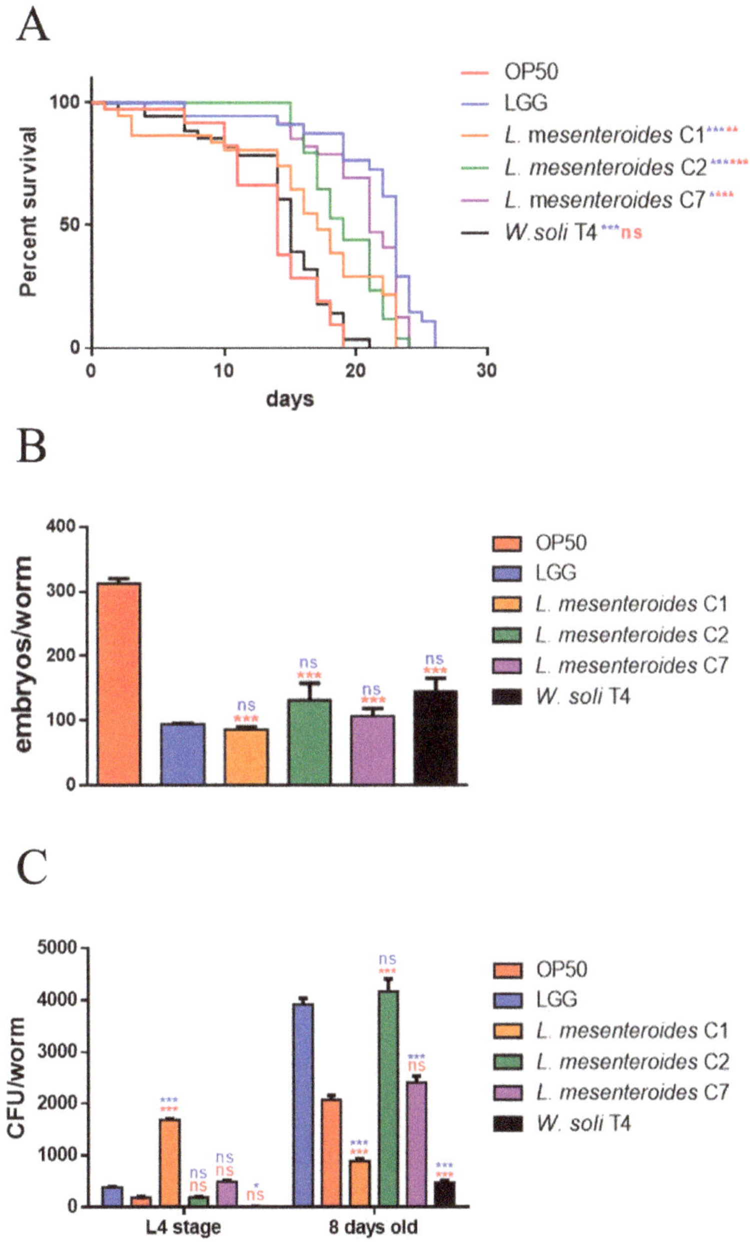

Figure 2. Effects of different isolates on *C. elegans* lifespan. (**A**) Kaplan–Meier survival plot of N2 worms fed *L. mesenteroides* and *W. soli* strains. Lifespans of OP50- and LGG-fed animals are reported as controls; $n = 60$ for each data point of single experiments. (**B**) Average embryos' production per worm of nematodes fed different bacterial isolates. (**C**) Bacterial colony-forming units (CFU) recovered from L4 larvae and 8-day-old adults fed the four isolates or LGG (LAB reference strain) and OP50 controls. Bars represent the mean of three independent experiments. Asterisks indicate significant differences (* $p < 0.05$, ** $p < 0.01$, *** $p < 0.001$) as compared to LGG (blue asterisks) or OP50 (red asterisks) controls, ns: not significant.

3.3.2. Effects on *C. elegans* Aging Processes

To investigate whether the pro-longevity effects exerted by LAB strains positively correlated to a delay in aging, age-related biomarkers, such as pumping, locomotion and lipofuscin accumulation, were analyzed. The pharyngeal pumping rate measures grinder contractions associated with food intake ability, normally declining with age. Figure 3A showed that *L. mesenteroides* C7-fed worms share a significantly high pumping rate, at 10 days of adulthood, similarly to LGG-fed worms. Moreover, nematodes fed with *L. mesenteroides* C2 and *W. soli* T4 showed an increase in grinder contraction of 12% and 8%, as compared to OP50-fed worms, respectively (Figure 3A). On the other hand, in *L. mesenteroides* C1-fed worms, a reduction in pumping rate of about 40% with respect to LGG was observed. Then, the locomotion rate of *C. elegans* was evaluated at day 10 of adulthood. In this case, nematodes fed different *L. mesenteroides* strains displayed a higher motility than OP50-fed worms, while locomotion of *W. soli* T4-fed worms was similar to the control (Figure 3B). Furthermore, accumulation of auto-fluorescent lipofuscin is also an aging marker of cellular impairment. Indeed, 10-day-old adult nematodes, when fed *L. mesenteroides* isolates, showed a reduced fluorescence compared to OP50-fed adults, while *W. soli* T4-fed animals showed a higher accumulation of fluorescent granules along the intestine, usual in old animals (Figure 4A,B).

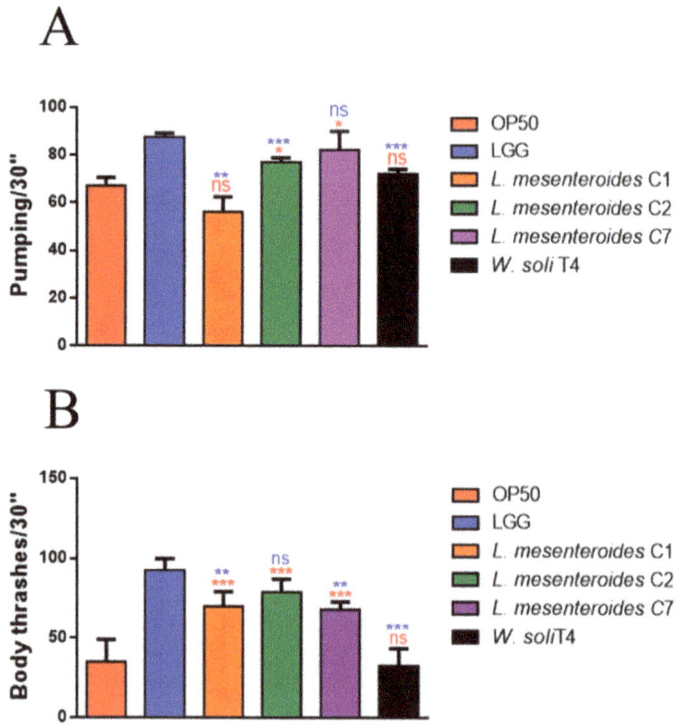

Figure 3. Impact of the different isolates on *C. elegans* aging markers. (**A**) Pumping rate of 10-day-old worms measured for 30 s. 10 worms were used for each condition. Worms fed OP50 or LGG were taken as controls. (**B**) Body bending of *C. elegans*-fed isolates as compared to LAB reference strain LGG or OP50, measured for 30 s. Bars represent the mean of three independent experiments. Statistical analysis was evaluated by one-way ANOVA with the Bonferroni post-test. Asterisks indicate significant differences (* $p < 0.05$, ** $p < 0.01$, *** $p < 0.001$) as compared to LGG (blue asterisks) or OP50 (red asterisks) controls, ns: not significant.

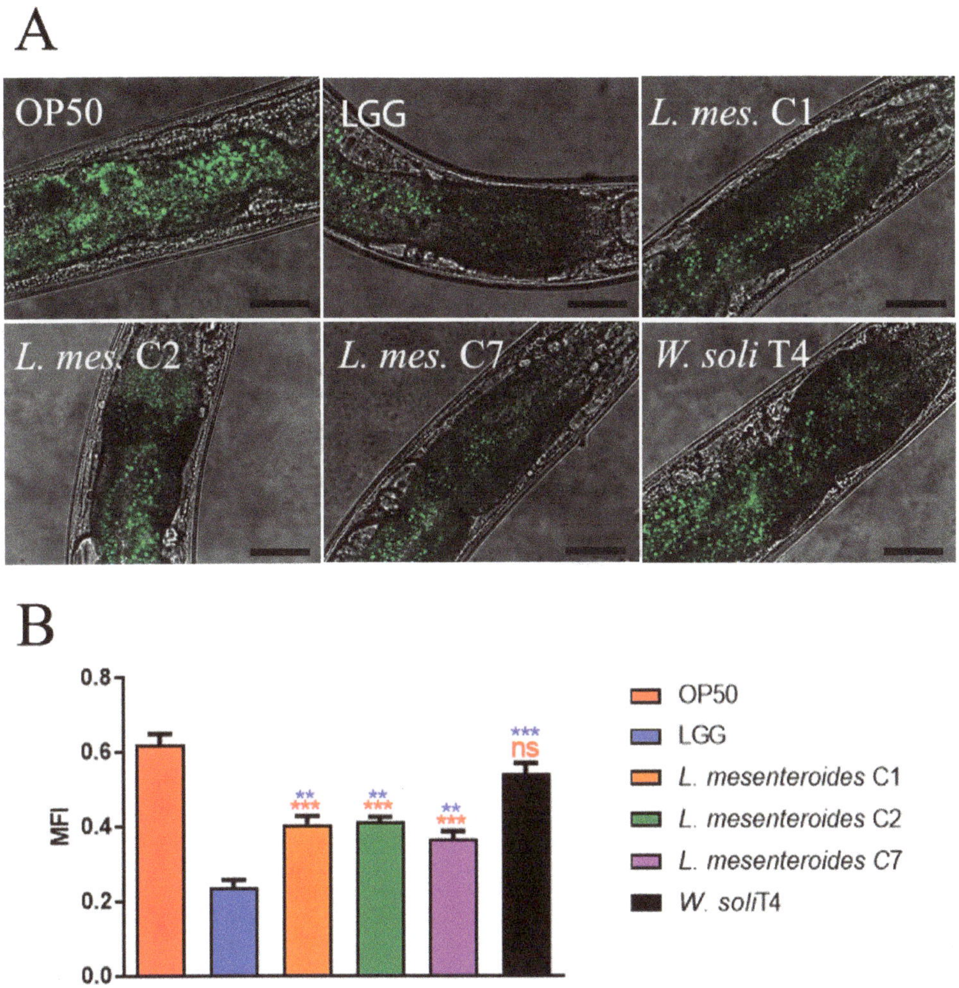

Figure 4. Evaluation of lipofuscin accumulation in *C. elegans*. (**A**) Autofluorescence of lipofuscin granules in *C. elegans* fed different LAB on day 10. Ten worms were used for each measurement. LGG- (LAB reference strain) and OP50-fed worms were used as controls. Scale bar = 50 μm. (**B**) Median fluorescence intensity of nematodes' lipofuscin. Statistical analysis was evaluated by one-way ANOVA with the Bonferroni post-test. Asterisks indicate significant differences (** $p < 0.01$, *** $p < 0.001$) as compared to LGG (blue asterisks) or OP50 (red asterisks) controls, ns: not significant. Bars represent the mean of three independent experiments.

3.3.3. Pathogen Resistance and Innate Immunity Stimulation

Among the four LAB isolates, *L. mesenteroides* C2 and *L. mesenteroides* C7 strains resulted as the most promising candidates in terms of beneficial features. Since a good probiotic is reported to compete with pathogens, protecting the host from infections, the *C. elegans* killing assay was performed to test possible protection from infection by the two isolates. *S. aureus* and *P. aeruginosa* were chosen as representatives of the Gram-positive and Gram-negative group, respectively. As shown in Table 4, nematodes displayed reduced survival when fed pathogens alone, as compared to nematodes fed co-cultures of the same pathogen with *L. mesenteroides* C2 and *L. mesenteroides* C7. Interestingly, 50% of viability was recorded at day 6 in nematodes fed co-cultures of *L. mesenteroides* strain *P. aeruginosa*, in comparison with day 3 of nematodes fed the pathogen alone. Similarly, *L. mesenteroides* C2 and *L. mesenteroides* C7 were able to protect worms from *S. aureus* infection, with 50% of viability recorded at day 7, with respect to day 5 in worms fed *S. aureus* alone.

To study whether *L. mesenteroides* C2 and C7 could stimulate nematodes' innate immunity, transcript levels were analyzed for *pmk-1*, *skn-1*, *sod-3*, *daf-16* and *hsf-1* genes, whose activation after probiotic feeding have been described [15,25]. Interestingly, in *L. mesenteroides* C2- and C7-fed nematodes, significantly increased expressions of *pmk*-1 and *hsf-1* transcripts were observed, similarly to LGG (Figure 5). On the other hand, while in LGG-fed worms DAF-16 also seemed to be activated, the two isolates were not able to stimulate this pathway. The expression of genes involved in detoxification processes (*skn-1* and *sod-3*) were instead very low, as compared to the OP50 control, but similar to LGG.

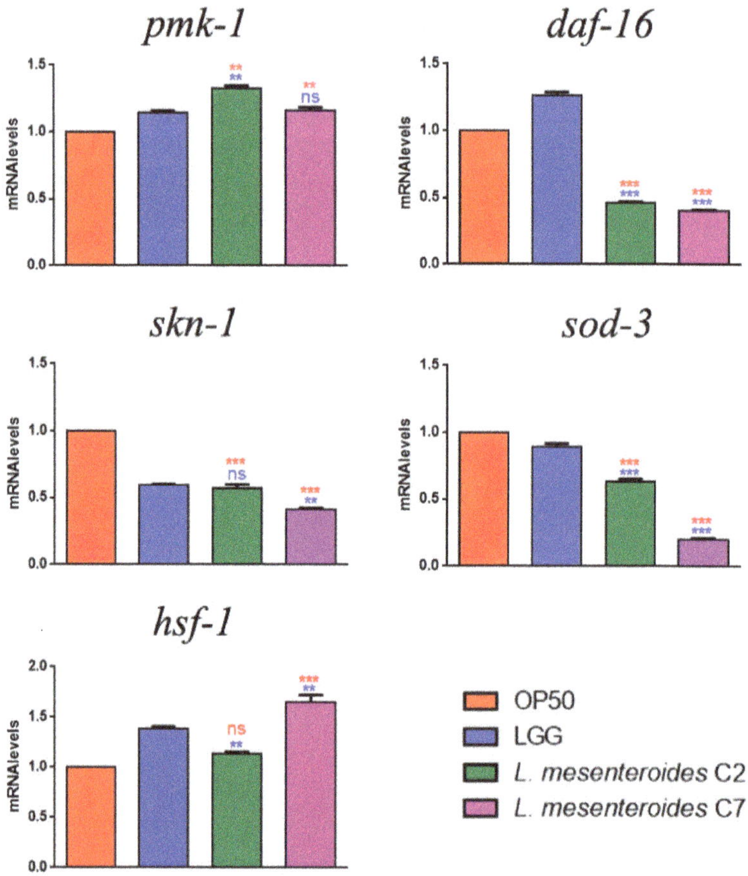

Figure 5. Transcript levels of genes involved in *C. elegans* immunity. Expression of *pmk-1*, *skn-1*, *sod-3*, *daf-16* and *hsf-1* genes in 1-day-old adults fed with different bacterial strains. Experiments were performed in triplicate. LGG (LAB reference strain) and OP50 were used as controls. Data are presented as mean ± SD. Asterisks indicate significant differences (** $p < 0.01$, *** $p < 0.001$) as compared to LGG (blue asterisks) or OP50 (red asterisks) controls; ns: not significant.

Table 4. In vivo resistance to pathogens. Survival assay of N2 worms fed co-cultures of LAB and *P. aeruginosa* or *S. aureus*. The lifespan of worms fed pathogens alone was reported as the control. Three experiments were performed in triplicate for each condition.

C. elegans Strain	Diet	Median Lifespan	Maximum Lifespan	Statistics
Wild-type N2	*P. aeruginosa*	3 ± 0.8	5 ± 0.8	-
	LGG + *P. aeruginosa*	5 ± 0.9	9 ± 0.4	$p < 0.001$
	L. mesenteroides C2 + *P. aeruginosa*	6 ± 1.2	12 ± 0.8	$p < 0.001$
	L. mesenteroides C7 + *P. aeruginosa*	6 ± 0.2	11 ± 0.3	$p < 0.001$
	S. aureus	5 ± 0.4	8 ± 0.9	-
	LGG + *S. aureus*	8 ± 0.5	13 ± 0.5	$p < 0.001$
	L. mesenteroides C2 + *S. aureus*	7 ± 0.6	11 ± 0.5	$p < 0.001$
	L. mesenteroides C7 + *S. aureus*	7 ± 0.9	11 ± 1.2	$p < 0.001$

4. Discussion

The human microbiota represents the first defense barrier against gut colonization by pathogens. This defense is due to microbiota competition with pathogenic bacteria, preventing their adhesion and subsequent internalization. Indeed, some commensal strains share adhesion receptors with higher affinity than pathogens, also competing for the availability of nutrients and trophic substances. Another strategy adopted by microbiota species is the production of antimicrobial substances called bacteriocins. Therefore, probiotics commonly share the ability to compete for receptor sites, nutrients and trophic substances, and they are able to synthesize bacteriocins. Recently, growing attention in pro-longevity effects exerted by different LAB strains allowed the development of several probiotic products [26]. In particular, fermented foods are the main source of LAB [14,19,27,28], but fruits have also been cited several times for the isolation of interesting LAB [29–31]. Due to the probiotics' relevance to human health, it is pivotal to characterize bacterial isolates to be used in alternative food products. In this context, the main objective of this work was the isolation of LAB from carrots and the selection of strains with potential to be used as probiotic microorganisms. The root carrot (*Daucus carota* L.) is one of the most important vegetables cultivated and consumed worldwide, rich in bioactive compounds, such as provitamin A [5,7]. It is also rich in dietary fiber, antioxidants and other nutrients, but especially in carotenoids. In this work, a combination of in vitro and in vivo methods was used to screen for new potential probiotic *Leuconostoc* and *Weissella* strains deriving from carrots grown in Fucino highland (Abruzzo, Italy). Characterization of LAB isolates at the species level identified three strains belonging to *Leuconostoc mesenteroides* and a *Weissella soli* strain. Indeed, a large percentage of probiotic microorganisms belongs to the LAB group. *Leuconostoc* and *Weissella*, together with *Lactobacillus* and *Pediococcus*, are important genera of LAB associated with foods and fermented products, such as meat, vegetables, dairy and bakery products, and also act as flavoring and texturizing agents [32]. This work showed that, among all tested strains, *L. mesenteroides* C2 and *L. mesenteroides* C7 could survive against the stress conditions assayed in this study. Survival to the adverse environment of the stomach is a key pre-requisite for effective colonization by a probiotic strain [33]. At first, tolerance to gastrointestinal conditions was carried out by evaluating the viability of each isolate in comparison with the commercial reference strain *L. rhamnosus* GG. Notably, *L. mesenteroides* C2 and *L. mesenteroides* C7 displayed survival rates equal to or higher than that of the commercial probiotic control LGG after the different treatments. Only in the case of long exposure to lysozyme and low pH resistance did *L. mesenteroides* C7 show a reduced recovery of viable cells. If this strain will result as positive in subsequent tests, this problem could be solved by encapsulating the bacteria cells, making them viable in the human gut. Furthermore, antibiotic resistance profiling of all isolates resulted similar to probiotic LGG, and this is an important trait to be verified for safety purposes [34]. In this study, different groups of antibiotics were used: cell wall inhibitors, inhibitors of protein synthesis and inhibitors of DNA and RNA synthesis. A recurrent ingestion of these types of antibiotics may cause imbalance in the intestinal sensitive microbiota. Moreover, the antibiotic resistance in probiotics usually does not constitute a safety issue, since resistance genes could be transferred to potential pathogens. It has been demonstrated that probiotics are able to prevent infections by foodborne pathogens, through different mechanisms, such as competitive exclusion or antimicrobial molecules' production [35,36]. Indeed, LAB show various antimicrobial features, such as the production of organic acids, cyclic dipeptides, phenylacetic acid, hydrogen peroxide, low molecular weight compounds, protein compounds, bacteriocins and fatty acids [37–40]. We therefore tested the antimicrobial activity exerted by *L. mesenteroides* C2 and *L. mesenteroides* C7 strains against common pathogens, such as *Pseudomonas aeruginosa* and *Staphylococcus aureus*. The isolates resulted to counteract the pathogens in vitro and in vivo. In parallel, the isolated strains were also analyzed evaluating possible beneficial properties in the in vivo model of *C. elegans*. Nematodes commonly feed on bacteria, but a significant number of bacterial cells escape the grinder contractions and can proceed to colonize the worm gut [41]. The *L. mesenteroides* C2 and C7 strains were able to significantly increase *C. elegans* lifespan as compared to the OP50 control strain, similar to the effect exerted by the reference strain LGG. The impact of *L. mesenteroides* C2 and C7 on *C. elegans* physiology could be due to the high gut colonization capacity of the bacterial strains. Moreover, the pro-longevity effects observed in lifespan experiments were associated with the anti-aging effects, highlighted by analyzing different aging markers, such as pharyngeal pumping rate, brood size and lipofuscin. These data further demonstrate the ability of specific LAB strains to prolong nematodes' lifespan, as described in previous studies [41–44]. As discussed above, the different isolates also displayed health-promoting activities in host defense against Gram-positive and Gram-negative pathogens in vivo, increasing the survival of infected worms. In *C. elegans*, host–pathogen interactions have been studied for a number of pathogens of human and animal origin [45], including *P. aeruginosa* and *S. aureus*, which colonize the worm gut and infect the nematode [46]. Moreover, innate immunity responses and lifespan were strongly correlated in nematodes [15]. Indeed, worms do not have adaptative immunity, but only innate immune defenses that have many aspects similar to human pathways [47]. Among them, the p38 MAPK and IIS pathways, the transforming growth factor-beta (TGF-beta) and the beta-catenin signaling pathways, are more conserved in humans and nematodes and they can be induced by probiotics [48]. In agreement with these works, real-time analysis highlighted the activation of *pmk-1* and *hsf-1* pathways, suggesting a stimulation of *C. elegans* immunity, which correlates with a reduction of oxidative stress, leading to the pro-longevity and anti-aging effects.

5. Conclusions

Among the different strains isolated from carrots, *L. mesenteroides* C2 and *L. mesenteroides C7* showed interesting probiotic characteristics, such as greater lysozyme, pH and bile tolerance, in vitro suppression of pathogen growth and in vivo beneficial effects, exerted on the *C. elegans* animal model. However, in vivo analysis on animal or human systems should be performed to further test their potential beneficial properties for human health.

Author Contributions: Conceptualization, A.M., A.T., W.A. and D.U.; methodology, E.S., J.M. and A.P.; software, E.S., J.M. and A.P.; investigation, E.S. and D.U.; data curation, E.S.; writing—original draft preparation, E.S.; writing—review and editing, D.U. and A.M.; visualization, E.S.; supervision, D.U., A.M. and P.M.; project administration, A.M. and D.U.; funding acquisition, W.A. All authors have read and agreed to the published version of the manuscript.

Funding: The present work has been carried out under the project "INNOPAQ", funded by POR-FESR Abruzzo 2014–2020, CUP number: C73D18000320007.

Institutional Review Board Statement: Not applicable.

Informed Consent Statement: Not applicable.

Data Availability Statement: Not applicable.

Conflicts of Interest: The authors declare no conflict of interest.

References

1. FAO/WHO Probiotics in Food. Health and Nutritional Properties and Guidelines for Evaluation 2006. Available online: http://www.fao.org/food/food-safety-quality/a-z-index/probiotics/en/ (accessed on 16 September 2021).
2. Markowiak-Kopeć, P.; Śliżewska, K. Effects of Probiotics, Prebiotics, and Synbiotics on Human Health. *Nutrients* **2017**, *9*, 1021. [CrossRef]
3. Roselli, M.; Natella, F.; Zinno, P.; Guantario, B.; Canali, R.; Schifano, E.; De Angelis, M.; Nikoloudaki, O.; Gobbetti, M.; Perozzi, G.; et al. Colonization Ability and Impact on Human Gut Microbiota of Foodborne Microbes from Traditional or Probiotic-Added Fermented Foods: A Systematic Review. *Front. Nutr.* **2021**, *8*, 689084. [CrossRef] [PubMed]
4. Sciubba, F.; Tomassini, A.; Giorgi, G.; Brasili, E.; Pasqua, G.; Capuani, G.; Aureli, W.; Miccheli, A. NMR-Based Metabolomic Study of Purple Carrot Optimal Harvest Time for Utilization as a Source of Bioactive Compounds. *Appl. Sci.* **2020**, *10*, 8493. [CrossRef]
5. Tomassini, A.; Sciubba, F.; Di Cocco, M.E.; Capuani, G.; Delfini, M.; Aureli, W.; Miccheli, A. 1H NMR-Based Metabolomics Reveals a Pedoclimatic Metabolic Imprinting in Ready-to-Drink Carrot Juices. *J. Agric. Food Chem.* **2016**, *64*, 5284–5291. [CrossRef] [PubMed]
6. Arscott, S.A.; Tanumihardjo, S.A. Carrots of Many Colors Provide Basic Nutrition and Bioavailable Phytochemicals Acting as a Functional Food. *Compr. Rev. Food Sci. Food Saf.* **2010**, *9*, 223–239. [CrossRef]
7. Que, F.; Hou, X.-L.; Wang, G.-L.; Xu, Z.-S.; Tan, G.-F.; Li, T.; Wang, Y.-H.; Khadr, A.; Xiong, A.-S. Advances in research on the carrot, an important root vegetable in the Apiaceae family. *Hortic. Res.* **2019**, *6*, 69. [CrossRef]
8. Da Silva, E.A.; Vieira, M.A.; Vieira, E.A.; Amboni, R.D.D.M.C.; Amante, E.R.; Teixeira, E. Chemical, Physical and Sensory Parameters of Different Carrot Varieties (*Daucus carota* L.). *J. Food Process. Eng.* **2007**, *30*, 746–756. [CrossRef]
9. Forssten, S.D.; Sindelar, C.W.; Ouwehand, A. Probiotics from an industrial perspective. *Anaerobe* **2011**, *17*, 410–413. [CrossRef]
10. Sornplang, P.; Piyadeatsoontorn, S. Probiotic isolates from unconventional sources: A review. *J. Anim. Sci. Technol.* **2016**, *58*, 26. [CrossRef]
11. Poupet, C.; Chassard, C.; Nivoliez, A.; Bornes, S. Caenorhabditis elegans, a Host to Investigate the Probiotic Properties of Beneficial Microorganisms. *Front. Nutr.* **2020**, *7*, 135. [CrossRef]
12. Kwon, G.; Lee, J.; Lim, Y.-H. Dairy Propionibacterium extends the mean lifespan of Caenorhabditis elegans via activation of the innate immune system. *Sci. Rep.* **2016**, *6*, 31713. [CrossRef] [PubMed]
13. Schifano, E.; Zinno, P.; Guantario, B.; Roselli, M.; Marcoccia, S.; Devirgiliis, C.; Uccelletti, D. The Foodborne Strain Lactobacillus fermentum MBC2 Triggers pept-1-Dependent Pro-Longevity Effects in Caenorhabditis elegans. *Microorganisms* **2019**, *7*, 45. [CrossRef] [PubMed]
14. Guantario, B.; Zinno, P.; Schifano, E.; Roselli, M.; Perozzi, G.; Palleschi, C.; Uccelletti, D.; Devirgiliis, C. In Vitro and In Vivo Selection of Potentially Probiotic Lactobacilli from Nocellara del Belice Table Olives. *Front. Microbiol.* **2018**, *9*, 595. [CrossRef] [PubMed]
15. Roselli, M.; Schifano, E.; Guantario, B.; Zinno, P.; Uccelletti, D.; Devirgiliis, C. Caenorhabditis Elegans and Probiotics Interactions from a Prolongevity Perspective. *Int. J. Mol. Sci.* **2019**, *20*, 5020. [CrossRef]
16. Schifano, E.; Cavallini, D.; De Bellis, G.; Bracciale, M.P.; Felici, A.C.; Santarelli, M.L.; Sarto, M.S.; Uccelletti, D. Antibacterial Effect of Zinc Oxide-Based Nanomaterials on Environmental Biodeteriogens Affecting Historical Buildings. *Nanomaterials* **2020**, *10*, 335. [CrossRef]
17. Samedi, L.; Charles, A.L. Isolation and characterization of potential probiotic Lactobacilli from leaves of food plants for possible additives in pellet feeding. *Ann. Agric. Sci.* **2019**, *64*, 55–62. [CrossRef]
18. Erkkilä, S.; Petäjä, E. Screening of commercial meat starter cultures at low pH and in the presence of bile salts for potential probiotic use. *Meat Sci.* **2000**, *55*, 297–300. [CrossRef]
19. Zanni, E.; Laudenzi, C.; Schifano, E.; Palleschi, C.; Perozzi, G.; Uccelletti, D.; Devirgiliis, C. Impact of a Complex Food Microbiota on Energy Metabolism in the Model Organism Caenorhabditis elegans. *BioMed Res. Int.* **2015**, *2015*, 621709. [CrossRef]
20. Zanni, E.; Schifano, E.; Motta, S.; Sciubba, F.; Palleschi, C.; Mauri, P.; Perozzi, G.; Uccelletti, D.; Devirgiliis, C.; Miccheli, A. Combination of Metabolomic and Proteomic Analysis Revealed Different Features among *Lactobacillus delbrueckii* Subspecies *bulgaricus* and *lactis* Strains While In Vivo Testing in the Model Organism Caenorhabditis elegans Highlighted Probiotic Properties. *Front. Microbiol.* **2017**, *8*, 1206. [CrossRef]
21. Schifano, E.; Marazzato, M.; Ammendolia, M.G.; Zanni, E.; Ricci, M.; Comanducci, A.; Goldoni, P.; Conte, M.P.; Uccelletti, D.; Longhi, C. Virulence behavior of uropathogenic Escherichia colistrains in the host model Caenorhabditis elegans. *Microbiologyopen* **2019**, *8*, e00756. [CrossRef]
22. Schifano, E.; Ficociello, G.; Vespa, S.; Ghosh, S.; Cipollo, J.F.; Talora, C.; Lotti, L.V.; Mancini, P.; Uccelletti, D. Pmr-1 gene affects susceptibility of Caenorhabditis elegans to Staphylococcus aureus infection through glycosylation and stress response pathways' alterations. *Virulence* **2019**, *10*, 1013–1025. [CrossRef]
23. Schifano, E.; Cicalini, I.; Pieragostino, D.; Heipieper, H.J.; Del Boccio, P.; Uccelletti, D. In vitro and in vivo lipidomics as a tool for probiotics evaluation. *Appl. Microbiol. Biotechnol.* **2020**, *104*, 8937–8948. [CrossRef]

24. Dunne, C.; O'Mahony, L.; Murphy, L.; Thornton, G.; Morrissey, D.; O'Halloran, S.; Feeney, M.; Flynn, S.; Fitzgerald, G.; Daly, C.; et al. In vitro selection criteria for probiotic bacteria of human origin: Correlation with in vivo findings. *Am. J. Clin. Nutr.* **2001**, *73*, 386s–392s. [CrossRef]
25. Bianchi, L.; Laghi, L.; Correani, V.; Schifano, E.; Landi, C.; Uccelletti, D.; Mattei, B. A Combined Proteomics, Metabolomics and In Vivo Analysis Approach for the Characterization of Probiotics in Large-Scale Production. *Biomolecules* **2020**, *10*, 157. [CrossRef]
26. Terpou, A.; Papadaki, A.; Lappa, I.K.; Kachrimanidou, V.; Bosnea, L.A.; Kopsahelis, N. Probiotics in Food Systems: Significance and Emerging Strategies Towards Improved Viability and Delivery of Enhanced Beneficial Value. *Nutrients* **2019**, *11*, 1591. [CrossRef]
27. Domingos-Lopes, M.; Stanton, C.; Ross, P.; Dapkevicius, M.; Silva, C. Genetic diversity, safety and technological characterization of lactic acid bacteria isolated from artisanal Pico cheese. *Food Microbiol.* **2017**, *63*, 178–190. [CrossRef] [PubMed]
28. Mokoena, M.P.; Mutanda, T.; Olaniran, A.O. Perspectives on the probiotic potential of lactic acid bacteria from African traditional fermented foods and beverages. *Food Nutr. Res.* **2016**, *60*, 29630. [CrossRef]
29. Rodríguez, L.G.R.; Gasga, V.M.Z.; Pescuma, M.; Van Nieuwenhove, C.; Mozzi, F.; Burgos, J.A.S. Fruits and fruit by-products as sources of bioactive compounds. Benefits and trends of lactic acid fermentation in the development of novel fruit-based functional beverages. *Food Res. Int.* **2021**, *140*, 109854. [CrossRef] [PubMed]
30. Di Cagno, R.; Filannino, P.; Gobbetti, M. Vegetable and Fruit Fermentation by Lactic Acid Bacteria. *Biotechnol. Lact. Acid Bact.* **2015**, 216–230. [CrossRef]
31. Ngea, G.L.N.; Yang, Q.; Tchabo, W.; Castoria, R.; Zhang, X.; Zhang, H. *Leuconostoc mesenteroides* subsp. *mesenteroides* LB7 isolated from apple surface inhibits P. *expansum* in vitro and reduces patulin in fruit juices. *Int. J. Food Microbiol.* **2021**, *339*, 109025. [CrossRef]
32. O'Sullivan, L.; Ross, R.; Hill, C. Potential of bacteriocin-producing lactic acid bacteria for improvements in food safety and quality. *Biochimie* **2002**, *84*, 593–604. [CrossRef]
33. Dicks, L.M.T.; Botes, M. Probiotic lactic acid bacteria in the gastro-intestinal tract: Health benefits, safety and mode of action. *Benef. Microbes* **2010**, *1*, 11–29. [CrossRef]
34. Imperial, I.C.V.; Ibana, J.A. Addressing the Antibiotic Resistance Problem with Probiotics: Reducing the Risk of Its Double-Edged Sword Effect. *Front. Microbiol.* **2016**, *7*, 1983. [CrossRef] [PubMed]
35. Rafieian-Kopaei, M.; Karami, S.; Roayaei, M.; Hamzavi, H.; Bahmani, M.; Hassanzad-Azar, H.; Leila, M. Isolation and identification of probiotic Lactobacillus from local dairy and evaluating their antagonistic effect on pathogens. *Int. J. Pharm. Investig.* **2017**, *7*, 137–141. [CrossRef]
36. Mathipa, M.G.; Thantsha, M.S. Probiotic engineering: Towards development of robust probiotic strains with enhanced functional properties and for targeted control of enteric pathogens. *Gut Pathog.* **2017**, *9*, 28. [CrossRef]
37. Pawlowska, A.M.; Zannini, E.; Coffey, A.; Arendt, E.K. "Green Preservatives": Combating Fungi in the Food and Feed Industry by Applying Antifungal Lactic Acid Bacteria. *Nutr. Benefits Kiwifruit* **2012**, *66*, 217–238. [CrossRef]
38. Peyer, L.C.; Zannini, E.; Arendt, E.K. Lactic acid bacteria as sensory biomodulators for fermented cereal-based beverages. *Trends Food Sci. Technol.* **2016**, *54*, 17–25. [CrossRef]
39. Sadiq, F.A.; Yan, B.; Tian, F.; Zhao, J.; Zhang, H.; Chen, W. Lactic Acid Bacteria as Antifungal and Anti-Mycotoxigenic Agents: A Comprehensive Review. *Compr. Rev. Food Sci. Food Saf.* **2019**, *18*, 1403–1436. [CrossRef]
40. Silva, H.; Balthazar, C.F.; Silva, R.; Vieira, A.; Costa, R.; Esmerino, E.; Freitas, M.; Cruz, A. Sodium reduction and flavor enhancer addition in probiotic prato cheese: Contributions of quantitative descriptive analysis and temporal dominance of sensations for sensory profiling. *J. Dairy Sci.* **2018**, *101*, 8837–8846. [CrossRef] [PubMed]
41. Nakagawa, H.; Shiozaki, T.; Kobatake, E.; Hosoya, T.; Moriya, T.; Sakai, F.; Taru, H.; Miyazaki, T. Effects and mechanisms of prolongevity induced by Lactobacillus gasseri SBT2055 in Caenorhabditis elegans. *Aging Cell* **2015**, *15*, 227–236. [CrossRef]
42. Ikeda, T.; Yasui, C.; Hoshino, K.; Arikawa, K.; Nishikawa, Y. Influence of Lactic Acid Bacteria on Longevity of Caenorhabditis elegans and Host Defense against Salmonella enterica Serovar Enteritidis. *Appl. Environ. Microbiol.* **2007**, *73*, 6404–6409. [CrossRef] [PubMed]
43. Komura, T.; Ikeda, T.; Yasui, C.; Saeki, S.; Nishikawa, Y. Mechanism underlying prolongevity induced by bifidobacteria in Caenorhabditis elegans. *Biogerontology* **2013**, *14*, 73–87. [CrossRef]
44. Oh, A.; Daliri, E.B.-M.; Oh, D.H. Screening for potential probiotic bacteria from Korean fermented soybean paste: In vitro and Caenorhabditis elegans model testing. *LWT* **2018**, *88*, 132–138. [CrossRef]
45. Clark, L.C.; Hodgkin, J. Commensals, probiotics and pathogens in the Caenorhabditis elegansmodel. *Cell. Microbiol.* **2013**, *16*, 27–38. [CrossRef] [PubMed]
46. Irazoqui, J.E.; Troemel, E.R.; Feinbaum, R.L.; Luhachack, L.G.; Cezairliyan, B.O.; Ausubel, F.M. Distinct Pathogenesis and Host Responses during Infection of *C. elegans* by *P. aeruginosa* and *S. aureus*. *PLoS Pathog.* **2010**, *6*, e1000982. [CrossRef]
47. Marsh, E.; May, R.C. Caenorhabditis elegans, a Model Organism for Investigating Immunity. *Appl. Environ. Microbiol.* **2012**, *78*, 2075–2081. [CrossRef]
48. Kwon, G.; Lee, J.; Koh, J.-H.; Lim, Y.-H. Lifespan Extension of Caenorhabditis elegans by Butyricicoccus pullicaecorum and Megasphaera elsdenii with Probiotic Potential. *Curr. Microbiol.* **2017**, *75*, 557–564. [CrossRef] [PubMed]

microorganisms

Article

Antimicrobial Effects of Potential Probiotics of *Bacillus* spp. Isolated from Human Microbiota: In Vitro and In Silico Methods

Alfonso Torres-Sánchez [1,2,†], Jesús Pardo-Cacho [1,†], Ana López-Moreno [1,2,3,*], Ángel Ruiz-Moreno [1,2], Klara Cerk [1] and Margarita Aguilera [1,2,3,*]

1 Department of Microbiology, Faculty of Pharmacy, University of Granada, Campus of Cartuja, 18071 Granada, Spain; alfons_ats@hotmail.com (A.T.-S.); jesuspar99@correo.ugr.es (J.P.-C.); angel_trm_@hotmail.com (Á.R.-M.); klara.cerk@gmail.com (K.C.)
2 Institute of Nutrition and Food Technology "José Mataix", CIBM, University of Granada, 18016 Granada, Spain
3 Instituto de Investigación Biosanitaria ibs (IBS), 18012 Granada, Spain
* Correspondence: alopezm@ugr.es (A.L.-M.); maguiler@ugr.es (M.A.); Tel.:+34-9-5824-5129 (A.L.-M.); Fax: +34-958-246235 (M.A.)
† These authors have contributed equally to this manuscript.

Citation: Torres-Sánchez, A.; Pardo-Cacho, J.; López-Moreno, A.; Ruiz-Moreno, Á.; Cerk, K.; Aguilera, M. Antimicrobial Effects of Potential Probiotics of *Bacillus* spp. Isolated from Human Microbiota: In Vitro and In Silico Methods. *Microorganisms* **2021**, *9*, 1615. https://doi.org/10.3390/microorganisms9081615

Academic Editor: Sabina Fijan

Received: 29 June 2021
Accepted: 26 July 2021
Published: 29 July 2021

Abstract: The variable taxa components of human gut microbiota seem to have an enormous biotechnological potential that is not yet well explored. To investigate the usefulness and applications of its biocompounds and/or bioactive substances would have a dual impact, allowing us to better understand the ecology of these microbiota consortia and to obtain resources for extended uses. Our research team has obtained a catalogue of isolated and typified strains from microbiota showing resistance to dietary contaminants and obesogens. Special attention was paid to cultivable *Bacillus* species as potential next-generation probiotics (NGP) together with their antimicrobial production and ecological impacts. The objective of the present work focused on bioinformatic genome data mining and phenotypic analyses for antimicrobial production. In silico methods were applied over the phylogenetically closest type strain genomes of the microbiota *Bacillus* spp. isolates and standardized antimicrobial production procedures were used. The main results showed partial and complete gene identification and presence of polyketide (PK) clusters on the whole genome sequences (WGS) analysed. Moreover, specific antimicrobial effects against *B. cereus, B. circulans, Staphylococcus aureus, Streptococcus pyogenes, Escherichia coli, Serratia marcescens, Klebsiella* spp., *Pseudomonas* spp., and *Salmonella* spp. confirmed their capacity of antimicrobial production. In conclusion, *Bacillus* strains isolated from human gut microbiota and taxonomic group, resistant to Bisphenols as xenobiotics type endocrine disruptors, showed parallel PKS biosynthesis and a phenotypic antimicrobial effect. This could modulate the composition of human gut microbiota and therefore its functionalities, becoming a predominant group when high contaminant exposure conditions are present.

Keywords: probiotics; *Bacillus*; antimicrobial effect; in vitro methods; in silico methods

1. Introduction

The human gut microbiota could be considered as a new source for the identification and isolation of multiple microorganisms producing bioactive compounds and enzymes of interest such as biopolymers, antimicrobials notably demanded by the food, health, and several biotechnological industries [1,2]. Identifying the composition of cultivable gut microbiota has always been a challenge due mainly to the requested anaerobic conditions [3]. Efforts in simulating these harsh culture conditions allow isolating potential NGP [4] and even a variety of taxonomy bacterial groups which were also tolerant to xenobiotics or obesogens [5] followed by characterization through 16S rRNA gene sequencing.

Microbiome compositional consortia are variable in each individual [6,7]. Culturing methods and directed-culturomics for isolating specific microorganisms deserve special attention. Thus, the genus *Bacillus* belonging to a predominant microbiota phylum, Firmicutes, is differentially present and its species are capable of synthesizing a wide variety of bioactive compounds and enzymes of interest for their potential technological applications in health and the modern food biotechnological sectors [8]. Several *Bacillus* species have also been considered as probiotics [9,10]. *Bacilli* taxa, concretely *Lactobacillus* and *Bacillus* genera in microbiota seem to play a role on the ecology of predominant groups present on individual microbiota in obesity and metabolic disorders as compiled in human clinical trials (Table 1). The potential impact on the other circumscribed taxa groups could be driven by antimicrobial substances released by the *Bacilli* taxa, such as bacteriocins, PKs, lipopeptides, etc. [11,12].

Table 1. *Bacilli taxa* modifications from clinical trials of metabolic related diseases.

Reference	Clinical Trials—Disease /Sample Size and Clinical Traits	Taxa Modifications
[13]	OB; n = 192; HC n = 25; OW n = 22; OB n = 145	↑ ***Bacillus*** in OW and OB
[14]	OB, AN; n = 49; HC n = 20; OB n = 20; AN n = 9	↑ ***Lactobacillus*** in OB
[15]	T2D; n = 36; HC n = 18; T2D n = 18	↑ ***Lactobacillus*** in T2D
[16]	T2D, OB; n = 60; HC n = 20; Obese-T2D n = 40	↑ ***Bacillus sporothermodurans*** in OB-T2D
[17]	T1D, T2D; n = 110; HC n = 40; T2D n = 49; T1D n = 21	↑ ***Lactobacillus*** in T1D and T2D
[18]	NAFLD; n = 126; HC n = 83; NAFLD n = 43	↓ ***Lactobacillus*** in NAFLD
[19]	NAFLD; n = 67; HC n = 37; NAFLD n = 30	↑ ***Lactobacillaceae*** in NAFLD
[20]	NAFLD; n = 60; HC n = 30; NAFLD n = 30	↑ ***Lactobacillus*** in NAFLD
[21]	NAFLD, OB; n = 73; HC n = 20; OB-NAFLD n = 36; OB-non-NAFLD n = 17	↑ ***Bacilli*** in OB-NAFLD ↑ ***Lactobacillus*** in non-NAFLD
[22]	MetS; n = 655; Monozygotic twins n = 306; Dizygotic twins n = 74; Siblings n = 275	↑ ***Lactobacillus*** in MetS

AN: anorexia nervosa; HC: healthy control; MetS: metabolic syndrome; NAFLD: non-alcoholic fatty liver disease; OB: obese; OW: overweight; T1D: type 1 diabetes; T2D: type 2 diabetes. ↑ Increasements.

Bisphenols are considered as microbiota disrupting chemicals (MDC) [5] and their presence in humans has been confirmed by detecting them in human biospecimens: feces, serum, urine, saliva, hair, tissue and blood [23,24]. Bisphenol A (BPA) is used in manufacturing polycarbonate and epoxy resins for food consumer products and packages. There is also cumulative exposure from contaminating soils, aquatic environments, drinking water, air and dust particles [25]. The estrogen activity alteration is the most widely studied effect of BPA and analogues, enhancing endocrine disruptor activities [26]. Moreover, some studies have shown obesogenic effects through microbiota dysbiosis [27], fat cell development, and lipid accumulation [28]. There are several regulations enforced concerning the hazards of Bisphenol A, as derivative of polycarbonates plastics and epoxy resins, used in food contact materials, toys, or other products. In order to protect the consumers from cumulative exposure, the tolerable daily intake (TDI) for BPA is permanently re-evaluated according to new toxicity data through specific international projects, such as U.S. National

Toxicology Program (CLARITY-BPA program) [29] or European Food Safety Authority (EFSA) comprehensive re-evaluation of BPA exposure and toxicity [30].

Moreover, commensal microorganisms isolated from human microbiota could in general fulfill the criteria of safety assessment and the status of Qualified Presumption of Safety (QPS) [31,32]. Similarly, most *Bacillus subtilis* cluster species are considered QPS [33] and they are increasingly marketed as products [34]. Conversely, *Bacillus cereus* cluster species can be also present in the gut microbiota, but they are not considered as QPS [34,35].

Next-generation sequencing (NGS) platforms and WGS of microorganisms have enlarged the molecular comparison knowledge on the gene collection for encoding enzymes, and better taxonomy has supported appropriate classification. Moreover, specific WGS gene description is needed to consider the food and feed safety aspects of microbiota cultivated strains [35].

Genome mining tools and phenotypic analysis are complementary approaches to predict and demonstrate the production of active secondary metabolites such as antimicrobial products from *Bacillus* species [36]. Genome mining revealed the potential for known and novel PKs extensively in *Bacillus* (Figure 1). Moreover, based on the prediction of the general architecture, novel clusters were identified in novel *Bacillus* spp. variants. In addition, more recent in silico and bioinformatics approaches seem to be successful to find and verify the microbial potential to produce valuable enzymes for biotechnological applications [36].

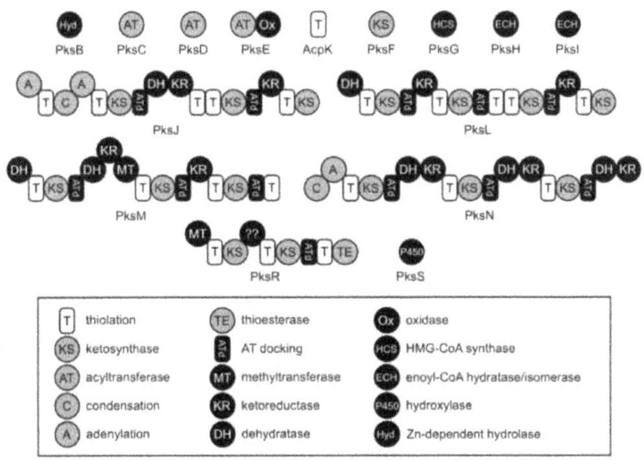

Figure 1. Conserved PKs proteins and functions in *Bacillus* modified from Straight et al. [37].

The main objective of the present study was to determine the antimicrobial effects of catalogue of microorganisms isolated from human gut, by applying directed-culturing methods after the addition of endocrine disruptor chemicals. Taxa groups of isolated bisphenol A (BPA)-degrading *Bacillus* spp. will be analyzed by with in vitro assays to demonstrate the bioactive substances released against commensals and critical pathogens according to the World Health Organization (WHO). Moreover, genome mining and in silico tests will be used for disclosing the genes responsible for antimicrobial production and its enzymatic pathways.

2. Materials and Methods

2.1. Microbiota Sampling Bank and Directed Culturing Approach

Ten isolates from fecal human microbiota collections of 0–1 year old infants (Isolates B-Project INFABIO) appropriately maintained at −80 °C underwent a directed culturing approach using 0.5 g of the fecal specimen in 1.5 mL of Brain Heart Infusion or Man Rogosa and Sharpe (BHI/MRS) broths, adding different concentrations of BPA (0.5, 10, 20, and 50 ppm), in order to search tolerant and/or potentially BPA biodegrading microorganisms, incubation for 72 h. Further serial dilutions and spreading onto BHI/MRS solid media plus incubation under aerobic and anaerobic conditions (anaerobic jars anaerocult®) at 37 °C over 72 h were applied. BPA-tolerant colonies with distinguishing features were isolated as pure culture for subsequent morphological, phenotypic, and genotypic identifications: bacterial cell counts, gram staining, spore staining, capsule staining, catalase activity, oxidase, and motility tests.

2.2. *BPA Microbiota Tolerance Testing*

BPA biodegradation microbiota capacity was tested directly adding BPA to the human fecal samples. The specimens were exposed to 25 ppm concentration of BPA at 30 °C during 72 h. BPA was measured in the extracts and supernatants through Liquid chromatography–mass spectrometry (LC-MS/MS) system for BPA quantification. Chemicals, reagents, instrumentation, and software for bisphenols determination were provided by CIC services under validated procedures previously described by García-Córcoles et al. [38].

2.3. *Culturing- Isolation of Bacillus Catalogue*

A common approach to isolate *Bacillus* strains from microbiota has been pursued in our research team [39]. For this study, ten isolates from fecal human microbiota collections of 0 to 1 year old infants (Isolates B-Project INFABIO) and 6–8 year-old children (Isolates C-Project OBEMIRISK) were obtained by a serial dilution method, with exposure to different BPA concentrations (0.5, 10, 20, and 50 ppm) over 72 h and further spreading in BHI/MRS media incubated under aerobic and anaerobic conditions (anaerobic jars anaerocult®) at 37 °C. The BPA-tolerant bacterial colonies with distinguishing features were isolated as pure culture for subsequent morphological, phenotypic, and genotypic identifications: bacterial cell counts, gram staining, spore staining, capsule staining, catalase activity, oxidase, and motility tests.

2.4. *Genomic DNA Extraction, Taxonomy Identification and Phylogenetic Analysis*

Genomic DNA was extracted using DNeasy columns (Qiagen®, Hilden, Germany) following the manufacturing instructions. The isolated DNA was quantified using Nanodrop (Thermo Scientific® Waltham, MA, USA) and biophotometer (Eppendorf® D30). The quality of DNA was monitored through gel electrophoreses. Complete 16S RNA gene sequencing of selected bacterial strains was done by Sanger method (Institute of Parasitology and Biomedicine "López-Neyra" IPBLN Service). Forward and reverse sequences were provided separately. Reverse sequence was converted to complementary sequence with Chromas Pro 2.0 software (Technelysium Pty Ltd., Tewantin, Australia). Sequences were examined for maximum homology against GenBank using National Center for Biotechnology Information NCBI's BLASTn program. The collection and comparison of complete 16S rRNA gene sequences were performed using the Ezbiocloud platform [40].

2.5. *Enzymes Tests*

Relevant enzymatic production assays were carried out to verify the potential of gut microbiota strains to synthetize relevant enzymes in the biotechnological and industrial context. Starch, carboxymethylcellulose, inulin, tween 20 and 80, and DNase supplemented media were used to determine the degradation of different substrates according to complementary methodologies [41–46].

2.6. *Antimicrobial In Vitro Tests*

Antimicrobial activity was tested by agar well diffusion method. Under Joint FAO/WHO Expert Committee on Food (JECFA) procedures [47] and the study carried out by Powthong & Suntornthiticharoen [48], nine different bacteria were used as indicators to verify the antimicrobial capacity of the *Bacillus* spp. isolated from the gut microbiota.

To determine the synthesis of antimicrobial compounds, several isolated strains were selected according to preliminary antimicrobial tests and the main taxonomy groups: strains close/represented by rB1 (*Bacillus* sp. AM1), strains close/represented by rB3 (*Bacillus siamensis* (KCTC 13613)), strains close/represented by rB7 (*Bacillus cereus* (AFS039342)). Plates with 20 mL of Müller-Hinton agar were prepared and test microorganisms used as indicators: *Bacillus cereus, Bacillus circulans, Staphylococcus aureus, Streptococcus pyogenes, Escherichia coli, Serratia marcescens, Klebsiella* spp., *Pseudomonas* spp., and *Salmonella* spp., were adjusted to a cell density of 0.5 on the McFarland scale in sterile 0.85% NaCl solution. The data were expressed as mean of the three replicates. Tests were done spreading the indicator microbial strains over the surface of the Müller-Hinton agar using sterile cotton swab. Inside six mm diameter oxford wells generated in agar, 20 µL of antibiotic producing bacteria extract was added. Standards appropriate positive controls (ampicillin, gentamycin, and streptomycin at 10 µg) and negative/blank (sterile media/ethanol) were used. The plates were incubated at 37 °C for 24 h and the inhibition zones were measured.

2.7. Genome Data Mining and Analysis –PKs Genes and Clusters

2.7.1. Genome Mining Tools for PKs Gene Searching

In order to discover the presence of secondary metabolites, several bioinformatics tools were used to perform genome mining. A data retrieving software has been specifically computed using Pascal programming language to obtain the PKs enzymes ID and the corresponding Loci from the genomes.

Type strain genomes from the closest species isolated were retrieved from NCBI Genome Data Bank in GenBank file format in order to list the proteins that they were able to potentially produce.

A more detailed prediction of the clusters was performed by checking the downstream and upstream genes of those involved in PKs synthesis using NCBI genome map viewer [49].

2.7.2. Prediction of Polyketides in WGS of *Bacillus* sp. AM1 Isolated from Microbiota

The identification of PKs gene cluster was carried out by the analysis of the WGS of *Bacillus* sp. AM1, GenBank CP047644.1, following the same approach explained above.

3. Results and Discussion

3.1. BPA-Tolerant Microorganisms Isolated from Human Gut Microbiota

3.1.1. BPA Microbiota Metabolization Capacities

The microbiota composition of each fecal sample was specific and contributed differentially to the biodegradation of BPA exposure levels (Figure 2). Each fecal sample (340, 349, and 437) showed a differential ability to eliminate BPA due to its taxa compositional and functional characteristics, showing sample 340 a maximum percentage of BPA degradation of 89.3% while sample 349 degraded 76% and 437 was able to eliminate 21% of the BPA concentration. Previous studies have shown the same effects in the environment [50], where they observed that different microbial communities presented a specific elimination rate dependent on their composition.

Cumulative exposure to a wide range of xenobiotics, such as BPA and its analogues, affects the microbiota diversity possessed by each individual, causing a selection of bacteria strains to populate the gut, and consequently modify its equilibrium through MDC [5]. This dysbiosis has been proven to be responsible for well-known diseases, such as obesity, diabetes, and even some hormonal-related cancers. Therefore, identification of the triggered main taxa variations and their functions remains a challenge. Moreover, the appropriate use of probiotics [50–52] or search for NGP to mitigate or reverse these dysbiosis are crucial [53,54]. A directed culturing approach allow us to select tolerant bacteria and mimic an ecological environment to understand better the impact of the specific enriched communities and their capacities to impact the taxa microbiota colonization.

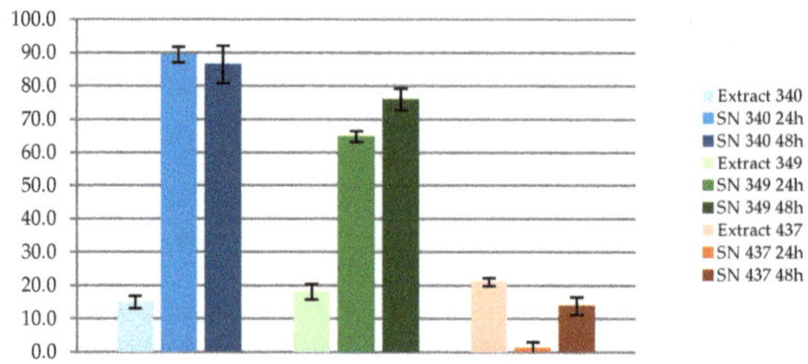

Figure 2. BPA relative percentage of degradation by human fecal specimens. (LC-MS/MS) system was used for BPA quantification; SN: Supernatant.

3.1.2. Catalogue of BPA-Tolerant *Bacillus* spp. Isolated from Human Microbiota

Isolation and identification of BPA-tolerant *Bacillus* spp. strains from microbiota samples were successfully performed with the different BPA concentrations plates (0.5; 10; 20 and 50 ppm). Out of these 11 isolates analyzed, the closest species by complete gene 16S rRNA sequence were *B. amyloliquefaciens, B. siamensis, B. velezensis, B. nematocida, B. cereus,* and *B. pacificus* (Table 2).

Table 2. *Bacillus* isolates from human microbiota and 16S rRNA complete gene homology description.

Microbiota Isolates	Closest Taxa—[Strain] Best Hit	bp Position 16S rRNA	Query Cover (%)	Identity (%)	Accession Number
B1	*Bacillus siamensis* [LRM10-3D]	15,030	100	100	MT645306.1
	Bacillus velezensis [XC1]		100	100	MT649755.1
B2	*Bacillus velezensis* [CR-502]	1483	95.4	99.14	AY603658
B3	*Bacillus siamensis* [KCTC 13613]	1490	100	98.00	AJVF01000043
B4	*Bacillus siamensis* [KCTC 13613]	1515	100	99.66	AJVF01000043
	Bacillus nematocida [B-16]		100	99.73	AY820954
	Bacillus amyloliquefaciens [DSM7]		100	99.52	FN597644
B5	*Bacillus siamensis* [KCTC 13613]	1516	100	98.91	AJVF01000043
	Bacillus nematocida [B-16]		100	98.98	AY820954
	Bacillus velezensis [CR-502]		95.4	99.22	AY603658 FN597644
	Bacillus amyloliquefaciens [DSM7]		100	98.78	
B6	*Bacillus velezensis* [CR-502]	1504	95.4	99.93	AY603658
B7	*Bacillus cereus* [AFS039342]	1510	100	99.39	NUMR01000072
	Bacillus pacificus [NCCP 15909]		100	99.34	CP041979.1
B8	*Bacillus velezensis* [CR-502]	1520	95.4	99.93	AY603658
B9	*Bacillus velezensis* [CR-502]	1499	95.4	99.22	AY603658
B9.2	*Bacillus siamensis* [KCTC 13613]	1499	100	99.52	AJVF01000043
	Bacillus nematocida [B-16]		100	99.59	AY820954
	Bacillus amyloliquefaciens [DSM 7]		100	99.39	FN597644
B12	*Bacillus cereus* [AFS039342]	1543	100	99.39	JMQC01000008
	Bacillus pacificus [NCCP 15909]		99.0	99.35	CP041979.1

Data obtained by parallel experimental work showed a BPA directed human fecal culturing catalogue that contained different BPA tolerant species from the following genera and percentages: *Enterococcus* 28%, *Bacillus* 27%, *Staphylococcus* 10%, *Escherichia* 8%, *Clostridium* 5%, and *Lactobacillus* 4% (data not shown). Representing *Bacilli* taxa (*Bacillus* and *Lactobacillus*) was a major taxa with approximately a 30% of BPA tolerant isolated strains from microbiota samples, which corroborates the predominant presence of these genera being able to overcome the impact of xenobiotics, such as BPA, as previous assays showed [39].

In line with these results, interesting properties and uses are specifically described for *Bacillus* spp. Recently, several *Bacilli* strains have been extensively proposed for use as human and animal probiotics [55,56]. Most of the species used belong to *Bacillus subtilis* and *Bacillus amyloliquefaciens* groups and special attention should be paid to the food and clinical studies with strains that showed special enzyme capacities [57] or those able to modulate and mitigate pathophysiological disorders [58].

3.1.3. Taxonomical and Phylogenetic Clustering

The phylogenetic tree based on complete 16S rRNA gene of *Bacillus* strains isolated from microbiota treated with BPA grouped the clusters to *B. subtilis, B. amyloliquefaciens, B. velezensis, B. siamensis, B.cereus,* and *B. pacificus* (Figure 3). The two main clustering of closely related *Bacillus* strains belong to *B. subtilis* and *B.amyloliquefaciens* taxonomic group (green) and *B. cereus* group (yellow). Three representative strains (rB1, rB3, and rB7) were further processed by bioactive compounds production tests. They were organized as follows: rB1 represented B1, B4, B5, B6, B7, B8, B9, and B9.2; rB3 represented B2 and B3; rB7 represented B7 and B12.

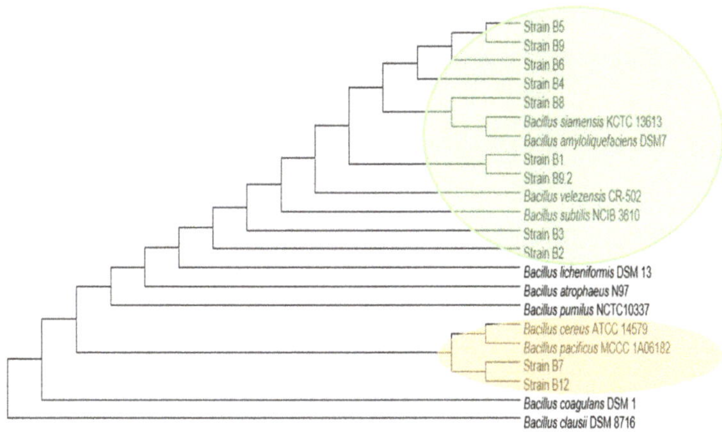

Figure 3. Phylogenetic tree based on gene sequences of isolated gut microbiota strains. The tree was obtained by applying Neighbor-Join and BioNJ algorithms to a matrix of pairwise distances estimated using the Maximum Composite Likelihood (MCL) approach and Kimura 2-parameter model. The species and strain names are shown. Bootstrap values shown after 1000 resamplings. Main clusters are highlighted: in green close to *B.subtilis* group and yellow close to *B.cereus* group.

The strains isolated in the present work were clustered in the two main groups: *B. subtilis*–like (non-pathogenic) [59] and *B. cereus*-like (pathogenic) [60], as shown in Figure 3, however the pathogenicity features are strain-specific dependent. The work approach is based on potential uses and predictive data analysis, but for further commercial uses, a safety assessment should be performed for each strain, to demonstrate that they do not pose any safety and/or pathogenicity concerns. The battery of tests usually requested is: antibiotic resistance test no greater than existing regulatory cutoffs against clinically important antibiotics, incapacity to induce hemolysis or produce surfactant factors, and the absence of virulence or toxigenic activity in vitro.

3.2. *Analysis of Bioactive Compounds Production Capacities*

3.2.1. Enzymatic Activity Tests

B. subtilis, B. amyloliquefaciens, and *B. licheniformis* have been used as bacterial resources in the industrial context for the production of a wide range of enzymes and bioactive compounds for decades. *Bacillus* sp. AM1 and other strains belonging to *Bacillus* genus have shown remarkable hydrolytic enzyme capacity (Table 3), being related to the performance of key roles in several biotechnological and many manufacturing processes [61–63].

Table 3. Enzymatic activity in gut microbiota isolates.

Enzyme Test	Microbiota Isolates		
	rB1	**rB3**	**rB7**
Starch	+	++	++
Carboxymethylcellulose	-	-	-
Inulin	+	-	+
Tween 80	-	-	-
DNase	++	-	-

3.2.2. Antimicrobial Activity Tests

The results obtained from antimicrobial experimental tests carried out with the representative isolated microorganisms from different taxonomic clusters confirmed the ability of the strains B1 and B3 to inhibit Gram-negative and Gram-positive bacteria (Table 4).

Preliminary results grouped the strains according to their capacity of antibiotic production with very similar inhibiting zone value, which were also in agreement with the main taxonomic clusters. rB1 represented B1, B4, B5, B6, B7, B8, B9, and B9.2; rB3 represented B2 and B3; rB7 represented B7 and B12.

Table 4. Antimicrobial activity of BPA-tolerant human gut microbiota isolated strains.

Target Indicator Bacteria	Strains rB1	Strains rB3	Strains rB7
	Diameter of inhibitory zone (mm) ± SD [1]		
Bacillus cereus	15 ± 0	17 ± 0	-
Bacillus circulans	13 ± 0	14.3 ± 1.2	-
Staphylococcus aureus	11.7 ± 0.6	10 ± 0	-
Streptococcus pyogenes	15 ± 0	13.3 ± 0.6	-
Serratia marcescens	17 ± 0	15.3 ± 1.5	-
E. coli	15 ± 0	13.3 ± 0.6	-
Salmonella	11 ± 0	10 ± 0	-
Klebsiella	20 ± 0 *	15 ± 0 *	-
Pseudomonas	-	-	-

[1] Values are mean diameter of inhibitory zone (mm) ± SD of three replicates. The diameter of well (6 mm) was included. (-) Diameter of inhibitory zone <7 mm considered as no antimicrobial activity. * Significant values compared to theroretical values from *B. subtilis* polyketides [64].

rB1 and rB3 strains were found to be antagonistic against Gram-positive *Bacillus cereus, Bacillus circulans, Staphylococcus aureus, Streptococcus pyogenes* (diameter of zone of growth inhibition 10–17 mm) and also against Gram-negative food-borne pathogenic bacteria *Serratia marcescens, Escherichia coli, Salmonella*, and *Klebsiella pneumoniae* (diameter of zone of growth inhibition 10–20 mm). Conversely, the strains rB7 did not show any production of antimicrobial effects.

Minimum inhibitory concentration (MIC) values were similar to those resultant of other polyketides antimicrobial effects previously described, being significant differential and higher the effects found against *Klebsiella* [64]. Therefore, the search for a putative biosynthetic pathway of the *pks* gene product proceeded after the validated molecular antimicrobial attributions.

3.3. WGS Data Mining and In Silico Analysis

3.3.1. WGS Mining in Type Strains

The bioinformatics analysis carried out on the type strains of closest species identified as cultivable *Bacillus* species from microbiota showed specific enzymes involved in PKs biosynthesis (Table 5). The genome mining identified the clusters with the genomes from closest homologue type strains available in the database. Bioinformatic tools and Pascal ad hoc software allowed the exhaustive analysis of genomes making it a powerful prediction tool.

According to the results, *Bacillus amyloliquefaciens, B. siamenensis, B. velezensis, B. subtilis* and *B. atrophaeus* harbor almost complete *pks* genetic macroclusters for the production of polyketides. While *B. licheniformis, B. cereus, B. pacificus,* and the *probiotics B. clausii, B. coagulans* did not contained the PKs loci. The antimicrobial effects of polyketides are site colonization specific and the strains are scarcely used for health biotechnological interests [65]. Moreover, the ecological impact of these antimicrobial substances on the gut microbiota composition may have a huge impact, beyond the modification and control of the colonization of commensals and pathogenic bacteria, e.g., to cause weight gain effects in humans as well as in animals [66].

3.3.2. WGS Representative *Bacillus* sp. AM1 from Microbiota: Genome Mining Data

From the analysis of the specific *Bacillus* sp. AM1 WGS, the cluster genes and enzymes related to PKs biosynthesis were identified (*bae, mln,* and *dfn*) and they were related to the production of bacillaene, and two other polyketides macrolactin and difficidin.

This complex microbial ecosystem seems to be enriched in new bacterial strains belonging to *Bacillus* genus that produce PKs with a wide range of applications in the current biotechnological context. Among these applications, PKs stand out for their antimicrobial capacity against certain bacterial species. Therefore, further identification through bioinformatics tools and experimental data will confirm the functionality of these bioactive substances.

Advances in NGS and in silico tools allow to perform an appropriate screening of genes of concern or interest in microbiota, such as antimicrobial resistance genes and the capacity of antimicrobial production of cultivable isolates WGS. A better understanding of the microbiota ecology, driven by the bioactive compounds released by its components, will lead to better clinical interventions. Antimicrobials naturally synthetized by gut microorganisms are mainly described as bacteriocins [12]. However, it is important to consider other molecules acting as antimicrobial as polyketides. Isolation and elucidation of PKs structures by nuclear magnetic resonance (NMR) methods are limited by the concentration needed for analysis [67]. Thus, it is possible to predict the types of PKs and their variants, as showed for Bacillales [37]. Genome mining performed in the present study allowed BLAST driven search for predicted PKs clusters. Pascal ad hoc software analysed the type strain genomes making it a powerful prediction tool. Similarly, another useful prediction tool could be used as nonribosomal peptide-synthetase NRPS/PKs substrate predictor [68].

Table 5. Gene-encoding and corresponding enzymes involved in Polyketide biosynthesis in WGS of Type strain of *Bacillus spp*.

Enzyme	Enzyme description EC number	B. amyloliquefaciens WF02T NZ_CP053376	B. siamenensis SCSIO 05746T NZ_CP025001	B. velezensis CBMB205T NZ_CP011937	B. subtilis 168T NC_000964	B. atrophaeus BSST NZ_CP007640	B. sp-AM1 B1T CP047644.1)
PksA	Hypothetical protein/EC:3.1.2.6	WP_024085315.1 174131..1741526	WP_060962748.1 2494188..2494397	WP_032874955.1 2222103..2222312	NP_000389590.1 1782713..1783390	WP_013390522.1 1165636..1167084	1787442..1787651 QHJ03379.1
-	Hypothetical protein/EC:3.1.2.6	WP_024085326.1 1816193..1816555	WP_016936035.1 2419160..2419522	WP_007410383.1 2146808..2147170	YP_0009513956.1 1783500..1783766	WP_003328852.1 1167393..1167932	-
Regulator	TetR family transcriptional regulator C terminal	-	-	-	NP_000389589.1 1781906..1782523	WP_003328851.1 1168054..1168644	-
PksB	MBL fold metallo hydrolase/ EC: 2.3.1.39	WP_024085316.1 1742160..1742837	WP_060962747.1 2492787..2493464	WP_032874957.1 2220496..2221173	YP_0009513956.1 1783500..1783766	WP_003328850.1 1168942..1169619	1788295..1788972 QHJ03380.1
PksC	ACP S malonyltransferase/ EC:2.3.1.51	WP_014305029.1 1743152..1744021	WP_060962746.1 2491603..2492472	WP_032874959.1 2219312..2220181	NP_000389591.1 1783763..1784629	WP_003328849.1 1170013..1170879	1789287..1790156 QHJ03381.1
PksD	Acyltransferase domain containing protein/EC: 2.3.1.39	WP_003154101.1 1744158..1745132	WP_060962745.1 2490494..2491468	WP_032874961.1 2218201..2219175	NP_000389592.2 1785133..1786107	WP_003328847.1 1171417..1172382	1790293..1791267 QHJ03382.1
PksE	ACP S malonyltransferase/ EC:1.3.1.9 and 1.3.1.10	WP_003154100.1 1745134..1747374	ID Not found 2488250..2490492	WP_032874963.1 2215959..2218199	NP_000389593.3 1786104..1788407	WP_003328846.1 1172389..1174752	1791269..1793509 QHJ03383.1
AcpK	Acyl carrier protein/EC:2.3.3.10	WP_003154099.1 1747440..1747688	WP_060962743.1 2487934..2488182	WP_012117592.1 2215645..2215893	NP_00570904.1 1788469..1788717	WP_003328845.1 1174891..1175139	1793575..1793823 QHJ03384.1
PksF	Polyketide beta ketoacyl:ACP synthase/EC: 4.2.1.17	-	-	-	NP_000389594.2 1788695..1789942	WP_003328844.1 1175117..1176364	-
PksG	Hydroxymethylglutaryl CoA synthase family/EC: 4.2.1.17	WP_003154098.1 1747740..1749002	WP_060962742.1 2486620..2487882	WP_032874965.1 2214331..2215593	NP_000389595.2 1789943..1791205	WP_010788667.1 1176364..1177626	1793875..1795137 QHJ03385.1
PksH	Enoyl CoA hydratase/isomerase	WP_024085319.1 1748999..1749772	WP_060962741.1 2485850..2486623	WP_032874967.1 2213561..2214334	NP_000389596.1 1791193..1791972	WP_087941777.1 1177614..1178390	1795134..1795907 QHJ03386.1
PksI	enoyl CoA hydratase/isomerase family protein	WP_003154094.1 1749782..1750531	WP_060962740.1 2485091..2485840	WP_003154094.1 2212802..2213551	NP_000389597.2 1792012..1792761	WP_003328841.1 1178438..1179184	1795917..1796666 QHJ03387.1
PksJ	Non ribosomal peptide synthetase	WP_024085320.1 1750571..1765525	WP_060962739.1 2470129..2485062	WP_032874969.1 2197814..2212762	NP_000389598.3 1792806..1807937	WP_013390525.1 1179247..1194429	1796706..1811657 QHJ03388.1
PksM	SDR family NAD(P) dependent oxidoreductase EC:1.6.5.2	WP_165869029.1 1765509..1778951	WP_167388675.1 2456724..2470145	WP_162859398.1 2184400..2197830	NP_000389601.3 1821553..1834341	WP_013390526.1 1194431..1208248	1811659..1825086 QHJ03389.1
PksM	SDR family NAD(P) dependent oxidoreductase/EC:1.6.5.2	WP_024085322.1 1778969..1789513	WP_101605493.1 2446202..2456707	WP_032874973.1 2173847..2184382	NP_000389602.3 1834409..1850875	WP_013390527.1 1208267..1221238	1825104..1835639 QHJ03390.1
PksN	Non ribosomal peptide synthetase	-	WP_101605492.1 2429908 2446212	WP_032874975.1 2157559 2173857	NP_000389604.2 1850890 1858521	WP_087941783.1 1221318..1237793	1835629..1851930 QHJ03391.1
PksR	Polyketide synthase dehydratase domain/EC:2.1.1.-	WP_024085324.1 1805818..1813275	WP_060962735.1 2422440..2429894	WP_032874977.1 2150088..2157545	NP_000389600.3 1807921..1821537	WP_003328830.1 1237809..1245533	1851944..1859401 QHJ03392.1
PksS	Cytochrome P450/EC:1.14.14.-	WP_024085325.1 1813410..1814621	WP_060962734.1 2421090..2422301	WP_032875233.1 2148742..2149953	NP_000389605.2 1858566..1859783	WP_003328829.1 1245647..1246888	1859536..1860747 QHJ03393.1

B. licheniformis (strain ATCC 14580)^{T}; NC_006270 PKs Loci was not found; *B. cereus (strain B4264)* NC_011725 PKs Loci was not found; *B. pacificus* (strain R1) NC_NJQG01000001 Loci was not found; *B. clausii* (strain 7520-2 contig00001)^{T} NZ_NPBN01000001 PKs Loci was not found; *B. coagulans* (B4099 NODE_1)^{T} NZ_LQYI01000001 PKs Loci was not found; *B. nematocida* (strain B-16^{T}) No WGS is available—Analysis PKS Loci was not applicable [69].

Importantly, *Bacillus* and specific WGS genes description is needed to verify the safety assessment of different strains if they are proposed to be used in food or feed chain [70]. Moreover, the safety of a beneficial microbe or probiotic strain must be sufficiently characterized by high-throughput technologies, safe for the intended use, and assessed through pathogenicity, immunotoxicity, and colonization, in addition to its antibiotic resistance profile [71]. However currently, there is no consensus or standardization for the interventional use of probiotics [72]. In addition to general guidelines for the qualification of the QPS, European Food Safety Authority (EFSA) made a supplementary requirement for *Bacillus* species other than the *Bacillus cereus* group, where a cytotoxicity test should be performed to determine whether the strain produces high levels of non-ribosomal synthesised peptides. One of the criteria for strains to fulfill and meet the requirements for QPS and generally recognized as safe (GRAS) standards is antimicrobial activity and the absence of antimicrobial resistance genes as a possible safety concern against critically important antimicrobials (CIAs) or highly important antimicrobials (HIAs), which might eventually be transferred via horizontal gene transfer to pathogenic bacteria during food manufacture or after consumption [33,73]. According to the general guidelines for the qualifications of the QPS, unless the strain qualifies for the QPS approach or belongs to a taxonomic unit, known not to produce antimicrobials relevant to use in humans and animals, assessment should be made to determine the inhibitory activity of culture supernatants against reference strains, known to be susceptible to a range of antibiotics and the inhibitory substance [47]. A slight adjustment has been made for the production strains, which have to demonstrate the absence of carry-over into the final product together with the exact phase of the industrial scale manufacturing process, and whether any CIAs or HIAs are used during the manufacturing of the product, to determine compatibility with other additives showing antimicrobial activity and, furthermore, possible co-/cross-resistance [35].

4. Conclusions

Bacillus strains isolated from human gut microbiota, and taxonomically closest to the safely qualified *B. subtilis* and *B. amyloliquefaciens* groups, became cultivable predominant taxa when high bisphenol exposure conditions were tested. In parallel, these strains harbored PKS molecular gene biosynthetic loci and showed phenotypic antimicrobial effects. Therefore, they might be proposed as beneficial microorganisms with molecular features that would contribute to modulate the ecological taxa composition and functionality of human gut microbiota. Intervention studies will be further needed to demonstrate the ability to recover from microbiota dysbiosis, triggered by high MDC exposure diets and lifestyles, towards eubiosis and healthier status.

5. Patents

IPR-823 Application in progress.

Author Contributions: Conceptualization, M.A.; methodology, A.T.-S. and J.P.-C.; writing—original draft preparation, A.T.-S., A.L.-M., and J.P.-C.; writing—review and editing: A.T.-S., J.P.-C., A.L.-M., Á.R.-M., K.C. and M.A.; supervision, M.A.; project administration, M.A.; funding acquisition, M.A. All authors have read and agreed to the published version of the manuscript.

Funding: This work was carried out within the frame of FEDER-Infrastructure Project: IE_2019-198 and OBEMIRISK EFSA-Partnering Grant Project GP/EFSA/ENCO/2018/03-GA04. A.T.-S. and Á.R.-M. "Colaboración Investigación—Master". J.P.-C. "ICARO-Extracurriculares Prácticas". A.L.-M. "Incentivación de la Investigación". Plan Propio-UGR. K.C. is collaborating with UGR under the EU-FORA Programme (2020/2021).

Institutional Review Board Statement: Ethical review and approval were waived for this specific study, due to the specimens collection belonged to our previous approved project INFABIO.

Informed Consent Statement: Not applicable.

Data Availability Statement: Not applicable.

Acknowledgments: Part of the results are from Alfonso Torres-Sánchez doctoral thesis, Nutrition and Food Technology Doctorate Program of the University of Granada and Ana López-Moreno doctoral thesis, Biomedicine Doctorate Program.

Conflicts of Interest: The authors declare no conflict of interest.

Abbreviations

AN	Anorexia Nervosa
BHI	Brain-Hearth Infusion
BPA	Bisphenol A
CIAs	Critically Important Antimicrobials
CIC	Centro de Instrumentación Científica
EC	Enzyme Commission number
EFSA	European Food Safety Authority
GRAS	Generally Recognized as Safe
HC	Healthy control
HIAs	Highly Important Antimicrobials
IPBLN	Institute of Parasitology and Biomedicine "López-Neyra"
JECFA	Joint FAO/WHO Expert Committee on Food AdditivesLiquid
LC-MS/MS	Liquid Chromatography-Mass Spectrometry
MCL	Maximum Composite Likelihood
MDC	Microbiota Disrupting Chemicals
MetS	Metabolic syndrome
MIC	Minimum Inhibitory Concentration
MRS	Man-Rogosa-Sharpe
NAFLD	Nonalcoholic fatty liver disease
NCBI	National Center for Biotechnology Information
NGP	Next Generation Probiotics
NGS	Next Generation Sequencing
NMR	Nuclear Magnetic Resonance
NRPS	Nonribosomal Peptide-synthetase
OB	Obesity
OW	Over-weigth
PKs	Poliketides
QPS	Qualified Presumption of Safety
SN	Supernatant
T1D	Type 1 Diabetes
T2D	Type 2 Diabetes
WGS	Whole Genome Sequences
WHO	World Health Organization

References

1. Casado, V.; Martín, D.; Torres, C.; Reglero, G. Phospholipases in food industry: A review. *Methods Mol. Biol.* **2012**, *861*, 495–523. [PubMed]
2. Moran, K.; de Lange, C.F.M.; Ferket, P.; Fellner, V.; Wilcock, P.; van Heugten, E. Enzyme supplementation to improve the nutritional value of fibrous feed ingredients in swine diets fed in dry or liquid form. *J. Anim. Sci.* **2016**, *94*, 1031–1040. [CrossRef]
3. Lagier, J.-C.; Armougom, F.; Million, M.; Hugon, P.; Pagnier, I.; Robert, C.; Bittar, F.; Fournous, G.; Gimenez, G.; Maraninchi, M.; et al. Microbial culturomics: Paradigm shift in the human gut microbiome study. *Clin. Microbiol. Infect.* **2012**, *18*, 1185–1193. [CrossRef] [PubMed]
4. López-Moreno, A.; Acuña, I.; Torres-Sánchez, A.; Ruiz-Moreno, Á.; Cerk, K.; Rivas, A.; Suárez, A.; Monteoliva-Sánchez, M.; Aguilera, M. Next Generation Probiotics for Neutralizing Obesogenic Effects: Taxa Culturing Searching Strategies. *Nutrients* **2021**, *13*, 1617. [CrossRef]
5. Aguilera, M.; Gálvez-Ontiveros, Y.; Rivas, A. Endobolome, a New Concept for Determining the Influence of Microbiota Disrupting Chemicals (MDC) in Relation to Specific Endocrine Pathogenesis. *Front. Microbiol.* **2020**, *11*, 578007. [CrossRef] [PubMed]
6. Adair, K.L.; Douglas, A.E. Making a microbiome: The many determinants of host-associated microbial community composition. *Curr. Opin. Microbiol.* **2017**, *35*, 23–29. [CrossRef]
7. Diakite, A.; Dubourg, G.; Dione, N.; Afouda, P.; Bellali, S.; Ngom, I.I.; Valles, C.; Million, M.; Levasseur, A.; Cadoret, F.; et al. Extensive culturomics of 8 healthy samples enhances metagenomics efficiency. *PLoS ONE* **2019**, *14*, e0223543. [CrossRef]

8. WoldemariamYohannes, K.; Wan, Z.; Yu, Q.; Li, H.; Wei, X.; Liu, Y.; Wang, J.; Sun, B. Prebiotic, Probiotic, Antimicrobial, and Functional Food Applications of Bacillus amyloliquefaciens. *J. Agric. Food Chem.* **2020**, *68*, 14709–14727. [CrossRef]
9. Konuray, G.; Erginkaya, Z. Potential Use of Bacillus coagulans in the Food Industry. *Foods* **2018**, *7*, 92. [CrossRef]
10. Cutting, S.M. Bacillus probiotics. *Food Microbiol.* **2011**, *28*, 214–220. [CrossRef] [PubMed]
11. Caulier, S.; Nannan, C.; Gillis, A.; Licciardi, F.; Bragard, C.; Mahillon, J. Overview of the Antimicrobial Compounds Produced by Members of the Bacillus subtilis Group. *Front. Microbiol.* **2019**, *10*, 302. [CrossRef]
12. Garcia-Gutierrez, E.; Mayer, M.J.; Cotter, P.D.; Narbad, A. Gut microbiota as a source of novel antimicrobials. *Gut Microbes* **2019**, *10*, 1–21. [CrossRef] [PubMed]
13. Gao, R.; Zhu, C.; Li, H.; Yin, M.; Pan, C.; Huang, L.; Kong, C.; Wang, X.; Zhang, Y.; Qu, S.; et al. Dysbiosis Signatures of Gut Microbiota Along the Sequence from Healthy, Young Patients to Those with Overweight and Obesity. *Obesity* **2018**, *26*, 351–361. [CrossRef] [PubMed]
14. Armougom, F.; Henry, M.; Vialettes, B.; Raccah, D.; Raoult, D. Monitoring bacterial community of human gut microbiota reveals an increase in Lactobacillus in obese patients and Methanogens in anorexic patients. *PLoS ONE* **2009**, *4*, e7125. [CrossRef] [PubMed]
15. Sedighi, M.; Razavi, S.; Navab-Moghadam, F.; Khamseh, M.E.; Alaei-Shahmiri, F.; Mehrtash, A.; Amirmozafari, N. Comparison of gut microbiota in adult patients with type 2 diabetes and healthy individuals. *Microb. Pathog.* **2017**, *111*, 362–369. [CrossRef]
16. Ahmad, A.; Yang, W.; Chen, G.; Shafiq, M.; Javed, S.; Zaidi, S.S.A.; Shahid, R.; Liu, C.; Bokhari, H. Analysis of gut microbiota of obese individuals with type 2 diabetes and healthy individuals. *PLoS ONE* **2019**, *14*, e0226372. [CrossRef] [PubMed]
17. Ejtahed, H.-S.; Hoseini-Tavassol, Z.; Khatami, S.; Zangeneh, M.; Behrouzi, A.; Ahmadi Badi, S.; Moshiri, A.; Hasani-Ranjbar, S.; Soroush, A.-R.; Vaziri, F.; et al. Main gut bacterial composition differs between patients with type 1 and type 2 diabetes and non-diabetic adults. *J. Diabetes Metab. Disord.* **2020**, *19*, 265–271. [CrossRef]
18. Wang, B.; Jiang, X.; Cao, M.; Ge, J.; Bao, Q.; Tang, L.; Chen, Y.; Li, L. Altered Fecal Microbiota Correlates with Liver Biochemistry in Nonobese Patients with Non-alcoholic Fatty Liver Disease. *Sci. Rep.* **2016**, *6*, 32002. [CrossRef]
19. Li, F.; Sun, G.; Wang, Z.; Wu, W.; Guo, H.; Peng, L.; Wu, L.; Guo, X.; Yang, Y. Characteristics of fecal microbiota in non-alcoholic fatty liver disease patients. *Sci. China Life Sci.* **2018**, *61*, 770–778. [CrossRef]
20. Raman, M.; Ahmed, I.; Gillevet, P.M.; Probert, C.S.; Ratcliffe, N.M.; Smith, S.; Greenwood, R.; Sikaroodi, M.; Lam, V.; Crotty, P.; et al. Fecal Microbiome and Volatile Organic Compound Metabolome in Obese Humans With Nonalcoholic Fatty Liver Disease. *Clin. Gastroenterol. Hepatol.* **2013**, *11*, 868–875.e3. [CrossRef]
21. Nistal, E.; Sáenz de Miera, L.E.; Ballesteros Pomar, M.; Sánchez-Campos, S.; García-Mediavilla, M.V.; Álvarez-Cuenllas, B.; Linares, P.; Olcoz, J.L.; Arias-Loste, M.T.; García-Lobo, J.M.; et al. An altered fecal microbiota profile in patients with non-alcoholic fatty liver disease (NAFLD) associated with obesity. *Rev. Española Enferm. Dig.* **2019**, *111*, 275–282. [CrossRef] [PubMed]
22. Lim, M.Y.; You, H.J.; Yoon, H.S.; Kwon, B.; Lee, J.Y.; Lee, S.; Song, Y.-M.; Lee, K.; Sung, J.; Ko, G. The effect of heritability and host genetics on the gut microbiota and metabolic syndrome. *Gut* **2017**, *66*, 1031–1038. [CrossRef] [PubMed]
23. Vandenberg, L.N.; Hauser, R.; Marcus, M.; Olea, N.; Welshons, W.V. Human exposure to bisphenol A (BPA). *Reprod. Toxicol.* **2007**, *24*, 139–177. [CrossRef]
24. Vandenberg, L.N.; Chahoud, I.; Heindel, J.J.; Padmanabhan, V.; Paumgartten, F.J.R.; Schoenfelder, G. Urinary, circulating, and tissue biomonitoring studies indicate widespread exposure to bisphenol A. *Cienc. Saude Coletiva* **2012**, *17*, 407–434. [CrossRef]
25. Louati, I.; Dammak, M.; Nasri, R.; Belbahri, L.; Nasri, M.; Abdelkafi, S.; Mechichi, T. Biodegradation and detoxification of bisphenol A by bacteria isolated from desert soils. *3 Biotech* **2019**, *9*, 228. [CrossRef]
26. Gramec Skledar, D.; Peterlin Mašič, L. Bisphenol A and its analogs: Do their metabolites have endocrine activity? *Environ. Toxicol. Pharmacol.* **2016**, *47*, 182–199. [CrossRef] [PubMed]
27. Gálvez-Ontiveros, Y.; Moscoso-Ruiz, I.; Rodrigo, L.; Aguilera, M.; Rivas, A.; Zafra-Gómez, A. Presence of parabens and bisphenols in food commonly consumed in spain. *Foods* **2021**, *10*, 92. [CrossRef]
28. Cohen, I.C.; Cohenour, E.R.; Harnett, K.G.; Schuh, S.M. BPA, BPAF and TMBPF Alter Adipogenesis and Fat Accumulation in Human Mesenchymal Stem Cells, with Implications for Obesity. *Int. J. Mol. Sci.* **2021**, *22*, 5363. [CrossRef]
29. Camacho, L.; Lewis, S.M.; Vanlandingham, M.M.; Olson, G.R.; Davis, K.J.; Patton, R.E.; Twaddle, N.C.; Doerge, D.R.; Churchwell, M.I.; Bryant, M.S.; et al. A two-year toxicology study of bisphenol A (BPA) in Sprague-Dawley rats: CLARITY-BPA core study results. *Food Chem. Toxicol.* **2019**, *132*, 110728. [CrossRef]
30. Gundert-Remy, U.; Bodin, J.; Bosetti, C.; FitzGerald, R.; Hanberg, A.; Hass, U.; Hooijmans, C.; Rooney, A.A.; Rousselle, C.; van Loveren, H.; et al. Bisphenol A (BPA) hazard assessment protocol. *EFSA Support. Publ.* **2017**, *14*, 1354E.
31. Gómez-Gallego, C.; Pohl, S.; Salminen, S.; De Vos, W.M.; Kneifel, W. Akkermansia muciniphila: A novel functional microbe with probiotic properties. *Benef. Microbes* **2016**, *7*, 571–584. [CrossRef] [PubMed]
32. Brodmann, T.; Endo, A.; Gueimonde, M.; Vinderola, G.; Kneifel, W.; de Vos, W.M.; Salminen, S.; Gómez-Gallego, C. Safety of Novel Microbes for Human Consumption: Practical Examples of Assessment in the European Union. *Front. Microbiol.* **2017**, *8*, 1725. [CrossRef]
33. Koutsoumanis, K.; Allende, A.; Alvarez-Ordóñez, A.; Bolton, D.; Bover-Cid, S.; Chemaly, M.; Davies, R.; Cesare, A.D.; Hilbert, F.; Lindqvist, R.; et al. Scientific Opinion on the update of the list of QPS-recommended biological agents intentionally added to food or feed as notified to EFSA (2017–2019). *EFSA J.* **2020**, *18*, e05966. [PubMed]

34. Bampidis, V.; Azimonti, G.; Bastos, M.d.L.; Christensen, H.; Dusemund, B.; Kouba, M.; Durjava, M.F.; López-Alonso, M.; Puente, S.L.; Marcon, F.; et al. Safety and efficacy of Bacillus subtilisPB6 (Bacillus velezensisATCC PTA-6737) as a feed additive for chickens for fattening, chickens reared for laying, minor poultry species (except for laying purposes), ornamental, sporting and game birds. *EFSA J.* **2020**, *18*, e06280. [PubMed]
35. Rychen, G.; Aquilina, G.; Azimonti, G.; Bampidis, V.; Bastos, M.d.L.; Bories, G.; Chesson, A.; Cocconcelli, P.S.; Flachowsky, G.; Gropp, J.; et al. Guidance on the characterisation of microorganisms used as feed additives or as production organisms. *EFSA J.* **2018**, *16*, e05206. [PubMed]
36. Aleti, G.; Sessitsch, A.; Brader, G. Genome mining: Prediction of lipopeptides and polyketides from Bacillus and related Firmicutes. *Comput. Struct. Biotechnol. J.* **2015**, *13*, 192–203. [CrossRef]
37. Straight, P.D.; Fischbach, M.A.; Walsh, C.T.; Rudner, D.Z.; Kolter, R. A singular enzymatic megacomplex from Bacillus subtilis. *Proc. Natl. Acad. Sci. USA* **2007**, *104*, 305–310. [CrossRef]
38. García-Córcoles, M.T.; Cipa, M.; Rodríguez-Gómez, R.; Rivas, A.; Olea-Serrano, F.; Vílchez, J.L.; Zafra-Gómez, A. Determination of bisphenols with estrogenic activity in plastic packaged baby food samples using solid-liquid extraction and clean-up with dispersive sorbents followed by gas chromatography tandem mass spectrometry analysis. *Talanta* **2018**, *178*, 441–448. [CrossRef]
39. López-Moreno, A.; Torres-Sánchez, A.; Acuña, I.; Suárez, A.; Aguilera, M. Representative Bacillus sp. AM1 from Gut Microbiota Harbor Versatile Molecular Pathways for Bisphenol A Biodegradation. *Int. J. Mol. Sci.* **2021**, *22*, 4952. [CrossRef]
40. Yoon, S.-H.; Ha, S.-M.; Kwon, S.; Lim, J.; Kim, Y.; Seo, H.; Chun, J. Introducing EzBioCloud: A taxonomically united database of 16S rRNA gene sequences and whole-genome assemblies. *Int. J. Syst. Evol. Microbiol.* **2017**, *67*, 1613–1617. [CrossRef]
41. Menasria, T.; Aguilera, M.; Hocine, H.; Benammar, L.; Ayachi, A.; Si Bachir, A.; Dekak, A.; Monteoliva-Sánchez, M. Diversity and bioprospecting of extremely halophilic archaea isolated from Algerian arid and semi-arid wetland ecosystems for halophilic-active hydrolytic enzymes. *Microbiol. Res.* **2018**, *207*, 289–298. [CrossRef] [PubMed]
42. Montalvo-Rodríguez, R.; Vreeland, R.H.; Oren, A.; Kessel, M.; Betancourt, C.; López-Garriga, J. *Halogeometricum borinquense* gen. nov., sp. nov., a novel halophilic archaeon from Puerto Rico. *Int. J. Syst. Bacteriol.* **1998**, *48 Pt 4*, 1305–1312. [CrossRef] [PubMed]
43. Kasana, R.C.; Salwan, R.; Dhar, H.; Dutt, S.; Gulati, A. A rapid and easy method for the detection of microbial cellulases on agar plates using gram's iodine. *Curr. Microbiol.* **2008**, *57*, 503–507. [CrossRef] [PubMed]
44. Allais, J.J.; Hoyos-Lopez, G.; Kammoun, S.; Baratti, J.C. Isolation and characterization of thermophilic bacterial strains with inulinase activity. *Appl. Environ. Microbiol.* **1987**, *53*, 942–945. [CrossRef]
45. Sierra, G. A simple method for the detection of lipolytic activity of micro-organisms and some observations on the influence of the contact between cells and fatty substrates. *Antonie Van Leeuwenhoek* **1957**, *23*, 15–22. [CrossRef]
46. Jeffries, C.D.; Holtman, D.F.; Guse, D.G. Rapid method for determining the activity of microorganisms on nucleic acids. *J. Bacteriol.* **1957**, *73*, 590–591. [CrossRef]
47. Combined Compendium of Food Additive Specifications. Available online: http://www.fao.org/3/a0691e/a0691e00.htm (accessed on 24 June 2021).
48. Powthong, P.; Suntornthiticharoen, P. Antimicrobial and enzyme activity produced by *Bacillus* spp. Isolated from soil. *Int. J. Pharm. Pharm. Sci.* **2017**, *9*, 205–210. [CrossRef]
49. Rangwala, S.H.; Kuznetsov, A.; Ananiev, V.; Asztalos, A.; Borodin, E.; Evgeniev, V.; Joukov, V.; Lotov, V.; Pannu, R.; Rudnev, D.; et al. Accessing NCBI data using the NCBI Sequence Viewer and Genome Data Viewer (GDV). *Genome Res.* **2020**, *31*, 159–169. [CrossRef]
50. Yu, K.; Yi, S.; Li, B.; Guo, F.; Peng, X.; Wang, Z.; Wu, Y.; Alvarez-Cohen, L.; Zhang, T. An integrated meta-omics approach reveals substrates involved in synergistic interactions in a bisphenol A (BPA)-degrading microbial community. *Microbiome* **2019**, *7*, 16. [CrossRef] [PubMed]
51. Jiménez-Pranteda, M.L.; Pérez-Davó, A.; Monteoliva-Sánchez, M.; Ramos-Cormenzana, A.; Aguilera, M. Food Omics Validation: Towards Understanding Key Features for Gut Microbiota, Probiotics and Human Health. *Food Anal. Methods* **2015**, *8*, 272–289. [CrossRef]
52. O'Toole, P.W.; Marchesi, J.R.; Hill, C. Next-generation probiotics: The spectrum from probiotics to live biotherapeutics. *Nat. Microbiol.* **2017**, *2*, 17057. [CrossRef]
53. Everard, A.; Belzer, C.; Geurts, L.; Ouwerkerk, J.P.; Druart, C.; Bindels, L.B.; Guiot, Y.; Derrien, M.; Muccioli, G.G.; Delzenne, N.M.; et al. Cross-talk between Akkermansia muciniphila and intestinal epithelium controls diet-induced obesity. *Proc. Natl. Acad. Sci. USA* **2013**, *110*, 9066–9071. [CrossRef] [PubMed]
54. Cani, P.D.; de Vos, W.M. Next-Generation Beneficial Microbes: The Case of Akkermansia muciniphila. *Front. Microbiol.* **2017**, *8*, 1765. [CrossRef] [PubMed]
55. Szlufman, C.; Shemesh, M. Role of Probiotic Bacilli in Developing Synbiotic Food: Challenges and Opportunities. *Front. Microbiol.* **2021**, *12*, 811. [CrossRef] [PubMed]
56. Khalid, F.; Khalid, A.; Fu, Y.; Hu, Q.; Zheng, Y.; Khan, S.; Wang, Z. Potential of Bacillus velezensis as a probiotic in animal feed: A review. *J. Microbiol.* **2021**, *59*, 627–633. [CrossRef]
57. Sultana, O.; Lee, S.; Seo, H.; Mahmud, H.A.; Kim, S.; Seo, A.; Kim, M.; Song, H.-Y. Biodegradation and removal of PAH by Bacillus velezensis isolated from fermented food. *J. Microbiol. Biotechnol.* **2021**, *31*. [CrossRef]
58. Kang, M.; Choi, H.J.; Yun, B.; Lee, J.; Yoo, J.; Yang, H.-J.; Jeong, D.-Y.; Kim, Y.; Oh, S. Bacillus amyloliquefaciens SCGB1 Alleviates Dextran Sulfate Sodium-Induced Colitis in Mice Through Immune Regulation. *J. Med. Food* **2021**, *24*, 709–719. [CrossRef]

59. Harwood, C.R.; Mouillon, J.-M.; Pohl, S.; Arnau, J. Secondary metabolite production and the safety of industrially important members of the Bacillus subtilis group. *FEMS Microbiol. Rev.* **2018**, *42*, 721–738. [CrossRef]
60. Bianco, A.; Capozzi, L.; Monno, M.R.; Del Sambro, L.; Manzulli, V.; Pesole, G.; Loconsole, D.; Parisi, A. Characterization of Bacillus cereus Group Isolates From Human Bacteremia by Whole-Genome Sequencing. *Front. Microbiol.* **2021**, *11*, 599524. [CrossRef]
61. Devaraj, K.; Aathika, S.; Periyasamy, K.; Periyaraman, P.M.; Palaniyandi, S.; Subramanian, S. Production of thermostable multiple enzymes from Bacillus amyloliquefaciens KUB29. *Nat. Prod. Res.* **2019**, *33*, 1674–1677. [CrossRef]
62. Deb, P.; Talukdar, S.A.; Mohsina, K.; Sarker, P.K.; Sayem, S.A. Production and partial characterization of extracellular amylase enzyme from Bacillus amyloliquefaciens P-001. *SpringerPlus* **2013**, *2*, 154. [CrossRef]
63. Latorre, J.D.; Hernandez-Velasco, X.; Wolfenden, R.E.; Vicente, J.L.; Wolfenden, A.D.; Menconi, A.; Bielke, L.R.; Hargis, B.M.; Tellez, G. Evaluation and Selection of Bacillus Species Based on Enzyme Production, Antimicrobial Activity, and Biofilm Synthesis as Direct-Fed Microbial Candidates for Poultry. *Front. Vet. Sci.* **2016**, *3*, 95. [CrossRef]
64. Chakraborty, K.; Thilakan, B.; Kizhakkekalam, V.K. Antibacterial aryl-crowned polyketide from Bacillus subtilis associated with seaweed Anthophycus longifolius. *J. Appl. Microbiol.* **2018**, *124*, 108–125. [CrossRef] [PubMed]
65. Piel, J.; Butzke, D.; Fusetani, N.; Hui, D.; Platzer, M.; Wen, G.; Matsunaga, S. Exploring the Chemistry of Uncultivated Bacterial Symbionts: Antitumor Polyketides of the Pederin Family. *J. Nat. Prod.* **2005**, *68*, 472–479. [CrossRef] [PubMed]
66. Angelakis, E. Weight gain by gut microbiota manipulation in productive animals. *Microb. Pathog.* **2017**, *106*, 162–170. [CrossRef] [PubMed]
67. Hoover, D.G.; Harlander, S.K. CHAPTER 2—Screening Methods for Detecting Bacteriocin Activity. In *Bacteriocins of Lactic Acid Bacteria*; Hoover, D.G., Steenson, L.R., Eds.; Academic Press: Cambridge, MA, USA, 1993; pp. 23–39. ISBN 978-0-12-355510-6.
68. Tang, G.-L.; Cheng, Y.-Q.; Shen, B. Leinamycin biosynthesis revealing unprecedented architectural complexity for a hybrid polyketide synthase and nonribosomal peptide synthetase. *Chem. Biol.* **2004**, *11*, 33–45. [CrossRef] [PubMed]
69. Huang, X.-W.; Niu, Q.-H.; Zhou, W.; Zhang, K.-Q. Bacillus nematocida sp. nov., a novel bacterial strain with nematotoxic activity isolated from soil in Yunnan, China. *Syst. Appl. Microbiol.* **2005**, *28*, 323–327. [CrossRef]
70. EFSA (European Food Safety Authority). Statement on the Requirements for Whole Genome Sequence Analysis of Microorganisms Intentionally Used in the Food Chain, 2021. Available online: https://www.efsa.europa.eu/sites/default/files/2021-03/EFSA-statement-EFSA-Q-2019-00434.pdf (accessed on 30 May 2021).
71. Swann, J.R.; Rajilic-Stojanovic, M.; Salonen, A.; Sakwinska, O.; Gill, C.; Meynier, A.; Fança-Berthon, P.; Schelkle, B.; Segata, N.; Shortt, C.; et al. Considerations for the design and conduct of human gut microbiota intervention studies relating to foods. *Eur. J. Nutr.* **2020**, *59*, 3347–3368. [CrossRef]
72. Silva, D.R.; Sardi, J.d.C.O.; Pitangui, N.d.S.; Roque, S.M.; Silva, A.C.B.d.; Rosalen, P.L. Probiotics as an alternative antimicrobial therapy: Current reality and future directions. *J. Funct. Foods* **2020**, *73*, 104080. [CrossRef]
73. World Health Organization. WHO Advisory Group on Integrated Surveillance of Antimicrobial Resistance. In *Critically Important Antimicrobials for Human Medicine: Ranking of Antimicrobial Agents for Risk Management of Antimicrobial Resistance Due to Non-Human Use*; World Health Organization: Geneva, Switzerland, 2017; ISBN 978-92-4-151222-0.

microorganisms

MDPI

Review

Gut Dysbiosis during COVID-19 and Potential Effect of Probiotics

Yuan-Pin Hung [1,2,†], Ching-Chi Lee [2,3,4,†], Jen-Chieh Lee [2], Pei-Jane Tsai [5,6,7] and Wen-Chien Ko [2,8,*]

1 Department of Internal Medicine, Tainan Hospital, Ministry of Health and Welfare, Tainan 700, Taiwan; yuebin16@yahoo.com.tw
2 Department of Internal Medicine, College of Medicine, National Cheng Kung University Hospital, National Cheng Kung University, Tainan 704, Taiwan; chichingbm85@yahoo.com.tw (C.-C.L.); jclee.eric@msa.hinet.net (J.-C.L.)
3 Clinical Medicine Research Center, College of Medicine, National Cheng Kung University Hospital, National Cheng Kung University, Tainan 705, Taiwan
4 Graduate Institute of Medical Sciences, College of Health Sciences, Chang Jung Christian University, Tainan 711, Taiwan
5 Department of Medical Laboratory Science and Biotechnology, College of Medicine, National Cheng Kung University, Tainan 705, Taiwan; peijtsai@mail.ncku.edu.tw
6 Institute of Basic Medical Sciences, College of Medicine, National Cheng Kung University, Tainan 705, Taiwan
7 Department of Pathology, National Cheng Kung University Hospital, Tainan 705, Taiwan
8 Department of Medicine, College of Medicine, National Cheng Kung University, Tainan 705, Taiwan
* Correspondence: winston3415@gmail.com
† These authors contribute equally.

Abstract: Severe acute respiratory syndrome coronavirus 2 (SARS-CoV-2), an RNA virus of the family *Coronaviridae*, causes coronavirus disease 2019 (COVID-19), an influenza-like disease that chiefly infects the lungs through respiratory transmission. The spike protein of SARS-CoV-2, a transmembrane protein in its outer portion, targets angiotensin-converting enzyme 2 (ACE2) as the binding receptor for the cell entry. As ACE2 is highly expressed in the gut and pulmonary tissues, SARS-CoV-2 infections frequently result in gastrointestinal inflammation, with presentations ordinarily ranging from intestinal cramps to complications with intestinal perforations. However, the evidence detailing successful therapy for gastrointestinal involvement in COVID-19 patients is currently limited. A significant change in fecal microbiomes, namely dysbiosis, was characterized by the enrichment of opportunistic pathogens and the depletion of beneficial commensals and their crucial association to COVID-19 severity has been evidenced. Oral probiotics had been evidenced to improve gut health in achieving homeostasis by exhibiting their antiviral effects via the gut–lung axis. Although numerous commercial probiotics have been effective against coronavirus, their efficacies in treating COVID-19 patients remain debated. In ClinicalTrials.gov, 19 clinical trials regarding the dietary supplement of probiotics, in terms of *Lactobacillus* and mixtures of *Bifidobacteria* and *Lactobacillus*, for treating COVID-19 cases are ongoing. Accordingly, the preventive or therapeutic role of probiotics for COVID-19 patients can be elucidated in the near future.

Keywords: SARS-CoV-2; COVID-19; gut microbiome; probiotics; *Lactobacillus*; *Bifidobacteria*

Citation: Hung, Y.-P.; Lee, C.-C.; Lee, J.-C.; Tsai, P.-J.; Ko, W.-C. Gut Dysbiosis during COVID-19 and Potential Effect of Probiotics. *Microorganisms* **2021**, *9*, 1605. https://doi.org/10.3390/microorganisms9081605

Academic Editor: Sabina Fijan

Received: 11 July 2021
Accepted: 26 July 2021
Published: 28 July 2021

1. Introduction

Severe acute respiratory syndrome coronavirus 2 (SARS-CoV-2), a new RNA virus of the family *Coronaviridae*, can cause coronavirus disease 2019 (COVID-19), majorly affecting pulmonary tissues by respiratory transmission [1,2]. Clinical presentations of COVID-19 vary greatly, ranging from no or mild symptoms often in young patients without comorbidities, moderate diseases with pneumonia, to severe diseases complicated by hypoxia, respiratory or multi-organ failure, and even death [2]. SARS-CoV-2 is composed of four structure proteins, including spike glycoproteins (S), small envelope glycoproteins

(E), glycoproteins membrane (M), nucleocapsid (N), and other accessory proteins [3]. The spike protein of SARS-CoV-2, a transmembrane protein, uses angiotensin-converting enzyme 2 (ACE2) as the receptor of the cell entry [3,4]. In addition to extensive existence in pulmonary tissue, ACE2 is highly expressed in the gut [3,4]; therefore, in the human small intestinal organoids model, enterocytes are easily infected by SARS-CoV-2, as demonstrated by confocal and electron microscopy [1,5]. In the gut, ACE2 is not only a key regulator of dietary amino acid homeostasis, innate immunity, gut microbial ecology, and transmissible susceptibility to colitis [6], but also is linked to the activation of intestinal inflammation [6]. Accordingly, SARS-CoV-2 infections frequently result in gastrointestinal inflammation, with clinical presentations ranging from intestinal cramps and diarrhea to intestinal perforations (Figure 1) [7,8]. Additionally, its abdominal presentation was more frequent in critically ill patients requiring intensive care than those who did not require intensive care, and 10% of patients presented with diarrhea and nausea within 1–2 days before the development of fever and respiratory symptoms [9]. However, the evidence detailing successful therapy for gastrointestinal involvement in COVID-19 patients is currently limited.

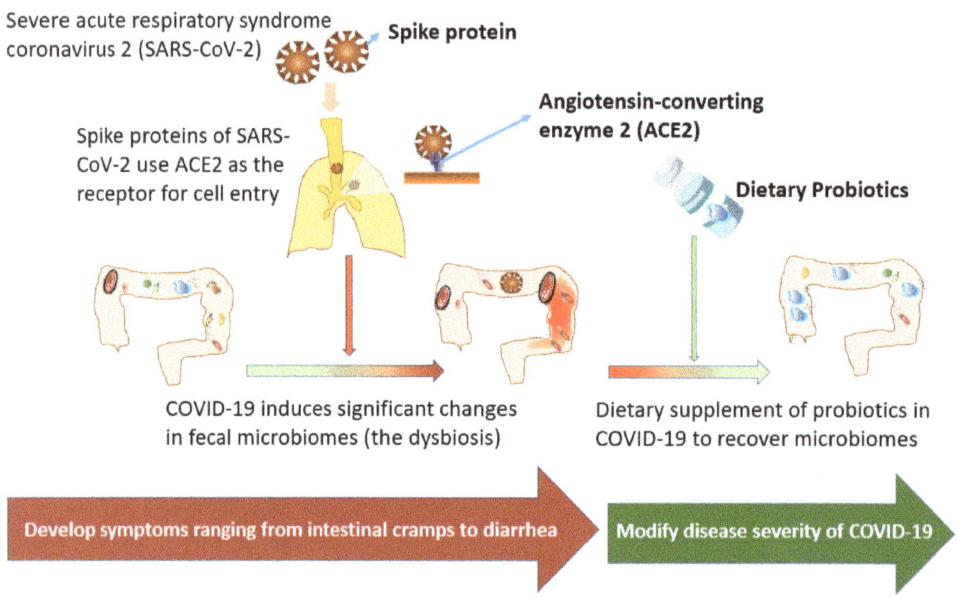

Figure 1. Gastrointestinal involvement and disturbance of gut microbiota during COVID-19 and recovery by dietary supplement of probiotics.

One possible mechanism linked to gut presentations in COVID-19 is the downregulation of ACE2, followed by the decreased activation of mechanistic targets of rapamycin and increased autophagy, further leading to dysbiosis [7]. Another theory is that the blockage of ACE2 induces the increased levels of angiotensinogen by the hyperactivation of the renin–angiotensin system, resulting in the shutdown of the amino acid transporter BA0T1 and a lack of cellular tryptophan. These alterations cause the decreased secretion of antimicrobial peptides and disturbance in the gut microbiome [10]. Therefore, COVID-19 impacts the human gut microbiome, with a decline in microbial diversity and beneficial microbes [11].

2. The Interaction between Respiratory Tract Diseases and Gut Microbiota

A crucial association between a modified gut microbiome and the immune response to respiratory viral infections is evidenced. Taking respiratory syncytial virus and influenza as examples, gut microbiota was significantly altered by viral infections itself and multifactorial variables, such as inflammation-induced tumor necrosis factor-alpha (TNF-α) [12]. Intact microbiota provides signals leading to inflammasome activation, expression of pro-interleukin (IL)-1β and pro-IL-18, and the migration of dendritic cells (DCs) from the lung to the draining lymph node and T-cells, which are critical for protective immunity following influenza virus infection [13]. Disturbed gut microbiota directly or indirectly affects innate and adaptive immune signals and cells in the pulmonary tissue, such as the increased susceptibility to asthma, pulmonary allergic diseases, and chronic obstructive pulmonary diseases [14–17]. More importantly, the severity of influenza infections has been vastly related to the heterogeneous responses of the gut microbiota, as noted by the finding that *Bifidobacterium* species in the gut can expand to enhance host resistance to influenza [18].

In addition, gut microorganisms regulate innate memory by eliciting pattern recognition receptors (PRRs) on monocytes/macrophages and natural killer cells to recognize microbe- or pathogen-associated molecular patterns on microbes [19]. Toll-like receptors (TLRs) and nucleotide-binding oligomerization domain (NOD)-like receptors, recognizable on the host's cells through PRRs, evoke different immunological reactions depending on the types of cells, ligands, or receptors [20]. The fine alteration of the regulatory balance of pro-inflammatory responses and inflammatory regulatory T cells (Tregs) ultimately controlled by the commensal microorganisms is critical in coordinating gut immune homeostasis [20,21]. For example, polysaccharide A, an immunomodulatory molecule, secreted by *Bacteroides fragilis,* can mediate the conversion of CD4+ T cells into IL-10-producing Foxp3(+) Treg cells, and may be considered for the prevention and treatment of experimental colitis in mice [21].

3. Gut Dysbiosis during COVID-19

Patients with COVID-19 had significant changes in fecal microbiomes, characterized by the enrichment of opportunistic pathogens and the depletion of beneficial commensals [22]. Dysbiosis has been vastly associated with COVID-19 severity [22–25], because the microbial diversity is regarded as a critical determinant of microbial ecosystem stability [26]. Among short-chain fatty acids (SCFAs), butyrate is not only responsible for energy requirements of the colonic epithelium, but also preserves tissues by mitigating chronic inflammatory responses through the regulation of pro- and anti-inflammatory cytokines [27]. Accordingly, decreases in the abundance of butyrate-producing bacteria (such as *Faecalibacterium prausnitzii* and *Clostridium* species), and the subsequent decline in SCFA availability have been correlated with severe COVID-19 [22–25,28,29]. Additionally, an increase in common pathogens in gut microbiota, such as *Prevotella*, *Enterococcus*, Enterobacteriaceae, or *Campylobacter*, were consistently associated with high infectivity, disease deterioration, or poor prognosis in COVID-19 patients [23–25,28]. The *Prevotella* species, for example, is associated with augmented T helper type 17 (Th17)-mediated mucosal inflammation, including activating TLR2 and Th17-polarizing cytokine production (such as IL-23 and IL-1), stimulating epithelial cells to produce IL-8, IL-6, and CCL20, and thus promoting neutrophil recruitment and inflammation [30]. The deterioration of the clinical course of patients with COVID-19 infection might be in part due to the activation of severe inflammation through disruption in gut microbiota and the out-growth of pathogenic bacteria.

Patients with COVID-19 also had the increased proportion of opportunistic fungal pathogens, such as *Aspergillus flavus* and *Aspergillus niger,* detected in fecal samples [31]. In metagenomic sequencing analyses of fecal samples from COVID-19 patients, the baseline abundance of *Coprobacillus*, *Clostridium ramosum*, and *Clostridium hathewayi* was correlated with disease severity, and an inverse correlation between abundance of *F. prausnitzii* (an anti-inflammatory bacterium) and disease severity was disclosed [22]. Furthermore, *Bacteroides dorei*, *Bacteroides thetaiotaomicron*, *Bacteroides massiliensis*, and *Bacteroides ovatus*, which downregulated the expression of ACE2 in the gut, were correlated inversely with

SARS-CoV-2 load [22]. The same study team also indicated that, in the cases of active SARS-CoV-2 infections, the gut microbiota presented a higher abundance of opportunistic pathogens, while increased nucleotide and amino acid biosynthesis, as well as carbohydrate metabolism, were evidenced [24]. In summary, these findings reasonably suggest that the development of therapeutic agents able to neutralize the SARS-CoV-2 activity in the gut, as well as to restore the physiological gut microbiota composition, may be warranted.

A crucial association between the predominance of opportunistic pathogens in gut microbiomes and unfavorable outcomes of COVID-19 patients has been comprehensively reported [23]. In a Chinese cohort of COVID-19 patients with different disease severity, the abundance of butyrate-producing bacteria decreased significantly, which may help discriminate critically ill patients from general and severe patients. The increased proportion of opportunistic pathogens, such as *Enterococcus* and Enterobacteriaceae, in critically ill patients might be associated with a poor prognosis [23]. In another study, a higher abundance of opportunistic pathogens, such as *Streptococcus, Rothia, Veillonella,* and *Actinomyces* species, and a lower abundance of beneficial symbionts, could be noted in the gut microbiota of COVID-19 patients [25]. In the American cohort, the specific alteration in the gut microbiome, particularly *Peptoniphilus, Corynebacterium,* and *Campylobacter,* was also noticed [28]. Nevertheless, opportunistic pathogens were prevalent in the COVID-19 cases, particularly among critically individuals, but the causal effect of the predominance of opportunistic pathogens, and a grave outcome remains to be determined.

The recovery of dysbiosis after active SARS-CoV-2 infections exhibited geographical and demographic differences [22,28,32]. After the clearance of SARS-CoV-2 and resolution of respiratory symptoms, depleted symbionts and gut dysbiosis were usually persistent among recovered COVID-19 patients, because microbiota richness did not yield to normal levels after 6-month recovery [22]. In contrast, in an American cohort including recovered COVID-19 cases, the dysbiosis could rapidly recover with a return of the human gut microbiota to an uninfected status [28]. Although the great diversity in the ability of the microbiota return was disclosed, it was evident that the recovery of gut microbiota could be regarded as an indicator of the favorable prognosis among patients with COVID-19.

4. Therapeutic Effects of Dietary Supplement of Probiotics for COVID-19

Oral probiotics had been proven to exhibit antiviral effects and thereby to improve gut health for achieving homeostasis [33,34]. To take the influenza infection as an example, *Lactococcus lactis* JCM 5805 demonstrated the activity against influenza virus through the activation of anti-viral immunity [34]. The oral administration of *Bacteroides breve* YIT4064 can enhance antigen-specific IgG against influenza virus [33]. Moreover, a meta-analysis report indicated the administration of these probiotics significantly reduced the incidence of ventilator-associated pneumonia, possibly through reducing the overgrowth of potentially opportunistic pathogens and stimulating immune responses [35]. However, such a promotion of oral probiotics in treating critically ill patients experiencing COVID-19 should be further explored.

In COVID-19 patients, the excessive production of pro-inflammatory cytokines, a so-called "cytokine storm", is pathologically related to acute respiratory distress syndrome and extensive tissue injury, multi-organ failure, or eventually death [36]. With COVID-19 progression, critically ill patients had higher plasma levels of many cytokines, in terms of IL-2, IL-7, IL-10, granulocyte colony-stimulating factor, IFN-γ-inducible protein-10, monocyte chemoattractant protein-1, macrophage inflammatory protein-1A, and TNF-α [37]. Therefore, therapeutic targeting on cytokines in COVID-19 treatment was evidenced to increase survival [36]. Fecal levels of IL-8 and IL-23 and intestinal specific IgA responses were vastly associated with severe COVID-19 disease, which indicated the co-existence of systemic and local intestine inflammation in critically ill patients [38]. One of the commercial probiotics, *Lactobacillus rhamnosus* HDB1258, might be effective in treating COVID-19 by modulating both microbiota-mediated immunity in gut and systemic inflammation induced by lipopolysaccharide [39]. Accordingly, concomitant targeting on local and systemic inflammatory responses by probiotics is reasonably believed to be valuable to counteract COVID-19-related gut and systemic inflammation.

Numerous probiotics and by-probiotic products exhibiting direct and indirect antiviral effects have been reported in the scientific literature. Lactic acid-producing bacteria such as *lactobacilli* can exert their antiviral activity by direct probiotic–virus interaction, the production of antiviral inhibitory metabolites, preventing secondary infection, and eliciting anti-viral immunity [40–47]. Nisin, one of the well-characterized bacteriocins from probiotics, contributes to probiotic antiviral effects against influenza A virus and other respiratory viruses [41,43]. A peptide, P18, produced by the probiotic *Bacillus subtilis* strain, was regarded as an antiviral compound against influenza virus [42]. Probiotics capsules containing live *B. subtilis* and *E. faecalis* (Medilac-S) can lower the acquisition of the gut colonization of potentially pathogenic microorganisms [44]. *L. rhamnosus* GG have been reported to prevent ventilator-associated pneumonia [45]. The heat-killed *L. casei* DK128 strain has been active against different subtypes of influenza viruses by an increasing proportion of alveolar macrophages in lungs and airways, the early induction of virus-specific antibodies, and reduced levels of pro-inflammatory cytokines and innate immune cells [46]. *S. salivarius* 24SMB and *S. oralis* 89a were able to inhibit the biofilm formation capacity of airway bacterial pathogens and even to disperse their pre-formed biofilms [47]. The *S. salivarius* strain K12 may stimulate IFN-γ release and suppress bronchial inflammation, and its colonization in the oral cavity and upper respiratory tract will actively interfere with the growth of pathogenic microbes [48]. Although these probiotics and their products provide the favorable antiviral interaction with immune composition in the gut, the feasibility and health effect of dietary probiotics to improve the dysbiosis in COVID-19 patients remains to be studied.

Numerous probiotics had been proposed to be beneficial in coronaviral infections, but the evidence detailing their efficacies in treating COVID-19 infection is limited [49]. *L. plantarum* Probio-38 and *L. salivarius* Probio-37 could inhibit transmissible gastroenteritis coronavirus [50]. The probiotic, *E. faecium* NCIMB 10415, has been approved as a feed additive for young piglets in the European Union for treating the transmissible coronavirus gastroenteritis [51]. The recombinant IFN-λ3-anchored *L. plantarum* can in vitro inhibit porcine gastroenteritis caused by coronavirus [52]. However, the clinical utility of probiotics in human infections caused by SARS-CoV-2 warrants further evaluations [53–57].

Another important issue regarding probiotics for COVID-19 cases is the patient safety. For an example, *B. longum* bacteremia had been reported in preterm infants receiving probiotics [58,59]. Since gastrointestinal SARS-CoV-2 involvement has been reported, the possibility of increased intestinal permeability should be expected and the risk of secondary bacterial infections in the gut is substantial if high-dosage steroid and other immunomodulation agents are administrated to treat the cytokine storm associated with COVID-19 [60,61]. The oral formulation Sivomixx®, which was a mixture of probiotics, was independently associated with a reduced risk for death in a retrospective, observational cohort study that included 200 adults with severe COVID-19 pneumonia [62]. In another study, nearly all COVID-19 patients treated with Sivomixx®showed remission of diarrhea and other symptoms within 72 h, in contrast to less than half in the control group [63]. However, the clinical application of probiotics in COVID-19 patients requires more evidence.

In ClinicalTrials.gov, 22 trials of probiotics for the prevention or adjuvant therapy of COVID-19 were registered since April 2020, including one aiming to study the effect of oxygen-ozone therapy, one studying intranasal probiotics, and the other using throat spray-containing probiotic [64]. Of the remaining 19 trials, 8 common probiotic strains include *Lactobacillus* (7 trials), a mixture of *Bifidobacteria* and *Lactobacillus* (5), and *Saccharomyces* species (2) (Table 1). The major outcome was greatly diverse in these trials, including disease prevention, symptom relief, antibody titers, disease progression, changes of viral load, microbiome effects, and mortality. Based on these trials, the role of dietary supplement probiotics for COVID-19 can be more evident in the near future.

Table 1. Nineteen clinical trials of dietary supplement of probiotics in coronavirus disease 2019 (COVID-19) registered at *ClinicalTrials.gov* posted from April 2020 to June 2021.

ClinicalTrials.gov Identifier	Study Title	First Posted	Study Design	Probiotic Strain	Location	Outcome Measures	Status
NCT04366180	Evaluation of probiotic *Lactobacillus coryniformis* K8 on COVID-19 prevention in healthcare workers	28 April 2020	Randomized	*L. coryniformis* K8	Granada, Spain	Incidence of COVID-19 infection in healthcare workers	Recruiting
NCT04390477	Study to evaluate the effect of a probiotic in COVID-19	15 May 2020	Randomized	Not revealed	Alicante, Spain	ICU admission rate	Recruiting
NCT04399252	Effect of *Lactobacillus* on the microbiome of household contacts exposed to COVID-19	22 May 2020	Randomized	*L. rhamnosus* GG	North Carolina, United States	Incidence of symptoms of COVID-19	Active, not recruiting
NCT04420676	Synbiotic therapy of gastrointestinal symptoms during COVID-19 infection (SynCov)	9 June 2020	Randomized	Omni-Biotic®10 AAD (chiefly *Lactobacillus* and *Bifidobacterium*)	Graz, Austria	Stool calprotectin	Recruiting
NCT04462627	Reduction of COVID 19 transmission to health care professionals	8 July 2020	Non-randomized	Metagenics Probactiol plus (chiefly *Lactobacillus* and *Bifidobacterium*)	Brussels, Belgium	Antibody concentration	Recruiting
NCT04507867	Effect of a NSS to reduce complications in patients with COVID-19 and comorbidities in stage III	11 August 2020	Randomized	*Saccharomyces bourllardii* with nutritional support system (NSS)	Mexico	Oxygen saturation	Not yet recruiting
NCT04517422	Efficacy of *L. plantarum* and *P. acidilactici* in adults with SARS-CoV-2 and COVID-19	18 August 2020	RCT	*L. plantarum* and *P. acidilactici*	Mexico City, Mexico	Severity progression of COVID-19	Completed
NCT04621071	Efficacy of probiotics in reducing duration and symptoms of COVID-19 (PROVID-19)	9 November 2020	RCT	Not revealed	Canada, Quebec	Duration of symptoms of the COVID-19	Recruiting
NCT04666116	Changes in viral load in COVID-19 after probiotics	14 December 2020	Randomized, single blind	GASTEEL PLUS (mixture of *Bifidobacteria* and *Lactobacillus*)	Valencia, Spain	Viral load in nasopharyngeal smear	Recruiting
NCT04734886	The effect of probiotic supplementation on SARS-CoV-2 antibody response after COVID-19	2 February 2021	Randomized	*L. reuteri* DSM 17938 + vitamin D	Örebro Län, Sweden	SARS-CoV-2 specific antibodies	Recruiting

Table 1. *Cont.*

ClinicalTrials.gov Identifier	Study Title	First Posted	Study Design	Probiotic Strain	Location	Outcome Measures	Status
NCT04756466	Effect of the consumption of a *Lactobacillus* strain on the incidence of COVID-19 in the elderly	16 February 2021	RCT	*Lactobacillus* strain	A Coruña, Spain	Incidence of SARS CoV-2 infection	Active, not recruiting
NCT04798677	Efficacy and tolerability of ABBC1 in volunteers receiving the influenza or COVID-19 Vaccine	15 March 2021	Non-randomized	*S. cerevisiae*, rich in selenium and zinc	Barcelona, Spain	Change in acute immune response to influenza vaccine after supplementation	Recruiting
NCT04813718	Post COVID-19 syndrome and the gut-lung axis	24 March 2021	Randomized	Omni-Biotic Pro Vi 5 (chiefly Lactobacillus)	Graz, Austria	Microbiome composition	Recruiting
NCT04847349	Live microbials to boost anti-severe acute respiratory syndrome coronavirus-2 (SARS-CoV-2) immunity clinical trial	19 April 2021	RCT	OL-1 (Content not revealed)	New Jersey, United States	Change in serum titer of anti-SARS-CoV-2 IgG	Recruiting
NCT04854941	Efficacy of probiotics in the treatment of hospitalized patients with novel coronavirus infection	22 April 2021	Randomized	*L. rhamnosus*, *B. bifidum*, *B. longum* subsp. *infantis* and *B. longum*	Moscow, Russian	Mortality	Completed
NCT04877704	Symprove (Probiotic) as an add-on to COVID-19 management	7 May 2021	Randomized	Symprove (*L. rhamnosus*, *E. faecium*, *L. acidophilus* and *L. plantarum*)	London, United Kingdom	Length of hospital stay	Not yet recruiting
NCT04884776	Modulation of gut microbiota to enhance health and immunity	13 May 2021	RCT	Probiotics blend (3 *Bifidobacteria*)	Hong Kong	Restoration of gut dysbiosis	Not yet recruiting
NCT04907877	*Bifidobacteria* and *Lactobacillus* in symptomatic adult COVID-19 outpatients (ProCOVID)	1 June 2021	Randomized	NordBiotic ImmunoVir (mixture of *Bifidobacteria* and *Lactobacillus*)	Not revealed	Global symptom score	Not yet recruiting
NCT04922918	*Ligilactobacillus salivarius* MP101 for elderly in a nursing home (PROBELDERLY)	11 June 2021	Single group	*Ligilactobacillus salivarius* MP101	Madrid, Spain	Barthel index, functional status score	Recruiting

RCT: randomized controlled trial; ICU: intensive care unit; IgG: immunoglobulin G.

There are microbiome-targeting agents other than oral probiotics for patients with COVID-19 infection. A clinical trial of oral prebiotics, KB109, a novel synthetic glycan to modulate gut microbiome composition and to increase SCFA production in the gut, is ongoing (NCT04414124) [64]. Throat spray containing three *Lactobacillus* strains was implemented in a clinical trial to change the severity of COVID-19 and prevent transmission of SARS-COV-2 virus to household members (NCT04793997) [64]. Moreover, there are several next-generation probiotics identified by metagenomic approaches, such as *F. prausnitzii* and *Akkermansia muciniphila*, which can generate diffusible metabolites, including butyrate, desaminotyrosine, and SCFAs, and may improve pulmonary immunity and prevent viral respiratory infections [65]. It can be expected, in the future, microbiome-targeting therapy may decrease disease severity, relief symptoms, or prevent viral transmission, and play a role in the treatment of patients with COVID-19 infection

5. Conclusions

Patients with COVID-19 had significant changes in fecal microbiomes, characterized by the enrichment of opportunistic pathogens and the depletion of beneficial commensals, which is vastly associated with disease severity. Besides anti-viral agents or supportive treatment, microbiome-targeting therapy may provide an alternative to prevent COVID-19 deterioration. Oral probiotics may have antiviral effects via the gut–lung axis and improve gut health for achieving homeostasis. Although some commercial probiotics have been effective against coronavirus, the evidence detailing their efficacies in treating COVID-19 patients is limited. Registered clinical trials of probiotics in COVID-19, mainly *Lactobacillus* and mixtures of *Bifidobacteria* and *Lactobacillus*, are ongoing and thus the preventive or therapeutic role of probiotics for such patients can be elucidated in the near future.

Author Contributions: Y.-P.H., P.-J.T. and W.-C.K. designed the experiments, performed the experiments, analyzed the data, and participated in the writing of the manuscript. Y.-P.H., J.-C.L., C.-C.L., P.-J.T. and W.-C.K. All authors have read and agreed to the published version of the manuscript.

Funding: The present study was supported by research grants from the Ministry of Health and Welfare (MOHW105-CDC-C-114-122113), and the Ministry of Science and Technology (MOST 108-2321-B-006-004, 108-2320-B-006-043-MY3, 109-2314-B-006-089-MY3, 110-2314-B-675-001).

Institutional Review Board Statement: Not applicable for our review study.

Informed Consent Statement: Not applicable for our review study.

Data Availability Statement: Data available in a publicly accessible repository.

Conflicts of Interest: All authors report no conflicts of interest relevant to this article.

References

1. Lamers, M.M.; Beumer, J.; van der Vaart, J.; Knoops, K.; Puschhof, J.; Breugem, T.I.; Ravelli, R.B.G.; van Schayck, J.P.; Mykytyn, A.Z.; Duimel, H.Q.; et al. SARS-CoV-2 productively infects human gut enterocytes. *Science* **2020**, *369*, 50–54. [CrossRef]
2. Wu, F.; Zhao, S.; Yu, B.; Chen, Y.M.; Wang, W.; Song, Z.G.; Hu, Y.; Tao, Z.W.; Tian, J.H.; Pei, Y.Y.; et al. A new coronavirus associated with human respiratory disease In China. *Nature* **2020**, *579*, 265–269. [CrossRef]
3. Suryana, K.D.; Simadibrata, M.; Renaldi, K. Impact of COVID-19 on the Gut: A review of the manifestations, pathology, management, and challenges. *Acta Med. Indones.* **2021**, *53*, 96–104.
4. Letko, M.; Marzi, A.; Munster, V. Functional assessment of cell entry and receptor usage for SARS-CoV-2 and other lineage B Betacoronaviruses. *Nat. Microbiol.* **2020**, *5*, 562–569. [CrossRef]
5. Zhou, J.; Li, C.; Liu, X.; Chiu, M.C.; Zhao, X.; Wang, D.; Wei, Y.; Lee, A.; Zhang, A.J.; Chu, H.; et al. Infection of bat and human intestinal organoids by SARS-CoV-2. *Nat. Med.* **2020**, *26*, 1077–1083. [CrossRef] [PubMed]
6. Hashimoto, T.; Perlot, T.; Rehman, A.; Trichereau, J.; Ishiguro, H.; Paolino, M.; Sigl, V.; Hanada, T.; Hanada, R.; Lipinski, S.; et al. ACE2 links amino acid malnutrition to microbial ecology and intestinal inflammation. *Nature* **2012**, *487*, 477–481. [CrossRef]
7. de Oliveira, A.P.; Lopes, A.L.F.; Pacheco, G.; de Sa Guimaraes Noleto, I.R.; Nicolau, L.A.D.; Medeiros, J.V.R. Premises among SARS-CoV-2, dysbiosis and diarrhea: Walking through the ACE2/mTOR/autophagy route. *Med. Hypotheses* **2020**, *144*, 110243. [CrossRef] [PubMed]
8. Bas, S.; Zarbaliyev, E. The role of dual-energy computed tomography in locating gastrointestinal tract perforations. *Cureus* **2021**, *13*, e15265.

9. Wang, D.; Hu, B.; Hu, C.; Zhu, F.; Liu, X.; Zhang, J.; Wang, B.; Xiang, H.; Cheng, Z.; Xiong, Y.; et al. Clinical characteristics of 138 hospitalized patients with 2019 novel coronavirus-infected pneumonia in Wuhan, China. *JAMA* **2020**, *323*, 1061–1069. [CrossRef]
10. Monkemuller, K.; Fry, L.C.; Rickes, S. Systemic inflammatory response and thrombosis due to alterations in the gut microbiota in COVID-19. *Rev. Esp. Enferm. Dig.* **2020**, *112*, 584–585. [PubMed]
11. Finlay, B.B.; Amato, K.R.; Azad, M.; Blaser, M.J.; Bosch, T.C.G.; Chu, H.; Dominguez-Bello, M.G.; Ehrlich, S.D.; Elinav, E.; Geva-Zatorsky, N.; et al. The hygiene hypothesis, the COVID pandemic, and consequences for the human microbiome. *Proc. Natl. Acad. Sci. USA* **2021**, *118*, e2010217118. [CrossRef]
12. Groves, H.T.; Higham, S.L.; Moffatt, M.F.; Cox, M.J.; Tregoning, J.S. Respiratory viral infection alters the gut microbiota by inducing inappetence. *Mbio* **2020**, *11*, e03236-19. [CrossRef]
13. Ichinohe, T.; Pang, I.K.; Kumamoto, Y.; Peaper, D.R.; Ho, J.H.; Murray, T.S.; Iwasaki, A. Microbiota regulates immune defense against respiratory tract influenza A virus infection. *Proc. Natl. Acad. Sci. USA* **2011**, *108*, 5354–5359. [CrossRef]
14. Enaud, R.; Prevel, R.; Ciarlo, E.; Beaufils, F.; Wieers, G.; Guery, B.; Delhaes, L. The gut-lung axis in health and respiratory diseases: A place for inter-organ and inter-kingdom crosstalks. *Front. Cell Infect. Microbiol.* **2020**, *10*, 9. [CrossRef] [PubMed]
15. Dang, A.T.; Marsland, B.J. Microbes, metabolites, and the gut-lung axis. *Mucosal Immunol.* **2019**, *12*, 843–850. [CrossRef]
16. Kuss, S.K.; Best, G.T.; Etheredge, C.A.; Pruijssers, A.J.; Frierson, J.M.; Hooper, L.V.; Dermody, T.S.; Pfeiffer, J.K. Intestinal microbiota promote enteric virus replication and systemic pathogenesis. *Science* **2011**, *334*, 249–252. [CrossRef]
17. Wilks, J.; Golovkina, T. Influence of microbiota on viral infections. *PLoS Pathog.* **2012**, *8*, e1002681. [CrossRef] [PubMed]
18. Zhang, Q.; Hu, J.; Feng, J.W.; Hu, X.T.; Wang, T.; Gong, W.X.; Huang, K.; Guo, Y.X.; Zou, Z.; Lin, X.; et al. Influenza infection elicits an expansion of gut population of endogenous *Bifidobacterium animalis* which protects mice against infection. *Genome Biol.* **2020**, *21*, 99. [CrossRef] [PubMed]
19. Negi, S.; Das, D.K.; Pahari, S.; Nadeem, S.; Agrewala, J.N. Potential role of gut microbiota in induction and regulation of innate immune memory. *Front. Immunol.* **2019**, *10*, 2441. [CrossRef] [PubMed]
20. Dhar, D.; Mohanty, A. Gut microbiota and Covid-19- possible link and implications. *Virus Res.* **2020**, *285*, 198018. [CrossRef]
21. Round, J.L.; Mazmanian, S.K. Inducible Foxp3+ regulatory T-cell development by a commensal bacterium of the intestinal microbiota. *Proc. Natl. Acad. Sci. USA* **2010**, *107*, 12204–12209. [CrossRef]
22. Zuo, T.; Zhang, F.; Lui, G.C.Y.; Yeoh, Y.K.; Li, A.Y.L.; Zhan, H.; Wan, Y.; Chung, A.C.K.; Cheung, C.P.; Chen, N.; et al. Alterations in gut microbiota of patients with COVID-19 during time of hospitalization. *Gastroenterology* **2020**, *159*, 944–955.e8. [CrossRef]
23. Tang, L.; Gu, S.; Gong, Y.; Li, B.; Lu, H.; Li, Q.; Zhang, R.; Gao, X.; Wu, Z.; Zhang, J.; et al. Clinical significance of the correlation between changes in the major intestinal bacteria species and COVID-19 severity. *Eng. Beijing* **2020**, *6*, 1178–1184. [CrossRef] [PubMed]
24. Zuo, T.; Liu, Q.; Zhang, F.; Lui, G.C.; Tso, E.Y.; Yeoh, Y.K.; Chen, Z.; Boon, S.S.; Chan, F.K.; Chan, P.K.; et al. Depicting SARS-CoV-2 faecal viral activity in association with gut microbiota composition in patients with COVID-19. *Gut* **2021**, *70*, 276–284. [CrossRef]
25. Gu, S.; Chen, Y.; Wu, Z.; Chen, Y.; Gao, H.; Lv, L.; Guo, F.; Zhang, X.; Luo, R.; Huang, C.; et al. Alterations of the gut microbiota in patients with coronavirus disease 2019 or H1N1 influenza. *Clin. Infect. Dis.* **2020**, *71*, 2669–2678. [CrossRef]
26. Lahti, L.; Salojarvi, J.; Salonen, A.; Scheffer, M.; de Vos, W.M. Tipping elements in the human intestinal ecosystem. *Nat. Commun.* **2014**, *5*, 4344. [CrossRef]
27. McNabney, S.M.; Henagan, T.M. Short chain fatty acids in the colon and peripheral tissues: A focus on butyrate, colon cancer, obesity and insulin resistance. *Nutrients* **2017**, *9*, 1348. [CrossRef] [PubMed]
28. Newsome, R.C.; Gauthier, J.; Hernandez, M.C.; Abraham, G.E.; Robinson, T.O.; Williams, H.B.; Sloan, M.; Owings, A.; Laird, H.; Christian, T.; et al. The gut microbiome of COVID-19 recovered patients returns to uninfected status in a minority-dominated United States cohort. *Gut Microbes* **2021**, *13*, 1–15. [CrossRef] [PubMed]
29. Xu, K.; Cai, H.; Shen, Y.; Ni, Q.; Chen, Y.; Hu, S.; Li, J.; Wang, H.; Yu, L.; Huang, H.; et al. Management of corona virus disease-19 (COVID-19): The Zhejiang experience. *Zhejiang Exp. J. Zhejiang Univ. Med. Sci.* **2020**, *49*, 147–157.
30. Larsen, J.M. The immune response to *Prevotella* bacteria in chronic inflammatory disease. *Immunology* **2017**, *151*, 363–374. [CrossRef]
31. Zuo, T.; Zhan, H.; Zhang, F.; Liu, Q.; Tso, E.Y.K.; Lui, G.C.Y.; Chen, N.; Li, A.; Lu, W.; Chan, F.K.L.; et al. Alterations in fecal fungal microbiome of patients with COVID-19 during time of hospitalization until discharge. *Gastroenterology* **2020**, *159*, 1302–1310. [CrossRef]
32. Chen, Y.; Gu, S.; Chen, Y.; Lu, H.; Shi, D.; Guo, J.; Wu, W.R.; Yang, Y.; Li, Y.; Xu, K.J.; et al. Six-month follow-up of gut microbiota richness in patients with COVID-19. *Gut* **2021**. [CrossRef]
33. Yasui, H.; Kiyoshima, J.; Hori, T.; Shida, K. Protection against influenza virus infection of mice fed *Bifidobacterium breve* YIT4064. *Clin. Diagn. Lab. Immunol.* **1999**, *6*, 186–192. [CrossRef] [PubMed]
34. Xia, Y.; Cao, J.; Wang, M.; Lu, M.; Chen, G.; Gao, F.; Liu, Z.; Zhang, D.; Ke, X.; Yi, M. Effects of Lactococcus lactis subsp. lactis JCM5805 on colonization dynamics of gut microbiota and regulation of immunity in early ontogenetic stages of tilapia. *Fish Shellfish Immunol.* **2019**, *86*, 53–63. [CrossRef]
35. Su, M.; Jia, Y.; Li, Y.; Zhou, D.; Jia, J. Probiotics for the prevention of ventilator-associated pneumonia: A meta-analysis of randomized controlled trials. *Respir. Care* **2020**, *65*, 673–685. [CrossRef] [PubMed]
36. Ragab, D.; Salah Eldin, H.; Taeimah, M.; Khattab, R.; Salem, R. The COVID-19 cytokine storm; what we know so far. *Front. Immunol.* **2020**, *11*, 1446. [CrossRef] [PubMed]

37. Huang, C.; Wang, Y.; Li, X.; Ren, L.; Zhao, J.; Hu, Y.; Zhang, L.; Fan, G.; Xu, J.; Gu, X.; et al. Clinical features of patients infected with 2019 novel coronavirus in Wuhan, China. *Lancet* **2020**, *395*, 497–506. [CrossRef]
38. Britton, G.J.; Chen-Liaw, A.; Cossarini, F.; Livanos, A.E.; Spindler, M.P.; Plitt, T.; Eggers, J.; Mogno, I.; Gonzalez-Reiche, A.; Siu, S.; et al. SARS-CoV-2-specific IgA and limited inflammatory cytokines are present in the stool of select patients with acute COVID-19. *MedRxiv* **2020**. [CrossRef]
39. Han, S.K.; Shin, Y.J.; Lee, D.Y.; Kim, K.M.; Yang, S.J.; Kim, D.S.; Choi, J.W.; Lee, S.; Kim, D.H. *Lactobacillus rhamnosus* HDB1258 modulates gut microbiota-mediated immune response in mice with or without lipopolysaccharide-induced systemic inflammation. *BMC Microbiol.* **2021**, *21*, 146. [CrossRef]
40. Al Kassaa, I.; Hober, D.; Hamze, M.; Chihib, N.E.; Drider, D. Antiviral potential of lactic acid bacteria and their bacteriocins. *Probiotics Antimicrob. Proteins* **2014**, *6*, 177–185. [CrossRef]
41. Baindara, P.; Chakraborty, R.; Holliday, Z.M.; Mandal, S.M.; Schrum, A.G. Oral probiotics in coronavirus disease 2019: Connecting the gut-lung axis to viral pathogenesis, inflammation, secondary infection and clinical trials. *New Microbes New Infect.* **2021**, *40*, 100837. [CrossRef] [PubMed]
42. Starosila, D.; Rybalko, S.; Varbanetz, L.; Ivanskaya, N.; Sorokulova, I. Anti-influenza activity of a *Bacillus subtilis* probiotic strain. *Antimicrob. Agents Chemother.* **2017**, *61*, e00539-17. [CrossRef]
43. Malaczewska, J.; Kaczorek-Lukowska, E.; Wojcik, R.; Siwicki, A.K. Antiviral effects of nisin, lysozyme, lactoferrin and their mixtures against bovine viral diarrhoea virus. *BMC Vet. Res.* **2019**, *15*, 318. [CrossRef]
44. Zeng, J.; Wang, C.T.; Zhang, F.S.; Qi, F.; Wang, S.F.; Ma, S.; Wu, T.J.; Tian, H.; Tian, Z.T.; Zhang, S.L.; et al. Effect of probiotics on the incidence of ventilator-associated pneumonia in critically ill patients: A randomized controlled multicenter trial. *Intensive Care Med.* **2016**, *42*, 1018–1028. [CrossRef]
45. Morrow, L.E.; Kollef, M.H.; Casale, T.B. Probiotic prophylaxis of ventilator-associated pneumonia: A blinded, randomized, controlled trial. *Am. J. Respir. Crit. Care Med.* **2010**, *182*, 1058–1064. [CrossRef] [PubMed]
46. Jung, Y.J.; Lee, Y.T.; Ngo, V.L.; Cho, Y.H.; Ko, E.J.; Hong, S.M.; Kim, K.H.; Jang, J.H.; Oh, J.S.; Park, M.K.; et al. Heat-killed *Lactobacillus casei* confers broad protection against influenza A virus primary infection and develops heterosubtypic immunity against future secondary infection. *Sci. Rep.* **2017**, *7*, 17360. [CrossRef] [PubMed]
47. Bidossi, A.; De Grandi, R.; Toscano, M.; Bottagisio, M.; De Vecchi, E.; Gelardi, M.; Drago, L. Probiotics *Streptococcus salivarius* 24SMB and *Streptococcus oralis* 89a interfere with biofilm formation of pathogens of the upper respiratory tract. *BMC Infect. Dis.* **2018**, *18*, 653. [CrossRef] [PubMed]
48. Di Pierro, F. A possible probiotic (*S. salivarius* K12) approach to improve oral and lung microbiotas and raise defenses against SAR S-CoV-2. *Minerva Med.* **2020**, *111*, 281–283. [CrossRef] [PubMed]
49. Bottari, B.; Castellone, V.; Neviani, E. Probiotics and COVID-19. *Int. J. Food Sci. Nutr.* **2021**, *72*, 293–299. [CrossRef]
50. Kumar, R.; Seo, B.J.; Mun, M.R.; Kim, C.J.; Lee, I.; Kim, H.; Park, Y.H. Putative probiotic *Lactobacillus* spp. from porcine gastrointestinal tract inhibit transmissible gastroenteritis coronavirus and enteric bacterial pathogens. *Trop. Anim. Health Prod.* **2010**, *42*, 1855–1860.
51. Chai, W.; Burwinkel, M.; Wang, Z.; Palissa, C.; Esch, B.; Twardziok, S.; Rieger, J.; Wrede, P.; Schmidt, M.F. Antiviral effects of a probiotic *Enterococcus faecium* strain against transmissible gastroenteritis coronavirus. *Arch. Virol.* **2013**, *158*, 799–807. [CrossRef]
52. Liu, Y.S.; Liu, Q.; Jiang, Y.L.; Yang, W.T.; Huang, H.B.; Shi, C.W.; Yang, G.L.; Wang, C.F. Surface-displayed porcine IFN-lambda3 in *Lactobacillus plantarum* inhibits porcine enteric coronavirus infection of porcine intestinal epithelial cells. *J. Microbiol. Biotechnol.* **2020**, *30*, 515–525. [CrossRef] [PubMed]
53. Mak, J.W.Y.; Chan, F.K.L.; Ng, S.C. Probiotics and COVID-19: One size does not fit all. *Lancet Gastroenterol. Hepatol.* **2020**, *5*, 644–645. [CrossRef]
54. Kurian, S.J.; Unnikrishnan, M.K.; Miraj, S.S.; Bagchi, D.; Banerjee, M.; Reddy, B.S.; Rodrigues, G.S.; Manu, M.K.; Saravu, K.; Mukhopadhyay, C.; et al. Probiotics in prevention and treatment of COVID-19: Current perspective and future prospects. *Arch. Med. Res.* **2021**. [CrossRef] [PubMed]
55. Peng, J.; Zhang, M.; Yao, G.; Kwok, L.Y.; Zhang, W. Probiotics as adjunctive treatment for patients contracted COVID-19: Current understanding and future needs. *Front. Nutr.* **2021**, *8*, 669808. [CrossRef]
56. Sundararaman, A.; Ray, M.; Ravindra, P.V.; Halami, P.M. Role of probiotics to combat viral infections with emphasis on COVID-19. *Appl. Microbiol. Biotechnol.* **2020**, *104*, 8089–8104. [CrossRef] [PubMed]
57. Khaled, J.M.A. Probiotics, prebiotics, and COVID-19 infection: A review article. *Saudi J. Biol. Sci.* **2021**, *28*, 865–869. [CrossRef]
58. Esaiassen, E.; Cavanagh, P.; Hjerde, E.; Simonsen, G.S.; Stoen, R.; Klingenberg, C. *Bifidobacterium longum* subspecies infantis bacteremia in 3 extremely preterm infants receiving probiotics. *Emerg. Infect. Dis.* **2016**, *22*, 1664–1666. [CrossRef]
59. Bertelli, C.; Pillonel, T.; Torregrossa, A.; Prod'hom, G.; Fischer, C.J.; Greub, G.; Giannoni, E. *Bifidobacterium longum* bacteremia in preterm infants receiving probiotics. *Clin. Infect. Dis.* **2015**, *60*, 924–927. [CrossRef]
60. Alataby, H.; Atemnkeng, F.; Bains, S.S.; Kenne, F.M.; Diaz, K.; Nfonoyim, J. A COVID-19 case complicated by *Candida dubliniensis* and *Klebsiella pneumoniae*-carbapenem-resistant Enterobacteriaceae. *J. Med. Cases* **2020**, *11*, 403–406. [CrossRef]
61. Miao, Q.; Ma, Y.; Ling, Y.; Jin, W.; Su, Y.; Wang, Q.; Pan, J.; Zhang, Y.; Chen, H.; Yuan, J.; et al. Evaluation of superinfection, antimicrobial usage, and airway microbiome with metagenomic sequencing in COVID-19 patients: A cohort study in Shanghai. *J. Microbiol. Immunol. Infect.* **2021**. [CrossRef] [PubMed]

62. Ceccarelli, G.; Borrazzo, C.; Pinacchio, C.; Santinelli, L.; Innocenti, G.P.; Cavallari, E.N.; Celani, L.; Marazzato, M.; Alessandri, F.; Ruberto, F.; et al. Oral bacteriotherapy in patients with COVID-19: A retrospective cohort study. *Front. Nutr.* **2020**, *7*, 613928. [CrossRef]
63. d'Ettorre, G.; Ceccarelli, G.; Marazzato, M.; Campagna, G.; Pinacchio, C.; Alessandri, F.; Ruberto, F.; Rossi, G.; Celani, L.; Scagnolari, C.; et al. Challenges in the management of SARS-CoV2 Infection: The role of oral bacteriotherapy as complementary therapeutic strategy to avoid the progression of COVID-19. *Front. Med.* **2020**, *7*, 389. [CrossRef] [PubMed]
64. Clinicaltrials.gov. Available online: https://clinicaltrials.gov/ct2/home (accessed on 24 July 2021).
65. Gautier, T.; Gall, S.D.L.; Sweidan, A.; Tamanai-Shacoori, Z.; Jolivet-Gougeon, A.; Loréal, O.; Bousarghin, L. Next-generation probiotics and their metabolites in COVID-19. *Microorganisms* **2021**, *9*, 941. [CrossRef] [PubMed]

microorganisms

MDPI

Article

Potential Probiotic *Bacillus subtilis* Isolated from a Novel Niche Exhibits Broad Range Antibacterial Activity and Causes Virulence and Metabolic Dysregulation in Enterotoxic *E. coli*

Sudhanshu Sudan [1], Robert Flick [2], Linda Nong [3] and Julang Li [1,*]

[1] Department of Animal Biosciences, University of Guelph, Guelph, ON N1G 2W1, Canada; ssudan@uoguelph.ca
[2] Biozone, Mass Spectrometry and Metabolomics, Department of Chemical Engineering and Applied Chemistry, University of Toronto, Toronto, ON M5S 3E5, Canada; robert.flick@utoronto.ca
[3] Department of Integrative Biology, University of Guelph, Guelph, ON N1G 2W1, Canada; nongl@uoguelph.ca
* Correspondence: jli@uoguelph.ca

Abstract: Microbial life in extreme environments, such as deserts and deep oceans, is thought to have evolved to overcome constraints of nutrient availability, temperature, and suboptimal hygiene environments. Isolation of probiotic bacteria from such niche may provide a competitive edge over traditional probiotics. Here, we tested the survival, safety, and antimicrobial effect of a recently isolated and potential novel strain of *Bacillus subtilis* (CP9) from desert camel in vitro. Antimicrobial assays were performed via radial diffusion, agar spot, and co-culture assays. Cytotoxic analysis was performed using pig intestinal epithelial cells (IPEC-J2). Real time-PCR was performed for studying the effect on ETEC virulence genes and metabolomic analysis was performed using LC-MS. The results showed that CP9 cells were viable in varied bile salts and in low pH environments. CP9 showed no apparent cytotoxicity in IPEC-J2 cells. CP9 displayed significant bactericidal effect against Enterotoxic *E. coli* (ETEC), *Salmonella Typhimurium*, and *Methicillin-resistant Staphylococcus aureus* (MRSA) in a contact inhibitory fashion. CP9 reduced the expression of ETEC virulent genes during a 5 h co-culture. Additionally, a unique emergent metabolic signature in co-culture samples was observed by LC-MS analysis. Our findings indicate that CP9 exhibits a strong antibacterial property and reveals potential mechanisms behind.

Keywords: probiotic; bacillus subtilis; antimicrobial; contact inhibition; extreme environment

Citation: Sudan, S.; Flick, R.; Nong, L.; Li, J. Potential Probiotic *Bacillus subtilis* Isolated from a Novel Niche Exhibits Broad Range Antibacterial Activity and Causes Virulence and Metabolic Dysregulation in Enterotoxic *E. coli*. *Microorganisms* **2021**, *9*, 1483. https://doi.org/10.3390/microorganisms9071483

Academic Editor: Sabina Fijan

Received: 20 June 2021
Accepted: 7 July 2021
Published: 12 July 2021

1. Introduction

Probiotics have gained much interest for the past decade in animal and human health research due to their ability to interact with the host microbiome and modulate cellular functions within the host [1,2]. The International Scientific Association for Probiotics and Prebiotics (ISAPP) defines probiotics as 'Live microorganisms which, when administered in adequate amounts, confer a health benefit on the host' [3]. One of the most intriguing host-benefiting properties of probiotics is its antagonism against pathogenic bacteria, which may be attributed to competitive inhibition, promoting growth of commensal or beneficial bacterial, and secretion of antimicrobial secondary metabolites [4–6]. In the host, probiotic bacteria may either directly aid exclusion of pathogenic bacteria by production of small antimicrobial compounds [7], or indirectly by strengthening mucosal membranes and modulating the immune capabilities of the host [8]. Various probiotics have been researched in the past; however, their application is limited to the survival of the strains in the intestinal gut microenvironment, which may be influenced by low pH, bile acids, digestive enzymes, host diet, and the colonization-resistant microbiome [9–12].

Little is known regarding the colonization pattern of probiotics in the context of attachment to the gastro-intestinal tract walls and metabolic interaction with enteric pathogens [10]. In the intestinal microbial landscape, probiotic interaction and communication with the commensal population occurs through metabolic exchange. Previous studies have shown the importance of studying microbial metabolic potential [13,14]. In addition, probiotics are able to cause shifts within the microbiome [15] and thus influence the colonization of enteric pathogens [16,17]. Hence, a more in-depth understanding on the metabolic potential of the probiotics is important in developing and enhancing the efficacy of probiotic interventions.

Bacillus subtilis is a gram positive, rod shaped, aerobic or facultative anaerobic bacteria belonging to genus *Bacillus,* widely found in the environment [18]. It has been previously studied and used as a probiotic in fermented foods and also as a supplement [18,19]. One of the unique properties of this bacterium is that it can form spores when challenged with unfavourable conditions for growth [20]. This hardy behaviour may help this bacterial strain to cross the gastrointestinal tract (GIT) fluid barrier and establish itself in the gut. More recently, studies have shown that extreme environment regions, such as deserts and deep oceans, may provide an additional evolutionary benefit to the resident bacteria for survival within the host as well as host adaptation to the environment [21,22]. Therefore, isolation of probiotic bacteria from such niches may allow identification of more robust strains for animal and human supplementation. This notion takes into consideration the fact that microbes living in these extreme conditions are able to overcome the constraints of limited nutrients, desiccation, and extreme fluctuating temperatures. Studying their molecular mechanisms and metabolic interactions with targeted pathogens could further provide cues to predict efficiency for novel antimicrobial probiotic intervention.

We recently isolated a novel *Bacillus subtilis* strain from Sub-Saharan camel [23]. Initial assessment showed a high extracellular protease and cellulase activity of the strain. In the current study, we attempted to test the safety and survival of this novel *Bacillus subtilis* as potential probiotic strain, CP9, in the intestinal environment in vitro, as well as its antagonistic properties against pathogenic bacteria. Moreover, the potential mechanism behind its antimicrobial property was also investigated.

2. Materials and Methods

2.1. Microbial Strains and Growth Conditions

We previously isolated and characterized *Bacillus subtilis* (CP9; [23]). *Bacillus subtilis* (ATCC 6633) as a control strain was acquired from American Type Culture Collection (ATCC; Edinburg, VA, USA). Enterotoxic *E. coli* (ETEC), *Salmonella typhimurium,* and Methicillin-resistant *Staphylococcus aureus* (MRSA) were acquired from Animal Health lab at University of Guelph, ON, Canada. CP9, ATCC 6633, ETEC, and *Salmonella Typ.* were grown aerobically in LB (Luria-Bertani, St. Louis, MO, USA) medium with constant shaking (200 RPM) at 37 °C. Tryptic soy broth (TSB; Becton–Dickinson, Sparks, MD, USA) was used to grow MRSA with constant shaking (200 RPM) at 37 °C. ETEC strain was positive for virulence factors K88: fimbrial variant 4 (F4), heat-labile enterotoxin A (eltA), heat-labile enterotoxin B (eltB), heat-stable enterotoxin A (estA), and heat-stable enterotoxin B (estB).

2.2. Survivability in Gastrointestinal Tract (GIT) Environment

2.2.1. Tolerance to Acid and Bile Salts

The tolerance of CP9 in acidic and bile salts environment was studied by methodology previously described [24] with minor modifications. Briefly, for assessing the tolerance of CP9 to acidic environment, 30 µL of the overnight cultures of CP9 were incubated with 70 µL LB broth adjusted to pH 2, 3, and 6.6 (control) using 1 N hydrochloric acid (HCl) in a 96-well microplate for 2 and 5 h. For assessing the tolerance of CP9 to bile salts environments, 30 µL of the overnight cultures of CP9 were incubated with 70 µL LB broth adjusted with 0% (control), 0.3%, 0.5%, and 1% bile salt (Sigma-aldrich, St. Louis, MO, USA) in a 96-well microplate for 1, 3, and 5 h. After the end of each incubation, cell viability and growth were measured spectrophotometrically via the metabolic activity of the cells using Bacterial Counting Colorimetric Assay Kit (BioVision Technologies, Inc., Chester Springs, PA, USA) following manufacturers protocol. Zero time period in all experiments represented the cellular activity of the initial cell concentration at the time of addition of the substrate. Metabolic cell activity and growth were then compared relative to the zero time point within each treatment group.

2.2.2. Tolerance to Swine GIT Fluids

Swine GIT fluids were collected, as previously described [25] courtesy of Anna Maystrenko. Briefly, the porcine gastrointestinal tract was obtained from the Meat Science Laboratory (University of Guelph, Guelph, ON, Canada). The GIT dissections and collection of gastric, duodenum, and jejunum contents were performed at 4 °C. Digestive contents were centrifuged at 10,000 rpm (9600× *g*) for 10 min at 4 °C, and supernatant fluid was collected, filter-sterilized using Fisherbrand 0.22 μm nylon filter (Fisher Scientific, Waltham, MA, USA) and stored at −80 °C until its use in the Tolerance to GIT fluids experiment. The pH of the collected supernatants was 3.5 and 6.5 for gastric fluid and duodenum fluid, respectively. To assess the tolerance of CP9 in the extracted swine GIT fluids, 30 μL of the overnight cultures of CP9 were incubated with 70 μL of extracted gastric, duodenum, and jejunum fluids in a 96-well microplate for 1, 2, and 5 h. Cell viability and growth were measured using Bacterial Counting Colorimetric Assay Kit (BioVision Technologies, Inc., Chester Springs, PA, USA) following manufacturers protocol. Zero time period in this experiment represented the cellular activity of the initial cell concentration at the time of addition of the substrate. Metabolic cell activity and growth were then compared relative to the zero time point within each treatment group.

2.3. *Evaluation of Antagonistic Activity of CP9 against ETEC, Salmonella Typ., and MRSA*

2.3.1. Agar Radial Diffusion Assay

The inhibitory activity of the CP9 cell-free supernatant (CFS) was evaluated by radial diffusion assay as preciously described [26] with minor modifications. Briefly, 10^8 CFUs of ETEC, *Salmonella typ.*, and MRSA were mixed with 30 mL of respective nutrient media agar and poured into a 100 mm round Petri dish. With the help of a sterile 1 mL pipette tip, approximately 5 mm diameter holes were punched in the agar and 100 μL of the filter-sterilized cell-free supernatant of CP9 or LB (negative control) or Hygromycin B (10 mg/mL, positive control, Sigma-aldrich, St. Louis, MO, USA) was added to the holes. After the supernatants were fully absorbed, plates were incubated at 37 °C under aerobic conditions. After an incubation period of 24 h, the diameters of the zones of inhibition were observed.

2.3.2. Agar Spot Assay

The contact-dependent inhibitory effect of CP9 was assessed by agar spot assay as previously described [27] with minor modifications. Briefly, 10^8 CFUs of ETEC, *Salmonella typ.* and MRSA were mixed with 30 mL of respective nutrient media agar and poured into a 100 mm round Petri dish. Overnight cultures of CP9 were grown to log phase until 10^8 CFUs were achieved and 10 μL of that culture, or LB (negative control) or Hygromycin B (10 mg/mL, positive control) was added to the petri dish with test pathogens. After the spots were fully absorbed, plates were incubated at 37 °C under aerobic conditions. After an incubation period of 24 h, the diameters of the zones of inhibition were observed.

2.3.3. Bacterial Co-Culture Assay

Quantitative analysis of CP9's inhibitory effect on the test pathogenic strains in a contact-dependent manner was performed by bacterial co-culture assay as previously described [28] with minor modifications. Briefly, 10% of 10^8 cells of overnight cultures of CP9, ETEC, *Salmonella typ.*, and MRSA were inoculated in 5 mL of their fresh respective nutrient media in 15 mL Falcon™ Round-Bottom Polypropylene Test Tubes (Fisher Scientific,

Waltham, MA, USA) and vortexed for 10 s. These cultures were named culture A. One ml of the CP9 culture A was mixed with 1 mL of ETEC culture A or *Salmonella typ.* culture A or MRSA culture A in a fresh 15 mL test tube and incubated at 37 °C under aerobic conditions for 5 h. After the end of the incubation, viable cell number of test pathogenic strains were analyzed by performing serial dilutions and colony forming units per ml were counted using pathogen-specific agar plates. MacConkey agar (Thermo Fisher Scientific, Waltham, MA, USA) was used for ETEC and *Salmonella typ.*, Columbia blood agar plates with 5% sheep blood were used for MRSA by counting typical hemolytic colonies, and bacillus cereus agar (PEMBA) with egg yolk and polymyxin B supplement was used for enumeration of CP9.

2.3.4. Cell Line Culture Conditions

The porcine intestinal epithelial cell line, IPEC-J2, originally derived from jejunum of neonatal piglet [29] was acquired from the American Type Culture Collection (ATCC; Virginia, USA). IPEC-J2 cells were cultured in a 1:1 mixture of Dulbecco's modified Eagle's medium/Ham's Nutrient Mixture F-12 (DMEM/F12) supplemented with 10% fetal bovine serum (FBS) and 1% Pen-Strep (10,000 units/mL, Invitrogen, Waltham, MA, USA) under 5% CO2 in a 95% aerobic atmosphere with 90% humidity at 37 °C.

2.3.5. Cell Cytotoxicity Assay

The impact of CP9 on IPEC-J2 cell viability was determined by performing cell cytotoxic assays as described [30]. Briefly, 2×10^5 cells/mL were seeded per well of a 96-well tissue culture plate and grown in 37 °C for 24 h. Media was then replaced with fresh DMEM/F12 media without antibiotics. Cell free supernatant (20, 50, 75, and 100 µL/mL) and CP9 (10^8 cells/mL) were added to IPEC-J2 cells and incubated at 37 °C for 8 h. Final well volume was 200 µL/well. After the end of incubation, cell viability was analyzed by using alamarBlue™ Cell Viability Reagent (Thermo Fisher Scientific, Waltham, MA, USA) following manufacturer's instructions.

2.3.6. Cell Surface Adhesion Assay

To determine CP9′s ability to adhere to the IPEC-J2 cells, cell surface adhesion assay was performed as preciously described [30] with minor modifications. Briefly, IPEC-J2 cells were seeded in 12 well tissue plates with 2×10^5 cells/well and grown for 24 h. Cells were then washed two times with PBS to remove the antibiotics in the medium. Fresh DMEM/F12 media without antibiotics was added to all wells. Commercial strain CS and CP9 were grown to log phase and 1×10^8 cells/mL were pelleted, washed with PBS and resuspended in DMEM/F12 incomplete media before incubating with IPEC-J2 cells for 3 h at 37 °C aerobically. After end of incubation, media was removed, and all the wells were washed twice with PBS to remove unadhered bacterial cells. Cells were collected using trypsin-EDTA solution, and serial dilutions were plated on LB nutrient agar plates and incubated aerobically at 37 °C overnight for enumerating and counting adhered bacterial cells.

2.3.7. Gene Expression Analysis

To analyze the effects of CP9 on the expression of virulence-related genes in ETEC, a co-culture experiment was performed where equal volumes of 10^8 cells of CP9 and 10^8 cells ETEC or 10^8 cells of their monocultures were grown in LB nutrient broth aerobically at 37 °C for 5 h. Prior to RNA extraction, RNAprotect Bacteria Reagent (Qiagen 76506) was added to each culture (2:1) for RNA stabilization. Total RNA was then extracted using an RNeasy Protect Bacteria Mini Kit (Qiagen, ON, Canada) according to the manufacturer's protocol. RNA yield and quality were assessed spectrophotometrically via A230, A260, and A280 nm measurements using a NanoDrop™ 8000 Spectrophotometer (Thermo Fisher Scientific, Waltham, MA, USA). cDNA synthesis was performed as previously described by [31] using a QuantiNova Reverse Transcription Kit (200) (Qiagen, ON, Canada). Quantitative real time-PCR (qPCR) was used to measure the change in the expression levels of transcripts of seven different virulence genes in ETEC, namely motA (motility-flagellar), faeG (adherence-K88, F4, fimbrial protein), tnaA (Tryptophanase-energy metabolism), estA and estB, (heat-stable enterotoxin A and B, respectively), and eltA and eltB (heat-labile enterotoxin A and B respectively), as previously described by [32]. Primers were designed using a Primer-BLAST tool (NCBI; National Center for Biotechnology Information) and synthesized by Integrated DNA Technologies, Guelph, Canada. Primer information is listed in Table S1. The efficiencies of the primers were calculated using CFX Manager Software (Bio-Rad Laboratories Ltd., Hercules, CA, USA). Gene expression was normalized using two reference genes, i.e., the *E. coli* D-glyceraldehyde-3-phosphate dehydrogenase A subunit (gapA) and the *E. coli* 16S ribosomal RNA genes. After determining the threshold cycle (Ct) for each gene, the relative changes in gene expression of ETEC co-cultured with CP9 compared to virulence gene expression of ETEC alone were calculated using the $2^{-\Delta\Delta Ct}$ method in CFX Manager Software (Bio-Rad Laboratories Ltd., Hercules, CA, USA) [32].

2.4. Metabolomic Analyses

Sample Preparation and LC/MS Procedure

To determine and compare the extracellular metabolite secretions of CP9 in a co-culture with ETEC, a co-culture experiment was performed for five hours. After the end of co-culture incubation, CP9 and ETEC monocultures along with their co-culture samples were centrifuged, supernatant was collected, and filter sterilized using Fisherbrand 0.22 µm nylon filters. LB nutrient media was used as a negative control sample. The samples were immediately frozen in liquid nitrogen and stored in a −80 °C ultrafreezer. Samples were packed in dry ice and shipped to the BioZone Mass Spectrometry Facility in the Chemical Engineering Department at the University of Toronto for metabolite extraction and liquid chromatography-mass spectrometry analysis, courtesy of metabolomics specialist, Robert Flick. Briefly, protein from the samples was precipitated and metabolites were vacuum dried using a speedvac at ambient temperature, followed by resuspension in one tenth the original volume using the appropriate starting solvent for each chromatography method. Samples were then analysed using a Thermo Scientific Ultimate 3000 UHPLC (Thermo Fisher Scientific, Waltham, MA, USA) equipped with a Hypersil Gold C18 column (50 mm × 2.1 mm, 1.9 um) (Thermo Scientific, Waltham, MA, USA) or a Phenomenex Luna NH2 column (150 mm × 2 mm, 3 um), both with guard columns. The temperature of the column was set to 40 °C with a flow rate of 300 µL.min^{-1}. Water and acetonitrile containing 0.1% formic acid were used as eluents. The gradient for the C18 column was performed at 5% B for 1 min, linear gradient at 98% B for 6 min, maintained at 98% B for 3 min, returned to 5% B for 0.5 min, and finally a re-equilibration at 5% B for 4.5 min (total runtime 15 min). The gradient for the Luna NH2 column was performed at 90% B for 1 min, linear gradient at 5% B for 4 min, maintained at 5% B for 8 min, returned to 90% B over 1 min, and finally a re-equilibration at 90% B for 6 min (total runtime 20 min). The autosampler of the Thermo Scientific Ultimate 3000 UHPLC was loaded with 10 µL liquid samples. The autosampler temperature was kept at 10 °C. A Q-Exactive Orbitrap mass spectrometer (Thermo Fisher Scientific, Waltham, MA, USA) equipped with a Heated Electrospray Ionization (HESI II) probe was used for compound detection. The system was operated in negative and positive ionization modes for generating spectra. MS1 spectra were acquired over an m/z range from 80 to 1200 with the mass resolution set to 70 k, AGC Target of 3E6, max injection time 100 ms, spray voltage 3.5 kV, capillary temperature 320 °C, sheath gas 15, aux gas 5, spare gas 2 and s-lens RF level 50. Data-dependent MS2 spectra using a Top5 approach were acquired using a mass resolution of 17.5 k, AGC Target of 1e5, max injection time 50 ms, isolation window of 1.0 m/z and HCD collision energy of 30. After generating the raw peaks, the untargeted metabolomic data was processed (raw signals exacting, data baselines filtering, peak identification and integration) and metabolite detection (KEGG and BioCyc database) using the differential analysis software package Compound Discoverer 2.1 (Thermo Scientific, Waltham, MA, USA).

2.5. Statistical Analyses

All experiments were performed in three biological replicates and data are presented as mean ± standard error of the mean (SEM). For gene expression analysis, experiments were performed in triplicate ($n = 3$) and data are presented as mean ± standard error of the mean (SEM). Data were analyzed with GraphPad Prism v. 7.0 (GraphPad Software, Inc., San Diego, CA, USA) using one-way or two-way ANOVA with Tukey's post hoc test. $p < 0.05$ was considered significant for all statistical tests.

Metabolomic data was analyzed by performing multivariate statistical analysis and one-way analysis of variance using Metaboanalyst (version 5.0) online analysis software (www.metaboanalyst.ca, accessed 18 October 2020). Briefly, samples were first normalized to the internal control and LB media control. Processed data was filtered to identify and remove any variables followed by normalization and scaling. Principal Component Analysis (PCA) and Partial Least Squares-Discriminant Analysis (PLS-DA) combined with one-way ANOVA and post-hoc analysis were used to screen the significantly differential metabolites. $p < 0.05$ was considered significant for all statistical tests. The model was evaluated by cross validation method using Q^2 as a performance measure. Clustering and pathway analysis was performed by generating a heat map using Euclidean distances and complete linkage with ANOVA results.

3. Results

3.1. CP9 Survives Gastrointestinal Environment

We first examined tolerance of CP9 to different bile salts and pH environments by measuring the metabolically live cell activity as described in *Materials and Method* section. Figure 1A shows that, compared to the initial cell activity at time point zero, no significant change in the cell activity of CP9 was observed in the presence of 0.3%, 0.5% and 1% bile salts for up to 3 h. However, by the end of the 5 h incubation, CP9 metabolic activity increased more than double from the initial 0 h time period across all bile concentrations tested, suggesting significant cell growth of CP9 in the varied bile salt environment after initial adaptation.

Similarly, data from the low pH incubation analysis showed that CP9 maintained its initial metabolic cell activity for up to 2 h of incubation in pH 2 and pH 3 environments (Figure 1B). Metabolic activity of CP9 by the 5 h time period increased significantly ($p < 0.05$) in both pH environments tested, suggesting CP9 could survive in low pH environments after initial adaptation in lower pH.

Overall, results from these suggest that, compared to the untreated CP9 cells, the metabolic activity of CP9 cells in varied concentrations of bile and low pH environments showed a halted growth, and the recovery in the cellular activity in higher time points is indicative of CP9 survival and growth in the stressed environments tested.

To further assess survival of CP9 in the intestinal environment, we incubated CP9 with freshly collected fluid from gastro-intestinal tract (GIT). It was found that, in gastric fluid, CP9 cells maintained a similar metabolic cell activity up to 2 h of incubation, which significantly increased ($p < 0.05$) by the 5 h incubation period (Figure 1C). Similarly, compared to initial metabolic activity at the 0 h time point, CP9 cells incubated with duodenum and jejunum fluids showed enhanced metabolic activity in all time periods tested, suggesting that CP9 may survive and propagate in a GIT environment.

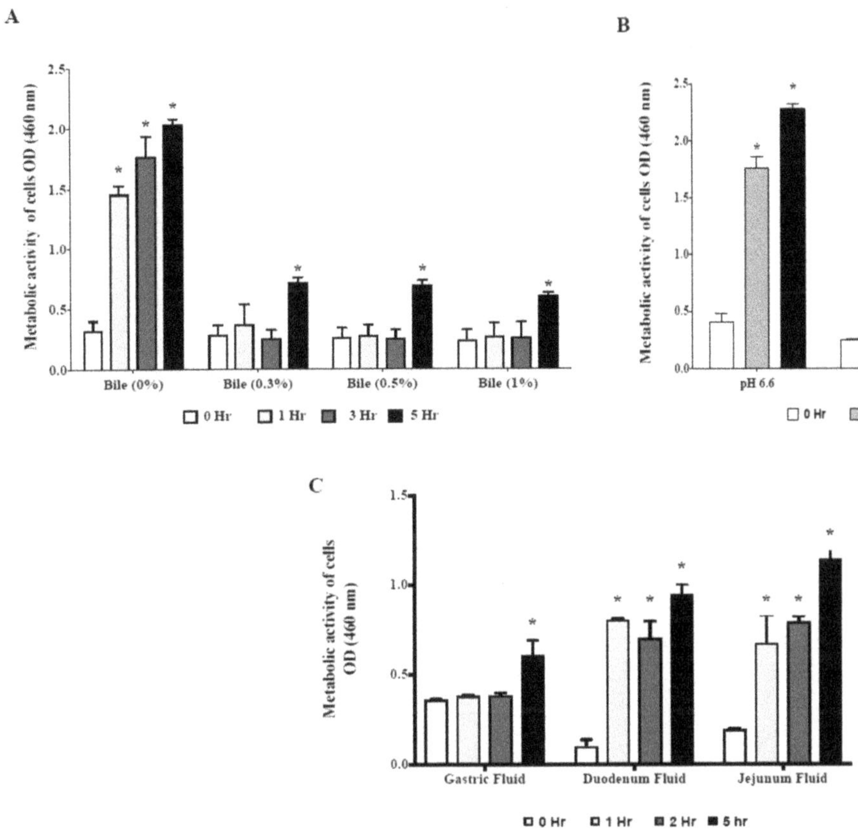

Figure 1. Metabolic activity and viability of CP9 cells in (**A**) bile environment, (**B**) low pH environment, and (**C**) swine gastro intestinal fluids. Data are presented as mean ± standard error of the mean (SEM). Bars with statistical significance denoted as * ($p \leq 0.05$), using Tukey's multiple comparison test in ANOVA. Significance in all tests is compared with the initial metabolic activity at time zero within each treatment group. The experiment was performed in triplicates and repeated thrice.

3.2. CP9 Adherence and Toxicity on IPEC Cells

As a potential probiotic, we next assessed the ability of CP9 to adhere to the pig intestinal epithelial cells, IPEC-J2. To compare the adhesion, we used a commercially available swine probiotic bacillus subtilis (CS) as a control. It was found that CP9 had a significantly higher ($p < 0.01$) adherence to the IPEC-J2 cells, which was 2.6 times higher than the than the commercially available strain CS (Figure 2A).

We next assessed if CP9 impacts intestine cell viability by incubating the CFS of CP9 and the bacterium itself with IPEC-J2 cells for 8 h. As shown in Figure 2B, no significant change in the relative metabolic activity of IPEC-J2 cells was observed when the IPEC-J2 was co-cultured with various concentrations of CP9 CFS or 10^8 CFU/mL CP9 cells. Similar results were observed for the commercial strain CS, where relative metabolic activity of the cells remained consistently well over 90%. Taken together, these results suggest that CP9 is not cytotoxic to the IPEC-J2 cells and shows better adhesion capacity to the cells compared to commercially available *B. subtilis* strain, CS, in vitro.

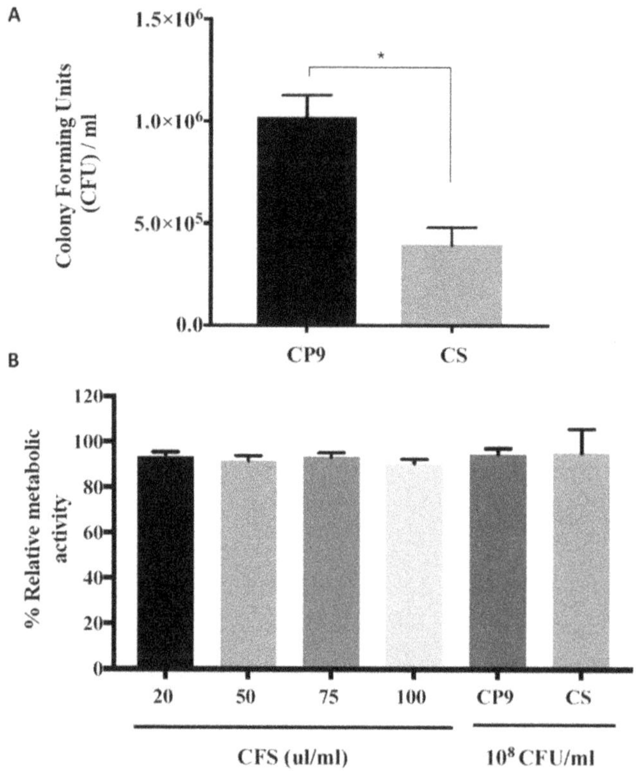

Figure 2. CP9 interaction with Swine intestinal cells, (**A**) CP9 cell surface adherence and (**B**) CP9 impact on swine intestinal epithelial (IPEC-J2) cell viability. Commercially available *Bacillus subtilis* strain (CS) was used as a comparative strain. Data are presented as mean ± standard error of the mean (SEM). Bars with statistical significance denoted as * ($p \leq 0.05$), using Tukey's multiple comparison test in ANOVA. The experiment was performed in triplicates and repeated thrice.

3.3. *CP9 Exhibits Anti-Pathogenic Activity via Contact Inhibition*

We first tested the inhibitory potential of CP9′s secretions against ETEC, *Salmonella Typ.* and MRSA. Cell-free supernatant from the log-phase CP9 culture was extracted and inoculated on agar plates containing ETEC, *Salmonella Typ.* and MRSA, separately, using radial diffusion assay. After 24 h of aerobic incubation, no inhibition zone on pathogen growth was observed by the CP9 CFS (Figure 3A), suggesting that CP9 did not secrete anti-pathogen substances in mono-cultures. Interestingly, when CP9 was spotted and grown on the pathogen-inoculated agar plates, clear inhibitory zones were observed in all the pathogens tested (Figure 3B). The results suggest that CP9 may act in a contact-dependent manner against ETEC, *Salmonella Typ.* and MRSA. To further evaluate the impact of CP9 on the viability of pathogenic bacterial strains, we performed a quantitative analysis, where the pathogenic bacterial strains were grown in a co-culture with CP9. Results from the 5 h co-culture experiment showed that, compared to the individual cultures, co-culture with CP9 significantly ($p < 0.05$) reduced the number of ETEC, *Salmonella Typ.* and MRSA (Figure 3C) by more than 75%, further confirming CP9′s ability to halt the growth of pathogenic bacteria when cultured together or in contact with the pathogen.

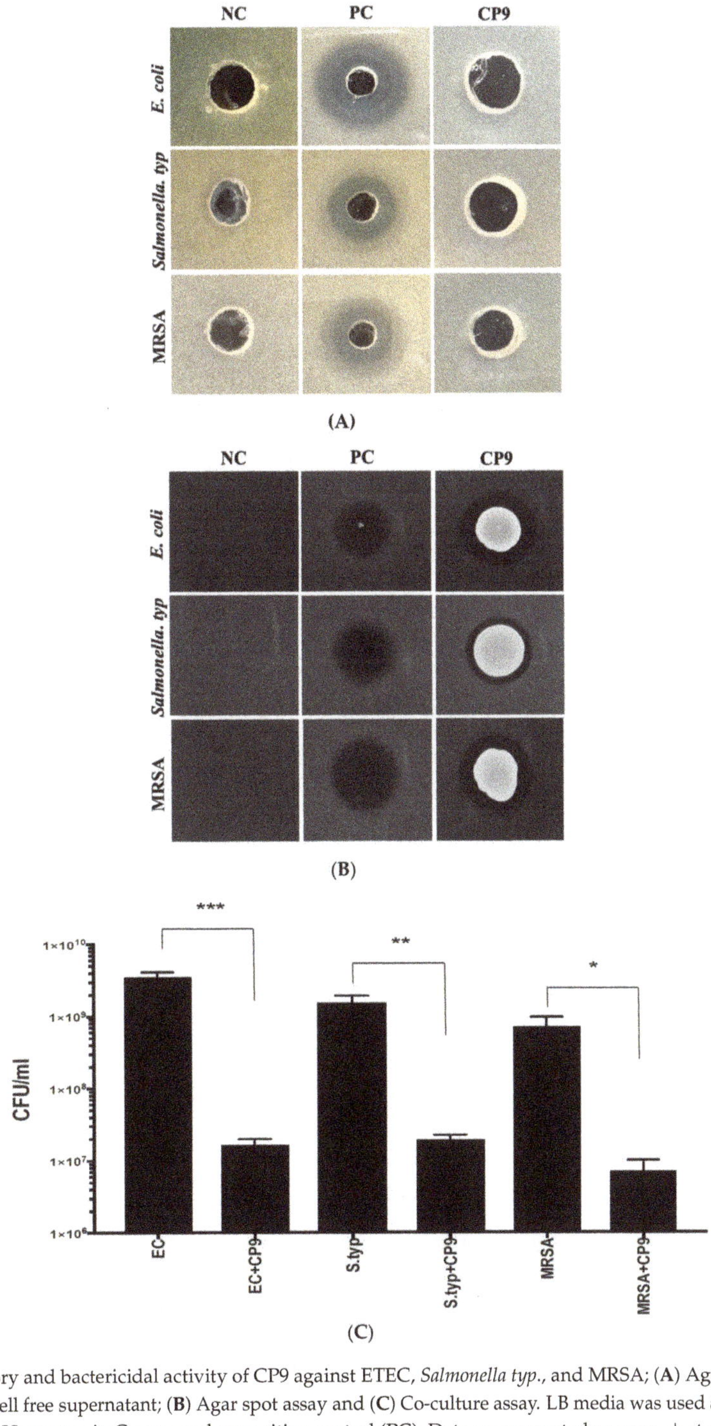

Figure 3. Inhibitory and bactericidal activity of CP9 against ETEC, *Salmonella typ.*, and MRSA; (**A**) Agar radial diffusion assay using CP9 cell free supernatant; (**B**) Agar spot assay and (**C**) Co-culture assay. LB media was used as negative control (NC), 10 mg/mL Hygromycin C was used as positive control (PC). Data are presented as mean ± standard error of the mean (SEM). Bars with statistical significance denoted as * ($p \leq 0.05$), ** ($p \leq 0.01$) and *** ($p \leq 0.001$) using Tukey's multiple comparison test in ANOVA. The experiment was performed in triplicates and repeated thrice.

3.4. CP9 Downregulates Virulence Genes Expression in ETEC

To evaluate if CP9 plays a role in attenuation of ETEC virulence, we analyzed the expression of the several virulence genes of ETEC such as motA (motility-flagellar), faeG (adherence-K88, F4, fimbrial protein), tnaA (Tryptophanase-energy metabolism), estA, estB, (heat-stable enterotoxin A and B, respectively) and eltA and eltB (heat-labile enterotoxin A and B respectively). As shown in Figure 4, while the expression of adherence gene faeG was downregulated upon co-culture with CP9, there was an increased expression of motA gene responsible for flagella motility. Expression of genes encoding ETEC toxins, estA, estB, and eltA, were significantly downregulated upon co-culture with CP9; however, there was no significant change observed in the expression of eltB gene. Finally, expression of tnaA gene was also seen significantly downregulated upon incubation with CP9. Overall, these data suggest that CP9 influenced and suppressed the expression of ETEC toxin genes and genes involved in pathogen adherence.

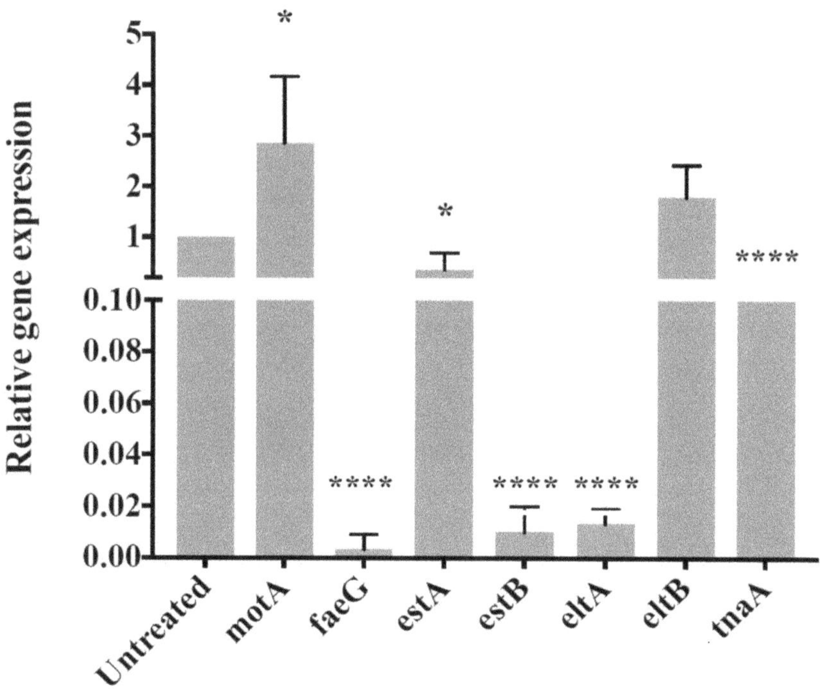

Figure 4. Relative gene expression of ETEC virulence-related genes in co-culture with CP9. Data are presented as mean ± standard error of the mean (SEM). Bars with statistical significance denoted as * ($p \leq 0.05$), **** ($p \leq 0.0001$), using Tukey's HSD test in ANOVA. All values are relative to untreated ETEC monoculture. This experiment was performed in triplicate.

3.5. Secreted Metabolic Repertoires of the CP9 and ETEC Co-Culture Vary Significantly Than Their Monocultures

In order to decipher the metabolic impact of bacteria-bacteria interaction and potential mechanisms on growth inhibition in a co-culture, we performed metabolomic analysis on secreted factors in the CFS of mono and co cultures of CP9 and ETEC, using liquid chromatography coupled to a mass spectrometer (LC-MS). An untargeted metabolomics approach was applied to capture a wide array of secreted metabolites in mono versus co-culture groups. The metabolomic features were first normalized and refined by using the same culture media as previously described [16]. The system successfully identified

199 metabolites (Table S2), which were then statistically analyzed through Metaboanalyst (version 5.0) online analysis software. Compared to the mono-cultures, the co-culture had substantially altered metabolomic profiling, as seen via heat map and clustering analysis (Figure 5A). To compare the metabolomic patterns in the secretions of CP9, ETEC and their co-culture, we first performed PCA and PLS-DA multivariate statistical analyses to evaluate the metabolic features that caused significant separation between the groups. Figure 5 shows clear separation in metabolomic profiles between the three groups. The R2 and Q2 values obtained from PLS-DA were >0.8 (Figure S1), suggesting that the models used were of reasonable and acceptable quality and could be further used for analyzing significant differences between the groups. Interestingly, samples from co-culture appeared to be located closer to the CP9 mono-culture, suggesting that the co-culture metabolome was less resembling to the negatively affected strain ETEC (Figure 5B,C). Importantly, the emergent separation of metabolomic profiles in the co-culture suggests that the interaction of CP9 and ETEC may have resulted in production of specific metabolites that may play a role in negatively affecting ETEC growth in the co-culture.

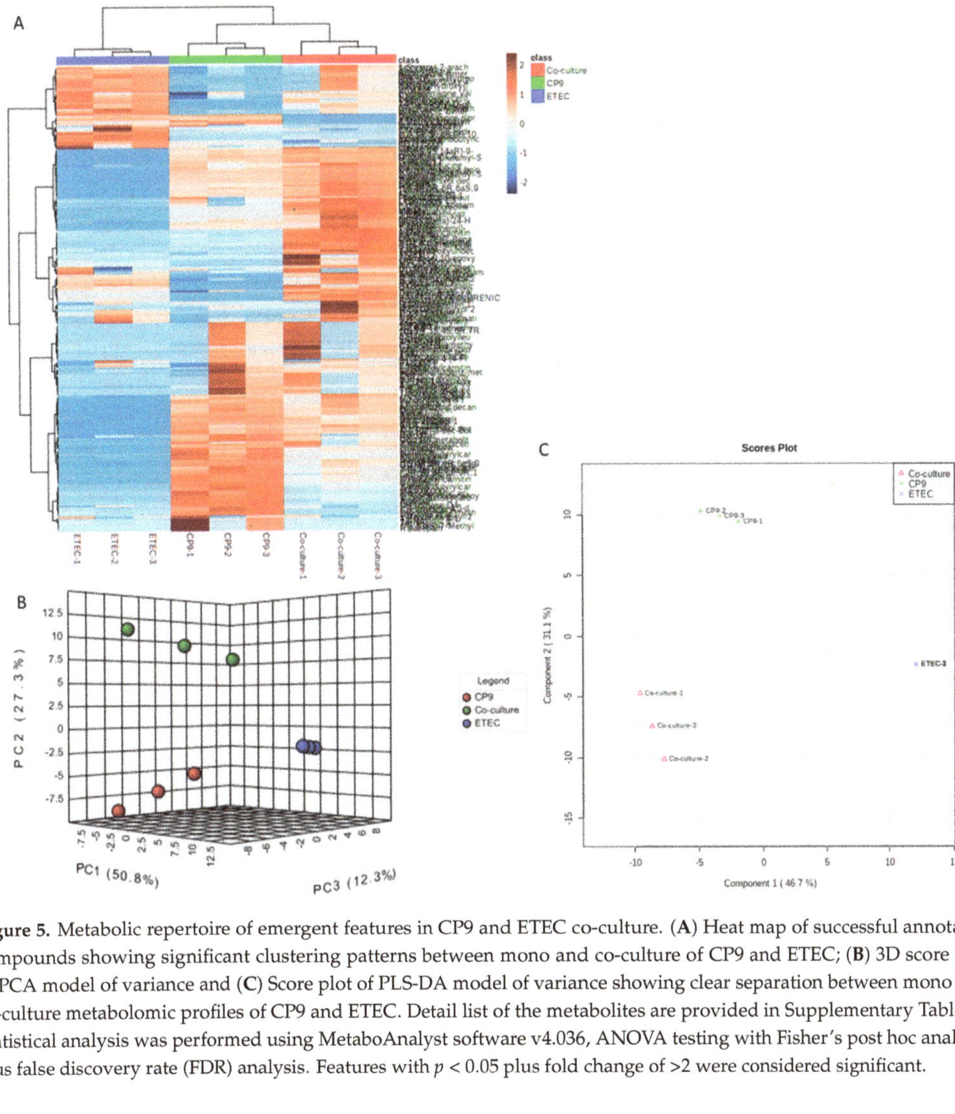

Figure 5. Metabolic repertoire of emergent features in CP9 and ETEC co-culture. (**A**) Heat map of successful annotated compounds showing significant clustering patterns between mono and co-culture of CP9 and ETEC; (**B**) 3D score plot of PCA model of variance and (**C**) Score plot of PLS-DA model of variance showing clear separation between mono and co-culture metabolomic profiles of CP9 and ETEC. Detail list of the metabolites are provided in Supplementary Table S2. Statistical analysis was performed using MetaboAnalyst software v4.036, ANOVA testing with Fisher's post hoc analysis plus false discovery rate (FDR) analysis. Features with $p < 0.05$ plus fold change of >2 were considered significant.

3.6. CP9 and ETEC Co-Culture Induces Emergence of New Metabolites with Antimicrobial Properties

To determine the significantly different metabolic features between the groups, we combined the Variable Importance for the Projection (VIP, VIP obtained by PLS-DA model) data with p-value obtained by performing one-way ANOVA analysis on the identified metabolites. For determining significant differential metabolites, the metabolite had to pass the screening criteria of VIP > 1 and $p < 0.05$ as previously described by [33]. A total of 143 metabolites were found to be significantly distinct between the mono- and CP9 and ETEC co-cultures (Table S3), and 82 differential metabolites were found to have a VIP score above 1 (Table S4). In order to look for the unique metabolites that may be secreted or influenced by CP9 in response to co-culture with ETEC, we focused on (i) the metabolites that emerged only as a result of co-culture and were absent in the monocultures, (ii) the metabolites that emerged in CP9 mono-cultures and were overexpressed in co-culture and (iii) the metabolites that had significantly higher concentrations in ETEC monoculture but were either suppressed or overexpressed in the co-culture (Table 1). Out of the 31 selected metabolites (Table 1), 11 metabolites presented only in the co-culture group. The unique profile consisted of metabolites belonging to fatty acid and energy metabolism, tryptophan metabolism, polyamine metabolism, nitrogen metabolism, and secondary metabolites with known antimicrobial properties. In the second group, 14 metabolites already observed in the CP9 monoculture group were seen to be increased significantly ($p < 0.05$) in the co-culture group. These unique metabolites belonged to fatty acid and energy metabolism, glutathione metabolism, polyamine metabolism and cell-cell signaling, nucleoside analogues, arachidonic acid metabolite, serine protease inhibitor and secondary bacterial bile acid. Additionally, three metabolites with structural similarities to commercial drugs/chemicals emerged in second group; however, no exact match in the metabolomic database or relevance to their role in microbial physiology could be determined. Lastly, the third group, reflecting metabolites that were present in ETEC and were significantly influenced by CP9 in co-culture, showed varied abundance of metabolites involved in tryptophan metabolism, secondary metabolites involved in ETEC virulence, purine metabolism, and cell growth.

Table 1. Differential metabolites uniquely emerging in co-culture and mono-cultures.

		VIP Scores			
Group	**Metabolite**	**Comp. 1**	**Comp. 2**	***p* Value**	**Pathway/Function**
	9-Decenoylcarnitine	1.3611	1.3578	5.69×10^{-8}	Fatty acid/energy Metabolism
	Carnosine	1.412	1.3994	1.47×10^{-6}	Fatty acid/energy Metabolism
	5-Methoxy-3-indoleaceate	1.3661	1.3571	2.08×10^{-6}	Tryptophan metabolism and antimicrobial metabolite
	Indole	1.5536	1.5057	1.70×10^{-6}	Tryptophan metabolism and antimicrobial metabolite
	Valclavam	1.3586	1.3543	4.26×10^{-6}	Antimicrobial metabolite
Co-culture only	3-[(3-Hydroxyundecanoyl)oxy]-4-(trimethylammonio) butanoate	1.3429	1.3343	0.00092045	Fatty acid/energy Metabolism

Table 1. *Cont.*

Group	Metabolite	VIP Scores Comp. 1	VIP Scores Comp. 2	*p* Value	Pathway/Function
	n-phenethyl acetamide	1.4326	1.4083	0.0011712	Antibacterial secondary metabolite
	LT9970000/ Furmecyclox	1.4475	1.4108	0.0029438	Drug
	Uric Acid	1.039	1.0652	0.0034036	Nitrogen metabolism/Amino acid and protein synthesis
	Putrescine	1.0175	1.0174	0.0071156	Polyamine/Cell growth and metabolism/ Virulence
	MFCD00059633/ 3-Hydroxymyristic Acid	1.0336	1.0203	0.023742	Bacterial metabolite/fatty acid metabolism
CP9 and overexpressed in Co-culture	C8-Carnitine	1.4194	1.4042	1.22×10^{-9}	Fatty acid/energy Metabolism
	L-Cysteinylglycine disulfide	1.5642	1.5136	2.15×10^{-9}	Di-peptides/ Glutathione metabolism
	N(1)-acetylspermidine	1.4443	1.4248	1.82×10^{-8}	Polyamine metabolite/Cell-Cell signalling/ Virulence
	5,6-Dihydrothymidine	1.4407	1.3837	4.66×10^{-8}	Nucleoside analogues
	Leukotriene C4	1.4209	1.3659	7.24×10^{-8}	Arachidonic Acid metabolite/ antimicrobial
	Naloxegol	1.4683	1.4096	3.51×10^{-7}	Drug
	gamma-Glu-gln	1.4715	1.4125	3.67×10^{-7}	Glutathione metabolism
	Aderbasib	1.5448	1.4852	1.11×10^{-6}	Drug
	Spironolactone	1.5097	1.4494	2.27×10^{-6}	Drug
	3-[(2,6-Dimethylheptanoyl)oxy]-4-(trimethylammonio) butanoate	1.4724	1.4402	5.38×10^{-6}	Fatty acid/energy Metabolism
	Leukotriene E3	1.262	1.247	6.88×10^{-6}	Arachidonic Acid metabolite/ antimicrobial
	Carnosine.1	1.4944	1.4362	0.00012389	Fatty acid/energy Metabolism
	Melagatran	1.5172	1.4729	0.00015017	Serine protease inhibitor
	(3beta,5beta)-24-Hydroxy-24-oxocholan-3-yl beta-D-glucopyranosiduronic acid	1.4674	1.4101	0.0010426	Secondary bacterial bile acid metabolite/ antibacterial metabolite

Table 1. *Cont.*

Group	Metabolite	VIP Scores Comp. 1	VIP Scores Comp. 2	*p* Value	Pathway/Function
ETEC and over/ under expressed in Co-culture	Kynurenic acid (↓)	1.2825	1.2397	7.61×10^{-8}	Tryptophan metabolism
	gamma-Aminobutyric acid (↓)	1.2656	1.2242	9.81×10^{-8}	Spore germination/bile and low pH resistance
	8-Methoxykynurenic acid (↑)	1.3368	1.3216	1.05×10^{-5}	Tryptophan metabolism
	Gln-Gln (↑)	1.2812	1.2803	0.00010342	L-Glutamine Dipeptide/acid resistance
	(1Z,3R,5E,8S,9S,10R)-N-[(Z)-2-(3-Chloro-4-hydroxyphenyl) vinyl]-3,9-dihydroxy-2,4-dimethoxy-6,8,10-trimethyl-7-oxo-5-tetradecenimidic acid (↑)	1.3569	1.3068	4.29×10^{-7}	Unknown
	Arabinosylhypoxanthine (↑)	1.1994	1.1915	1.07×10^{-7}	Purine metabolism/E. coli cellular growth and virulence in mixed culture

4. Discussion

Over the past couple decades, probiotics have been researched for their unique antagonistic properties to pathogenic bacteria. They may achieve this by various mechanisms such as competitive exclusion and secretion of small antibacterial molecules. Our study evaluated properties of a novel *Bacillus subtilis* strain, CP9, for its probiotic and antimicrobial potential in vitro and identified unique small molecules during CP9-ETEC interaction.

Gastrointestinal tract in vitro-mimicking models have been widely and successfully used for testing the passage survival and colonization of the probiotic strains [34]. In our study, CP9 showed significant resistance in the GIT environment conditions in vitro, which is a positive trait of a potential probiotic bacteria, since the colonization and persistence of probiotic bacteria in the GIT is an important factor for exerting a beneficial effect on the host [35]. We observed an initial adaptation of CP9 in the low pH and varied bile environments (Figure 1A,B), which was consistent to a previous study, where *Bacillus subtilis* cultures were seen to adapt initially to varied pH and alkaline stress before recovering growth rapidly [36].

Bacillus subtilis' existence is ubiquitous in the environment and has been shown to be found in symbiotic existence within plants and animal kingdom [37]. Due to its spore-forming practical edge over the other vegetative forms of probiotics, it has gained a substantial research interest in human and animal consumption and is generally considered safe due its long history of consumption [18]. However, due to the strain-specific properties, behavior, and interactions in the mammalian intestinal tract, the toxigenic potential of a novel probiotic strain is an inevitable check point [38]. In the current study, CP9 showed no cytotoxicity to the swine intestinal epithelial cells and was consistent with the previous studies performed on *Bacillus subtilis*-based probiotics [39,40]. Furthermore, CP9 showed a higher adherence to the IPEC-J2 intestinal cells than a commercially available probiotic

Bacillus subtilis. Intestinal adherence is an important determining factor for probiotics to modulate a host's immune system as well as competitively prevent the adhesion of opportunistic and pathogenic enteric bacteria [41,42].

It has been shown that CFS of *Bacillus subtilis* can inhibit enteric pathogens such as ETEC and *Salmonella typ*. [43,44]. However, in our study, CP9 appeared to display an antimicrobial effect via contact-dependent inhibition and, in parts, via metabolic influence. The absence of anti-pathogenic activity in the CP9 cell-free supernatant suggests the absence of toxins or antimicrobial molecules in the monocultures of CP9. Interestingly, upon contact with the pathogenic strains ETEC, *Salmonella*, and MRSA, a substantial decrease in the pathogenic cell growth was observed, which may suggest, in parts, activation of pathways for CP9′s cellular response to pathogens. Highly competitive bacteria survive by using their exploiting (nutrient depletion) and/or interfering (release of antagonistic factors) abilities to survive in heavily populated environments [45]. As part of the interference mechanism, contact-dependent inhibition (CDI) describes the bacterial adjustment of internal cellular responses and cell differentiation pathways in response to external cue [45,46]. Upon sensing interbacterial competition, members of the same microbial community can ramp up their cellular growth, activate the secretion system and deliver the regulatory factor across membrane upon contact with the competitor strain. These regulatory factors can influence the cellular processes of the competitor strains and inhibit their cell growth [47,48]. Both gram-positive and gram-negative bacteria have been shown to utilize their secretion system for CDI to influence cellular growth of competitor strains. For example, *Bacillus subtilis* have been shown to utilize the CDI secretion system to secrete and deliver toxic polymorphic protein regulatory factors to influence morphological changes and growth inhibition in target strains such as *E. coli* [49,50]. Contact-dependent growth inhibition is also profoundly used by gram-negative bacteria, such as *E. coli*, for delivering toxins to the neighbouring target cells [46,51], however, since we did not observe contact inhibition from the ETEC on CP9 in our study may suggest that CDI growth inhibition of ETEC was driven by CP9.

Enteric pathogens such as ETEC express various virulence factors that are regulated by the environment and help ETEC outcompete its rival commensals in the GIT and evade host defenses such as motility (*motA*, flagellar movement), adherence (*faeG*, F4, fimbrial protein), heat-stable enterotoxins (*estA*, *estB*), heat-labile enterotoxins (*eltA*, *eltB*), and tryptophanase (*tnaA*, virulence regulator and energy metabolism) [52–54]. Interestingly, when co-cultured with ETEC, CP9 significantly downregulated the expression of the virulence genes that are responsible for adherence, faeG and toxin-releasing genes *estA*, *estB*, *eltA*. The finding that flagellar motility gene motA, one of the ETEC virulence genes, was overexpressed during co-culture was surprising. The significance of the increase expression is currently unknown. Interestingly, it has been reported that overexpression of motA is associated with reduced ETEC cell growth [55].

Energy metabolism is vital for physiological processes and biochemical pathways for driving division and cell growth in microbes such as bacteria [56,57]. Interactions in mixed microbial cultures are driven by metabolite exchanges and are dependent on symbiotic and sometimes competitive behaviours [58,59]. Tryptophan and its metabolic derivatives such as indole, indole derivatives, and kynurenic acid are vital for bacterial protein synthesis and cell growth [60]. In ETEC, tryptophan metabolism is executed by enzyme tryptophanase (*tnaA*) and its expression is tightly regulated by external tryptophan availability [61,62]. Importantly, in ETEC, pathogenicity and virulence have also been shown to be regulated by the *tnaA* gene [32,52,63]. In our study, exposure to CP9 downregulated the ETEC *tnaA* gene in the co-culture, which was also reflected in the lower kynurenic acid levels observed in the ETEC group (Table 1). In addition to these results, higher abundance of its downstream metabolite, 8-Methoxy kynurenic acid, in the co-culture samples might suggest depletion of tryptophan from the media by CP9 as part of competitive exclusion, a typical strategy for survival in mixed microbial cultures for its own growth. This notion is further supported by reduced growth of ETEC in co-culture and abundance of downstream metabolites of tryptophan such as, indole and its derivatives seen in the co-culture, suggesting an external tryptophan utilization by CP9 for its rapid growth and production of survival proteins (Table 1). Interestingly, microbially derived indole and its derivatives, known for their antimicrobial effects [64], have been previously shown to negatively regulate virulence of GIT pathogens, such as enterohemorrhagic *Escherichia coli* (EHEC) and *Citrobacter rodentium* [65,66]. Consistent to our study, Singh et al. 2014 previously found that in the co-culture with *Bacillus subtilis,* there was a higher indole yield, the number of *E. coli* decreased dramatically compared to its monoculture, and *Bacillus subtilis* in co-culture [67]. This might suggest, in parts, the antibacterial effect of the *Bacillus subtilis* derived indole and its derivatives seen in our study. However, the study by Singh and colleagues was an experimental demonstration of indole production in the co-culture using mathematical modelling and function of time. Therefore, results from our study should be taken carefully as further experiments may be needed to model and quantify the depletion of tryptophan and production of its downstream metabolites such kynurenic acid and indole by individual strains in the co-culture, especially when *E. coli* is also shown to produce indole under stress conditions [68]. This could additively represent the indole production in the co-culture in our experiment.

In mixed microbial cultures, competitive exclusion is achieved by either rapid nutrient utilization for energy and protein production for cellular growth, by secreting antimicrobial metabolites, or by both [69]. Endogenous and exogenous fatty acid metabolisms play a critical role in energy derivation, protein synthesis, transport for cellular growth, and survival in bacterial physiology [70,71]. Our finding on the significant emergence of carnitine, acyl carnitines, and other fatty acid metabolites in the co-culture samples reflects a rapid metabolism of fatty acid for intracellular transport and energy production [72–74] in the co-culture samples. Additionally, there was significant emergence of polyamine putrescine and its intermediate, N(1)-acetylspermidine, which are responsible for regulating virulence factors for survival and cellular growth in stressful environment in eukaryotes including *Bacillus subtilis* and *E. coli* [75,76] in co-culture samples. This may further suggest stimulation of the stress response between CP9 and ETEC [75,76] in the co-culture samples. The decrease in ETEC cells in co-culture and emergence of the co-culture metabolic features appearing closer to the CP9 samples in PCA and PLS-DA plot are suggestive of the notion that production of these metabolites may be driven by CP9 for its defense, rapid cellular growth to outcompete and weaken ETEC. Furthermore, metabolites that appeared in high concentrations in ETEC mono-cultures were significantly regulated in the co-culture, suggesting that CP9 may influence ETEC cellular metabolism and growth. For example, secondary metabolite gamma-Aminobutyric acid (GABA), which is responsible for spore germination, bile and low pH resistance, and tight regulation of virulent factors in enterotoxic and enteropathogenic *E. coli* [77,78], was observed in significantly lower abundance in the co-culture compared to ETEC mono-culture. This data is further supported by the lower cell number of ETEC after the co-culture. Similarly, di-peptide gln-gln involved in ETEC acid resistance [79] and arabinosylhypoxanthine involved in the purine metabolism, *E. coli* cellular growth, and virulence in mixed culture [80,81] were seen in higher abundance in the co-culture group. This may reflect an initial defensive response of ETEC to CP9 in the co-culture. However, as explained above, it should be noted that our study did not analyze the emergent metabolomic profile as a measure of the production or consumption of metabolites by either of the strains. Our study is in agreement with previous research by Medlock and colleagues, where, through metabolic modelling, it was shown that in mixed culture pairings, co-culture metabolomic profiles were less similar to the negatively impacted strain than the other strain, and the emergent metabolic profile of co-culture was directly correlated to the abundance of the highly competitive strain in the culture [16]. This notion is further supported by the emergence of unique antimicrobial secondary metabolites in co-culture and CP9 samples (Table 1) respectively, that may have synergistically impacted the growth of ETEC in co-culture. For example, valclavam, which is a metabolite of clavam class of β-lactam antibiotics, has been shown to strongly inhibit pathogenic *E. coli* blocking methionine biosynthesis [82]. However, to our knowledge, these have only been shown to be produced by *Streptomyces antibioticus* spp. [83,84]. Hence, emergence of valclavam in the co-culture warrants further investigation to analyze if its biosynthesis was triggered by CP9. Similarly, we observed a unique presence of leukotriene C4 and leukotriene E3 in CP9 metabolome samples and their significant abundance in the co-culture (Table 1). Leukotrienes are inflammatory mediators and are formed by oxidation of arachidonic acid by lipoxygenase enzyme. They are traditionally known to be exclusively produced in mammalian leukocytes for defense against microbial infections [85]. Interestingly, lipoxygenase activity, which was historically thought to be of eukaryotic function, has recently been found in various bacterial species [86]. This opens the door for further investigation into the presence of lipoxygenase activity in CP9 that may have resulted in the biosynthesis of leukotrienes in our study. This will be particularly important, as none of the probiotic classes of bacteria have been shown to possess this enzyme activity, which could have a direct impact on host immune response towards pathogenic microbes. Another unique metabolite significantly expressed in the co-culture samples was 3-Hydroxymyristic Acid, which is the most common fatty acid constituent of the lipid A component of bacterial lipopolysaccharides (LPS) [87]. Its significant presence in the co-culture suggests that CP9 may have caused the lysis of ETEC. This notion is supported by a previous study where *Bacillus subtilis* was shown to sensitize and lyse *E. coli* cells, which was driven by its proteolytic activity [88]. Alternatively, ETEC may have released LPS as an initial stress response in co-culture with CP9; however, this has a low probability since we did not observe any growth inhibitory effect of the ETEC CFS or co-culture CFS on CP9 growth (Data not shown). Lastly, emergence of an increased expression of melagatran, a serine protease inhibitor in the co-culture, is intriguing, since in gram negative bacteria such as *E. coli,* serine protease is secreted via autotransporter pathway and are implicated in expression of virulence and direct pathogenicity of its infection [89]. This could be a possible way CP9 may have downregulated the expression of ETEC virulence factors observed in our study. Overall, these results suggest how these unique metabolites may regulate the interactions between CP9 and ETEC by influencing their metabolic pathways and secreting secondary metabolites in the mixed culture either to weaken the opponent or depleting the essential nutrients for cell growth. However, these results warrant further investigation into the biosynthesis and purification of these metabolites to determine the extent of the impact they might have on antagonistic potential of CP9.

In conclusion, our study showed a strong antibacterial effect of potential probiotic, *Bacillus subtilis* CP9, which was driven by a contact-dependent mechanism of inhibition. We also found a substantial survival rate of CP9 in GIT fluids. However, since GIT fluids can vary in composition and pH depending on the diet of the animal, more tests need to be performed by utilizing GIT fluids from pigs or animals in question fed varied diets. Our study further revealed that CP9 successfully downregulates the virulence factors in the ETEC on a molecular level upon direct contact, which may be one of the possible mechanisms of CP9′s antagonistic potential. However, whether this effect is translated on a protein level was not analyzed in this study and warrants further investigation. Interspecies interactions within the gut are highly complex and impacted by metabolic cooperation and competitiveness [59,90]. Therefore, a detailed understanding about mechanisms of interactions of novel probiotic strains with gut pathogens may likely improve the predictability of the biological effect of the probiotic. Our study used an untargeted, data-driven approach to identify metabolic patterns that may influence bacterial growth in ETEC and CP9 co-culture, and proposed mechanisms that may contribute to the appearance of these patterns. However, this study did not analyze the biosynthesis and substrate utilization by either of the strain in co-culture. Perhaps incorporating a metabolic model that analyzes biosynthesis and utilization of these metabolites over time could provide normalized behaviors of the CP9 and ETEC metabolic patterns in co-culture. Developing such a model and validating these experiments will require a much larger data set than used in the current study. Nonetheless, extension of our approach to time-coursed metabolic modelling will provide more specific insights into CP9-induced growth inhibition of ETEC as well as other pathogenic bacteria. There is an increasing interest in developing novel probiotic-based interventions for animal and human use. However, traditional methods have primarily been focused on characteristics based on survival and properties of the probiotic strains. We envision that our study may provide the basis of preliminary understanding into the complex interaction of probiotic bacteria with an enteric pathogen, laying the foundation for the potential application of the probiotic for animal and human use.

Supplementary Materials: The following are available online at https://www.mdpi.com/article/10.3390/microorganisms9071483/s1, Figure S1: Cross Validation analysis for predicting PLS-DA model accuracy, Table S1: Genes and primer sequences used for ETEC reference genes and virulence-related genes, Table S2: Metabolomic profiles of co-culture and mono-cultures of CP9 and ETEC, Table S3: Significant metabolites emerging in Co-culture and mono-cultures of CP9 and ETEC, Table S4: Variable Importance in projection (VIP) scores across PLS-DA components indicating scale of variable metabolite concentration in mono-cultures and co-culture of CP9 and ETEC.

Author Contributions: Conceptualization, S.S. and J.L.; formal analysis, S.S.; funding acquisition, J.L.; Methodology-Metabolomics, R.F., L.N. and S.S.; writing—original draft, S.S.; writing—review and editing, S.S., J.L., L.N. and R.F. All authors have read and agreed to the submitted version of the manuscript.

Funding: This work was supported by the Swine Innovation Porc, Canada and Ontario Ministry of Agriculture, Food, and Rural Affairs (OMAFRA), Canada.

Institutional Review Board Statement: Not applicable.

Informed Consent Statement: Not applicable.

Data Availability Statement: The data presented in this study are available on request from the corresponding author.

Conflicts of Interest: The authors declare no conflict of interest.

References

1. Valdes, A.; Walter, J.; Segal, E.; Spector, T.D. Role of the gut microbiota in nutrition and health. *BMJ* **2018**, *361*, k2179. [CrossRef]
2. Anadón, A.; Irma, A.; Martínez-Larrañaga, M.R.; Maria, A.M. Prebiotics and Probiotics in Feed and Animal Health. In *Nutraceuticals in Veterinary Medicine*; Springer International Publishing: Cham, Switzerland, 2019; pp. 261–285. [CrossRef]
3. Hill, C.; Guarner, F.; Reid, G.; Gibson, G.R.; Merenstein, D.J.; Pot, B.; Morelli, L.; Canani, R.B.; Flint, H.J.; Salminen, S.; et al. The International Scientific Association for Probiotics and Prebiotics consensus statement on the scope and appropriate use of the term probiotic. *Nat. Rev. Gastroenterol. Hepatol.* **2014**, *11*, 506–514. [CrossRef] [PubMed]
4. Woo, J.; Ahn, J. Probiotic-mediated competition, exclusion and displacement in biofilm formation by food-borne pathogens. *Lett. Appl. Microbiol.* **2013**, *56*, 307–313. [CrossRef]
5. Liu, Q.; Yu, Z.; Tian, F.; Zhao, J.; Zhang, H.; Zhai, Q.; Chen, W. Surface components and metabolites of probiotics for regulation of intestinal epithelial barrier. *Microb. Cell Factories* **2020**, *19*, 1–11. [CrossRef] [PubMed]
6. Silva, D.R.; Sardi, J.D.C.O.; Pitangui, N.D.S.; Roque, S.M.; da Silva, A.C.B.; Rosalen, P.L. Probiotics as an alternative antimicrobial therapy: Current reality and future directions. *J. Funct. Foods* **2020**, *73*, 104080. [CrossRef]
7. Bron, P.A.; van Baarlen, P.; Kleerebezem, M. Emerging molecular insights into the interaction between probiotics and the host intestinal mucosa. *Nat. Rev. Genet.* **2011**, *10*, 66–78. [CrossRef] [PubMed]
8. O'Flaherty, S.; Saulnier, D.M.; Pot, B.; Versalovic, J. How can probiotics and prebiotics impact mucosal immunity? *Gut Microbes* **2010**, *1*, 293–300. [CrossRef]
9. Trush, E.A.; Poluektova, E.A.; Beniashvilli, A.G.; Shifrin, O.S.; Poluektov, Y.M.; Ivashkin, V.T. The Evolution of Human Probiotics: Challenges and Prospects. *Probiotics Antimicrob. Proteins* **2020**, *12*, 1291–1299. [CrossRef]
10. Suez, J.; Zmora, N.; Segal, E.; Elinav, E. The pros, cons, and many unknowns of probiotics. *Nat. Med.* **2019**, *25*, 716–729. [CrossRef]
11. Veiga, P.; Suez, J.; Derrien, M.; Elinav, E. Moving from probiotics to precision probiotics. *Nat. Microbiol.* **2020**, *5*, 878–880. [CrossRef]
12. Evivie, S.E.; Huo, G.-C.; Igene, J.O.; Bian, X. Some current applications, limitations and future perspectives of lactic acid bacteria as probiotics. *Food Nutr. Res.* **2017**, *61*, 1318034. [CrossRef]
13. Visconti, A.; Le Roy, C.; Rosa, F.; Rossi, N.; Martin, T.; Mohney, R.P.; Li, W.; De Rinaldis, E.; Bell, J.T.; Venter, J.C.; et al. Interplay between the human gut microbiome and host metabolism. *Nat. Commun.* **2019**, *10*, 1–10. [CrossRef]

14. Koren, O.; Goodrich, J.K.; Cullender, T.C.; Spor, A.; Laitinen, K.; Bäckhed, H.K.; Gonzalez, A.; Werner, J.J.; Angenent, L.T.; Knight, R.; et al. Host Remodeling of the Gut Microbiome and Metabolic Changes during Pregnancy. *Cell* **2012**, *150*, 470–480. [CrossRef] [PubMed]
15. Wieërs, G.; Belkhir, L.; Enaud, R.; Leclercq, S.; De Foy, J.-M.P.; Dequenne, I.; De Timary, P.; Cani, P.D. How Probiotics Affect the Microbiota. *Front. Cell. Infect. Microbiol.* **2020**, *9*, 454. [CrossRef] [PubMed]
16. Medlock, G.L.; Carey, M.; McDuffie, D.G.; Mundy, M.; Giallourou, N.; Swann, J.R.; Kolling, G.L.; Papin, J.A. Inferring Metabolic Mechanisms of Interaction within a Defined Gut Microbiota. *Cell Syst.* **2018**, *7*, 245–257. [CrossRef]
17. Hove, P.R.; Nora, J.N.; Siu, H.J.C.; Shea, M.B.; Hannah, B.H.; Elizabeth, P.R. Metabolomics and Proteomics of L. Rhamnosus GG and E. Coli Nissle Probiotic Supernatants Identify Distinct Pathways That Mediate Growth Suppression of Antimicrobial-Resistant Pathogens. *BioRxiv* **2020**. [CrossRef]
18. Elshaghabee, F.M.F.; Rokana, N.; Gulhane, R.D.; Sharma, C.; Panwar, H. Bacillus As Potential Probiotics: Status, Concerns, and Future Perspectives. *Front. Microbiol.* **2017**, *8*, 1490. [CrossRef] [PubMed]
19. Kim, B.J.; Hong, J.-H.; Jeong, Y.S.; Jung, H.K. Evaluation of two Bacillus subtilis strains isolated from Korean fermented food as probiotics against loperamide-induced constipation in mice. *J. Korean Soc. Appl. Biol. Chem.* **2014**, *57*, 797–806. [CrossRef]
20. Mazkour, S.; Shekarforoush, S.S.; Basiri, S.; Nazifi, S.; Yektaseresht, A.; Honarmand, M. Effects of two probiotic spores of Bacillus species on hematological, biochemical, and inflammatory parameters in Salmonella Typhimurium infected rats. *Sci. Rep.* **2020**, *10*, 8035. [CrossRef]
21. Gaete, A.; Mandakovic, D.; González, M. Isolation and Identification of Soil Bacteria from Extreme Environments of Chile and Their Plant Beneficial Characteristics. *Microorganisms* **2020**, *8*, 1213. [CrossRef]
22. Bang, C.; Dagan, T.; Deines, P.; Dubilier, N.; Duschl, W.J.; Fraune, S.; Hentschel, U.; Hirt, H.; Hülter, N.F.; Lachnit, T.; et al. Metaorganisms in extreme environments: Do microbes play a role in organismal adaptation? *Zoology* **2018**, *127*, 1–19. [CrossRef]
23. Akhtar, N.; Cai, H.; Kiarie, E.G.; Li, J. A novel *Bacillus sp.* with rapid growth property and high enzyme activity that allows efficient fermentation of soybean meal for improving digestibility in growing pigs. *J. Appl. Microbiol.* **2021**, Submitted.
24. Ramos, C.; Thorsen, L.; Schwan, R.F.; Jespersen, L. Strain-specific probiotics properties of Lactobacillus fermentum, Lactobacillus plantarum and Lactobacillus brevis isolates from Brazilian food products. *Food Microbiol.* **2013**, *36*, 22–29. [CrossRef] [PubMed]
25. Maystrenko, A.; Feng, Y.; Akhtar, N.; Li, J. The Addition of a Synthetic LPS-Targeting Domain Improves Serum Stability While Maintaining Antimicrobial, Antibiofilm, and Cell Stimulating Properties of an Antimicrobial Peptide. *Biomolecules* **2020**, *10*, 1014. [CrossRef] [PubMed]
26. Taha-Abdelaziz, K.; Astill, J.; Kulkarni, R.R.; Read, L.R.; Najarian, A.; Farber, J.M.; Sharif, S. In vitro assessment of immunomodulatory and anti-Campylobacter activities of probiotic lactobacilli. *Sci. Rep.* **2019**, *9*, 17903. [CrossRef] [PubMed]
27. Kemme, M.; Heinzel-Wieland, R. Quantitative Assessment of Antimicrobial Activity of PLGA Films Loaded with 4-Hexylresorcinol. *J. Funct. Biomater.* **2018**, *9*, 4. [CrossRef]
28. Alcoforado, D.; Juliana, B.H.; Sarah, J.C. Quantitative Determination of Anti-Bacterial Activity during Bacterial Co-Culture. In *Methods in Molecular Biology*; Humana Press Inc.: New York, NY, USA, 2017; Volume 1615, pp. 517–524. [CrossRef]
29. Wang, J.; Zeng, Y.; Wang, S.; Liu, H.; Zhang, D.; Zhang, W.; Wang, Y.; Ji, H. Swine-Derived Probiotic Lactobacillus plantarum Inhibits Growth and Adhesion of Enterotoxigenic Escherichia coli and Mediates Host Defense. *Front. Microbiol.* **2018**, *9*, 1364. [CrossRef] [PubMed]
30. Ho, S.W.; El-Nezami, H.; Shah, N.P. The protective effects of enriched citrulline fermented milk with Lactobacillus helveticus on the intestinal epithelium integrity against Escherichia coli infection. *Sci. Rep.* **2020**, *10*, 1–15. [CrossRef] [PubMed]
31. Medellin-Peña, M.J.; Wang, H.; Johnson, R.; Anand, S.; Griffiths, M.W. Probiotics Affect Virulence-Related Gene Expression in Escherichia coli O157:H7. *Appl. Environ. Microbiol.* **2007**, *73*, 4259–4267. [CrossRef]
32. Bonetti, A.; Tugnoli, B.; Rossi, B.; Giovagnoni, G.; Piva, A.; Grilli, E. Nature-Identical Compounds and Organic Acids Reduce *E. coli* K88 Growth and Virulence Gene Expression In Vitro. *Toxins* **2020**, *12*, 468. [CrossRef]
33. Wang, W.; He, J.; Pan, D.; Wu, Z.; Guo, Y.; Zeng, X.; Lian, L. Metabolomics analysis of Lactobacillus plantarum ATCC 14917 adhesion activity under initial acid and alkali stress. *PLoS ONE* **2018**, *13*, e0196231. [CrossRef] [PubMed]
34. Derrien, M.; Vlieg, J.E.V.H. Fate, activity, and impact of ingested bacteria within the human gut microbiota. *Trends Microbiol.* **2015**, *23*, 354–366. [CrossRef] [PubMed]
35. Corcoran, B.M.; Stanton, C.; Fitzgerald, G.F.; Ross, R. Survival of Probiotic Lactobacilli in Acidic Environments Is Enhanced in the Presence of Metabolizable Sugars. *Appl. Environ. Microbiol.* **2005**, *71*, 3060–3067. [CrossRef]
36. Wilks, J.C.; Kitko, R.D.; Cleeton, S.H.; Lee, G.E.; Ugwu, C.S.; Jones, B.D.; BonDurant, S.S.; Slonczewski, J.L. Acid and Base Stress and Transcriptomic Responses in Bacillus subtilis. *Appl. Environ. Microbiol.* **2009**, *75*, 981–990. [CrossRef]
37. Lefevre, M.; Racedo, S.M.; Denayrolles, M.; Ripert, G.; Desfougères, T.; Lobach, A.R.; Simon, R.; Pélerin, F.; Jüsten, P.; Urdaci, M.C. Safety assessment of Bacillus subtilis CU1 for use as a probiotic in humans. *Regul. Toxicol. Pharmacol.* **2017**, *83*, 54–65. [CrossRef]
38. Alayande, K.A.; Aiyegoro, O.A.; Ateba, C.N. Probiotics in Animal Husbandry: Applicability and Associated Risk Factors. *Sustainability* **2020**, *12*, 1087. [CrossRef]
39. Rhayat, L.; Maresca, M.; Nicoletti, C.; Perrier, J.; Brinch, K.S.; Christian, S.; Devillard, E.; Eckhardt, E. Effect of Bacillus subtilis Strains on Intestinal Barrier Function and Inflammatory Response. *Front. Immunol.* **2019**, *10*. [CrossRef]
40. Gu, M.J.; Song, S.K.; Park, S.M.; Lee, I.K.; Yun, C.-H. Bacillus subtilis Protects Porcine Intestinal Barrier from Deoxynivalenol via Improved Zonula Occludens-1 Expression. *Asian Australas. J. Anim. Sci.* **2014**, *27*, 580–586. [CrossRef] [PubMed]

41. Monteagudo-Mera, A.; Rastall, R.A.; Gibson, G.R.; Charalampopoulos, D.; Chatzifragkou, A. Adhesion mechanisms mediated by probiotics and prebiotics and their potential impact on human health. *Appl. Microbiol. Biotechnol.* **2019**, *103*, 6463–6472. [CrossRef]
42. Macfarlane, G.T.; Cummings, J.H. Probiotics and prebiotics: Can regulating the activities of intestinal bacteria benefit health? *BMJ* **1999**, *318*, 999–1003. [CrossRef] [PubMed]
43. Tazehabadi, M.H.; Algburi, A.; Popov, I.V.; Ermakov, A.M.; Chistyakov, V.A.; Prazdnova, E.V.; Weeks, R.; Chikindas, M.L. Probiotic Bacilli Inhibit Salmonella Biofilm Formation Without Killing Planktonic Cells. *Front. Microbiol.* **2021**, *12*. [CrossRef] [PubMed]
44. Ramachandran, R.; Chalasani, A.G.; Lal, R.; Roy, U. A Broad-Spectrum Antimicrobial Activity ofBacillus subtilisRLID 12.1. *Sci. World J.* **2014**, *2014*, 968487. [CrossRef] [PubMed]
45. Stubbendieck, R.M.; Straight, P.D. Multifaceted Interfaces of Bacterial Competition. *J. Bacteriol.* **2016**, *198*, 2145–2155. [CrossRef]
46. Garcia, E.C. Contact-dependent interbacterial toxins deliver a message. *Curr. Opin. Microbiol.* **2018**, *42*, 40–46. [CrossRef]
47. Garcia, E.C.; Perault, A.I.; Marlatt, S.A.; Cotter, P.A. Interbacterial signaling via Burkholderia contact-dependent growth inhibition system proteins. *Proc. Natl. Acad. Sci. USA* **2016**, *113*, 8296–8301. [CrossRef]
48. Lopez, D.; Kolter, R. Extracellular signals that define distinct and coexisting cell fates inBacillus subtilis. *FEMS Microbiol. Rev.* **2010**, *34*, 134–149. [CrossRef]
49. Kaundal, S.; Deep, A.; Kaur, G.; Thakur, K.G. Molecular and Biochemical Characterization of YeeF/YezG, a Polymorphic Toxin-Immunity Protein Pair From Bacillus subtilis. *Front. Microbiol.* **2020**, *11*, 95. [CrossRef] [PubMed]
50. Holberger, L.E.; Garza-Sánchez, F.; Lamoureux, J.; Low, D.A.; Hayes, C.S. A novel family of toxin/antitoxin proteins inBacil-lusspecies. *FEBS Lett.* **2011**, *586*, 132–136. [CrossRef] [PubMed]
51. Ikryannikova, L.; Kurbatov, L.; Gorokhovets, N.; Zamyatnin, A. Contact-Dependent Growth Inhibition in Bacteria: Do Not Get Too Close! *Int. J. Mol. Sci.* **2020**, *21*, 7990. [CrossRef]
52. Anyanful, A.; Dolan-Livengood, J.M.; Lewis, T.; Sheth, S.; DeZalia, M.N.; Sherman, M.A.; Kalman, L.V.; Benian, G.M.; Kalman, D. Paralysis and killing of Caenorhabditis elegans by enteropathogenic Escherichia coli requires the bacterial tryptophanase gene. *Mol. Microbiol.* **2005**, *57*, 988–1007. [CrossRef]
53. Zhou, M.; Duan, Q.; Zhu, X.; Guo, Z.; Li, Y.; Hardwidge, P.R.; Zhu, G. Both flagella and F4 fimbriae from F4ac+ enterotoxigenic Escherichia coli contribute to attachment to IPEC-J2 cells in vitro. *Vet. Res.* **2013**, *44*, 30. [CrossRef] [PubMed]
54. Dubreuil, J.D.; Isaacson, R.E.; Schifferli, D.M. Animal Enterotoxigenic Escherichia coli. *EcoSal Plus* **2016**, *7*. [CrossRef]
55. Wilson, M.L.; Macnab, R.M. Overproduction of the MotA protein of Escherichia coli and estimation of its wild-type level. *J. Bacteriol.* **1988**, *170*, 588–597. [CrossRef]
56. Jurtshuk, P. Good info about different types and pathways: Bacterial Metabolism. In *Medical Microbiology*; Baron, S., Ed.; University of Texas Medical Branch at Galveston: Galveston, TX, USA, 1996. Available online: http://www.ncbi.nlm.nih.gov/pubmed/21413278 (accessed on 23 May 2021).
57. Ward, B. Bacterial Energy Metabolism. In *Molecular Medical Microbiology*, 2nd ed.; Elsevier Ltd.: Amsterdam, The Netherlands, 2014; Volume 1. [CrossRef]
58. Hansen, S.K.; Rainey, P.; Haagensen, J.A.J.; Molin, S. Evolution of species interactions in a biofilm community. *Nat. Cell Biol.* **2007**, *445*, 533–536. [CrossRef] [PubMed]
59. Freilich, S.; Zarecki, R.; Eilam, O.; Segal, E.S.; Henry, C.S.; Kupiec, M.; Gophna, U.; Sharan, R.; Ruppin, E. Competitive and cooperative metabolic interactions in bacterial communities. *Nat. Commun.* **2011**, *2*, 589. [CrossRef] [PubMed]
60. Zhang, Y.J.; Rubin, E.J. Feast or famine: The host-pathogen battle over amino acids. *Cell. Microbiol.* **2013**, *15*, 1079–1087. [CrossRef]
61. Li, G.; Young, K.D. A new suite of tnaA mutants suggests that Escherichia coli tryptophanase is regulated by intracellular sequestration and by occlusion of its active site. *BMC Microbiol.* **2015**, *15*, 1–17. [CrossRef]
62. Orozco-Gómez, D.I.; Sosa-Hernández, J.E.; Gallardo-Navarro, Ó.A.; Santana-Solano, J.; Santillán, M. Bistable behaviour and medium-dependent post-translational regulation of the tryptophanase operon regulatory pathway in Echerichia coli. *Sci. Rep.* **2019**, *9*, 5451. [CrossRef]
63. Di Martino, P.; Fursy, R.; Bret, L.; Sundararaju, B.; Phillips, R. Indole can act as an extracellular signal to regulate biofilm formation of Escherichia coli and other indole-producing bacteria. *Can. J. Microbiol.* **2003**, *49*, 443–449. [CrossRef]
64. Roager, H.M.; Licht, T.R. Microbial tryptophan catabolites in health and disease. *Nat. Commun.* **2018**, *9*, 3294. [CrossRef] [PubMed]
65. Kumar, A.; Sperandio, V. Indole Signaling at the Host-Microbiota-Pathogen Interface. *mBio* **2019**, *10*, e01031-19. [CrossRef] [PubMed]
66. Bommarius, B.; Anyanful, A.; Izrayelit, Y.; Bhatt, S.; Cartwright, E.; Wang, W.; Swimm, A.; Benian, G.M.; Schroeder, F.; Kalman, D. A Family of Indoles Regulate Virulence and Shiga Toxin Production in Pathogenic E. coli. *PLoS ONE* **2013**, *8*, e54456. [CrossRef]
67. Singh, R.; Sharma, S.; Kareenhalli, V.V.; Giri, L.; Mitra, K. Experimental investigation into indole production using passaging of E. coli and B. subtilis along with unstructured modeling and parameter estimation using dynamic optimization: An integrated framework. *Biochem. Eng. J.* **2020**, *163*, 107743. [CrossRef]
68. Melander, R.J.; Minvielle, M.J.; Melander, C. Controlling bacterial behavior with indole-containing natural products and derivatives. *Tetrahedron* **2014**, *70*, 6363–6372. [CrossRef]
69. Hibbing, M.E.; Fuqua, C.; Parsek, M.R.; Peterson, S.B. Bacterial competition: Surviving and thriving in the microbial jungle. *Nat. Rev. Microbiol.* **2010**, *8*, 15–25. [CrossRef]

70. Nunn, W.D.; Kelly, D.L.; Stumfall, M.Y. Regulation of fatty acid synthesis during the cessation of phospholipid biosynthesis in Escherichia coli. *J. Bacteriol.* **1977**, *132*, 526–531. [CrossRef]
71. Parsons, J.B.; Rock, C.O. Bacterial lipids: Metabolism and membrane homeostasis. *Prog. Lipid Res.* **2013**, *52*, 249–276. [CrossRef] [PubMed]
72. Meadows, J.A.; Wargo, M.J. Carnitine in bacterial physiology and metabolism. *Microbiology* **2015**, *161*, 1161–1174. [CrossRef]
73. Bjørndal, B.; Alterås, E.K.; Lindquist, C.; Svardal, A.; Skorve, J.; Berge, R.K. Associations between fatty acid oxidation, hepatic mitochondrial function, and plasma acylcarnitine levels in mice. *Nutr. Metab.* **2018**, *15*, 1–13. [CrossRef] [PubMed]
74. Li, S.; Gao, D.; Jiang, Y. Function, Detection and Alteration of Acylcarnitine Metabolism in Hepatocellular Carcinoma. *Metabolites* **2019**, *9*, 36. [CrossRef] [PubMed]
75. Tofalo, R.; Cocchi, S.; Suzzi, G. Polyamines and Gut Microbiota. *Front. Nutr.* **2019**, *6*, 16. [CrossRef]
76. Shah, P.; Swiatlo, E. A multifaceted role for polyamines in bacterial pathogens. *Mol. Microbiol.* **2008**, *68*, 4–16. [CrossRef] [PubMed]
77. Shin, S.; Castanie-Cornet, M.-P.; Foster, J.W.; Crawford, J.A.; Brinkley, C.; Kaper, J.B. An activator of glutamate decarboxylase genes regulates the expression of enteropathogenic Escherichia coli virulence genes through control of the plasmid-encoded regulator, Per. *Mol. Microbiol.* **2008**, *41*, 1133–1150. [CrossRef] [PubMed]
78. Braun, H.-S.; Sponder, G.; Aschenbach, J.R.; Kerner, K.; Bauerfeind, R.; Deiner, C. The GadX regulon affects virulence gene expression and adhesion of porcine enteropathogenic Escherichia coli in vitro. *Vet. Anim. Sci.* **2017**, *3*, 10–17. [CrossRef]
79. Lu, P.; Ma, D.; Chen, Y.; Guo, Y.; Chen, G.-Q.; Deng, H.; Shi, Y. L-glutamine provides acid resistance for Escherichia coli through enzymatic release of ammonia. *Cell Res.* **2013**, *23*, 635–644. [CrossRef]
80. Leung, H.B.; Doering, A.M.; Cohen, S.S. Effect of 9-β- d -Arabinofuranosyladenine on Polymer Synthesis in a Polyauxotrophic Strain of Escherichia coli. *J. Bacteriol.* **1966**, *92*, 558–564. [CrossRef]
81. Xi, H.; Schneider, B.L.; Reitzer, L. Purine Catabolism in Escherichia coli and Function of Xanthine Dehydrogenase in Purine Salvage. *J. Bacteriol.* **2000**, *182*, 5332–5341. [CrossRef] [PubMed]
82. Röhl, F.; Rabenhorst, J.; Zahner, H. Biological properties and mode of action of clavams. *Arch. Microbiol.* **1987**, *147*, 315–320. [CrossRef] [PubMed]
83. Baldwin, J.E.; Fujishima, Y.; Goh, K.-C.; Schofield, C.J. Enzymes of valclavam biosynthesis. *Tetrahedron Lett.* **1994**, *35*, 2783–2786. [CrossRef]
84. McGowan, S.J.; Barrie, W.B.; George, P.C.S. Bacterial production of carbapenems and clavams: Evolution of β-Lactam antibiotic pathways. *Trends Microbiol.* **1998**, *6*, 203–208. [CrossRef]
85. Nigam, P.K.; Sehgal, U. Leukotrienes. *Indian J. Dermatol. Venereol. Leprol.* **1989**, *55*, 155–163. [CrossRef] [PubMed]
86. Hansen, J.; Garreta, A.; Benincasa, M.; Fusté, M.C.; Busquets, M.; Manresa, A. Bacterial lipoxygenases, a new subfamily of enzymes? A phylogenetic approach. *Appl. Microbiol. Biotechnol.* **2013**, *97*, 4737–4747. [CrossRef] [PubMed]
87. Parlesak, A. Alcohol, Altered Gut Permeability, and Endotoxins. In *Comprehensive Handbook of Alcohol Related Pathology*; Elsevier Inc.: Amsterdam, The Netherlands, 2005; Volume 2. [CrossRef]
88. Dean, C.R.; Ward, O.P. Nature of Escherichia coli cell lysis by culture supernatants of Bacillus species. *Appl. Environ. Microbiol.* **1991**, *57*, 1893–1898. [CrossRef] [PubMed]
89. Ruiz-Perez, F.; Nataro, J.P. Bacterial serine proteases secreted by the autotransporter pathway: Classification, specificity, and role in virulence. *Cell. Mol. Life Sci.* **2014**, *71*, 745–770. [CrossRef]
90. Nicholson, J.K.; Holmes, E.; Kinross, J.; Burcelin, R.; Gibson, G.; Jia, W.; Pettersson, S. Host-Gut Microbiota Metabolic Interactions. *Science* **2012**, *336*, 1262–1267. [CrossRef]

MDPI
St. Alban-Anlage 66
4052 Basel
Switzerland
Tel. +41 61 683 77 34
Fax +41 61 302 89 18
www.mdpi.com

Microorganisms Editorial Office
E-mail: microorganisms@mdpi.com
www.mdpi.com/journal/microorganisms

www.ingramcontent.com/pod-product-compliance
Lightning Source LLC
Chambersburg PA
CBHW041948260926
51850CB00018B/1960
* 9 7 8 3 0 3 6 5 6 9 3 3 8 *